OUR WORLD TODAY SERIES

THE
EASTERN HEMISPHERE

OUR WORLD TODAY SERIES

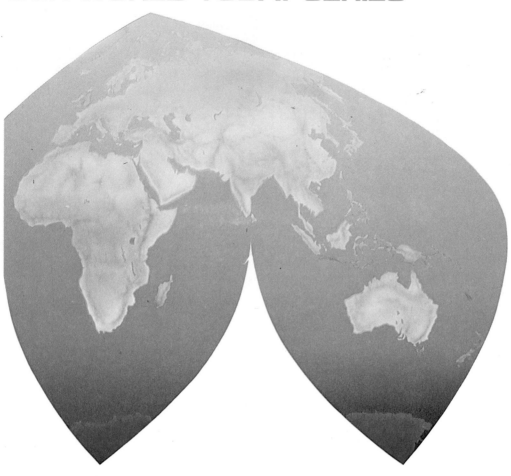

THE EASTERN HEMISPHERE

Harold D. Drummond

James W. Hughes

Allyn and Bacon, Inc.

Boston   Rockleigh, N.J.   Atlanta   Dallas   Belmont, Calif.

The Eastern Hemisphere was originally written by De Forest Stull and Roy W. Hatch. It has been completely rewritten by the present authors, Harold D. Drummond and James W. Hughes.

Harold D. Drummond is Professor Emeritus of Elementary Education at the University of New Mexico and the former Associate Dean for Curriculum and Instruction at the College of Education. He has been active in the field of education for many years, as teacher at the elementary and college levels, administrator, and as a member of numerous professional organizations. He is the author of many educational publications and textbooks, including *Our World Today Series*, Allyn and Bacon, Inc.

James W. Hughes is presently Professor of Education at Oakland University in Rochester, Michigan. In the past he has served as Director of Teacher Education and Area Chairperson in Elementary Education at Oakland University. As a supervision specialist for the National Education Association, he has assisted the governments of Kenya and Nepal, in improving their educational systems. In addition he is the author of eleven books on Africa and has written numerous articles for various educational publications.

Editor: Richard V. Foster
Designer: Beverly Fell
Photo Researcher: Marjorie H. Bishop
Buyer: Linda Card

Graphic material and maps prepared by Jeppesen and Co., I² Geographics, Lee Ames and Zak, Ltd.

Printed in the United States of America

ISBN 0–205–06627–5

Library of Congress Catalog Card Number 79–8701

1 2 3 4 5 6 7 8 9 8 8 8 7 8 6 8 5 8 4 8 3 8 2 8 1 8 0 79

PREFACE

The Eastern Hemisphere was written to help you learn more about the people and countries of Europe, Africa, Asia, and Australia. There are so many countries and peoples in the Eastern Hemisphere that some countries have been only briefly mentioned. Others have been chosen for emphasis.

The Importance of Geographic Knowledge

Geographic knowledge becomes more important each year. Improvements in transportation and communications have brought the people on this planet closer together. All people of the world, in one sense, are now closer neighbors. Knowing more about our world may help us understand other peoples better.

Aids to Study

Many sections of this book have been planned to help you study and learn about the Eastern Hemisphere. To help you find information quickly, this book includes near the front a table of contents and a list of maps, charts, graphs, and diagrams. At the back of the book is an Appendix, a Glossary, and an Index.

In each "Unit Introduction" there are suggestions of "Things You Might Like To Do." These activities include both group projects and things to do by yourself. To help you remember the information presented, every few pages there are review questions and activities called "REMEMBER, THINK, DO." You will also find that the text includes Sidelights here and there. These are interesting stories about people, places, and events. At the end of each chapter or unit is a "REVIEW" section with questions and some additional activities under the headings of "WHAT HAVE YOU LEARNED?" and "BE A GEOGRAPHER." Finally, each "REVIEW" section has some "QUESTIONS TO THINK ABOUT." All of these activities are designed to help you use what you have learned.

Many changes are rapidly taking place in the Eastern Hemisphere. Watch for articles in newspapers and magazines about the countries and peoples of *The Eastern Hemisphere*.

CONTENTS

MAPS, GRAPHS, CHARTS, DIAGRAMS

Unit

THE EASTERN HEMISPHERE

Unit Preview

LOOK

- Can you see on these globes:
    - a. Air that surrounds our Earth?
    - b. Water that we drink when we are thirsty?
    - c. Ice?
    - d. Mountains?
    - e. Deserts?
    - f. Oceans?

- Be prepared to tell where.

Globe 1

- What hemisphere is this?
    - a. Northern
    - b. Southern
    - c. Eastern
    - d. Western

- What *is* a hemisphere?

Globe 2

- What hemisphere is this?

    - a. Northern
    - b. Southern
    - c. Eastern
    - d. Western

- How many continents are there in this hemisphere?

DO YOU KNOW?

- On which hemisphere do most people live?

- Which hemisphere has the highest mountains?

- Which hemisphere has the largest river? The longest river?

- Which hemisphere has the country with the largest area?

- Which hemisphere has the country with the most people?

Unit Introduction

The Eastern Hemisphere

Look at the pictures of globes on page 2. They show different views of the planet Earth. Globe 1 shows the Western *Hemisphere*. (**Hemisphere** means "half a *sphere*.") A **sphere** is shaped like a ball. The Western Hemisphere is where most young persons using this book live. Can you find about where you live on the planet Earth? On what continent do you live? In what country? In what state or province? How many continents are there in the Western Hemisphere? What is a continent?

Now look at Globe 2. Can you see any part of the Western Hemisphere? No, this side of the globe shows the Eastern Hemisphere.

How many continents can you find on this hemisphere? Geographers do not agree about how many continents there are in the Eastern Hemisphere. Some of them count three continents, and some of them count four. Do you already know the names of the three or four continents shown on Globe 2? If you counted three continents, the answer would be: Eurasia, Africa, and Australia. If you counted four continents the answer would be: Europe, Asia, Africa, and Australia. Find all these continents on a real globe or a world map in your classroom. In this book we have chosen to use four continents.

3

Chapter

WHY STUDY
—THE EASTERN HEMISPHERE?—

What Do You Already Know About the Eastern Hemisphere?

What countries are on the four continents and the islands near them? What large cities are in these countries? How many people live in the Eastern Hemisphere? How do the people make a living? What kinds of homes do they have? Are there large rivers in the Eastern Hemisphere? High mountains? Deserts?

Here are some interesting facts about the Eastern Hemisphere:

● Most of the people on Earth live in the Eastern Hemisphere. For every person living in the Western Hemisphere there are six persons living in the Eastern Hemisphere.

● There are many islands near the four continents in the Eastern Hemisphere. Six of the world's ten largest islands are in the Eastern Hemisphere. You might like to try to locate them.

● In the Eastern Hemisphere there is more land north of the equator than south of it. All the lands of two continents are north of the equator. One continent has land both north and south of the equator. All of one continent is south of the equator.

● The highest mountain and the longest river on the planet Earth are in the Eastern Hemisphere. Do you know their names?

● The lowest spot on the surface of the planet Earth not covered by water is in the Eastern Hemisphere. Do you know where this lowest spot is?

● The largest desert on the planet Earth is in the Eastern Hemisphere. Several other large deserts are also found there. In these deserts there are places where it hardly ever rains.

● At some places in the Eastern Hemisphere it rains almost every day. In some places it rains several times almost every day. In such places tropical rain forests grow. (See page 370 and the Glossary for an explanation of tropical rain forests.)

● The country with the largest amount of land is in the Eastern

Mountain ranges, such as the Alps in Switzerland, are barriers. They slow the movement of people from region to region. They also affect the climate of the surrounding areas.

Hemisphere. So are the three countries with the most people. Can you name these countries?

● The Eastern Hemisphere has many large cities. Three of the world's five largest cities are there. Do you know their names?

● Most people in the Eastern Hemisphere still live in small villages, however. Most of them are farmers. They walk or ride an animal to the fields each day. They return to the village every evening after a long day of work in the fields. The tools they use are simple ones. Most of the power they use is their own muscle power or that of animals.

● Some parts of the Eastern Hemisphere have many earthquakes and volcanoes.

● Some parts of the Eastern Hemisphere near oceans often have huge rain and windstorms called typhoons. (See Glossary for definition.) Other areas have almost perfect weather every day.

Things You Might Like To Do As You Study About the Eastern Hemisphere

1. Ask around your neighborhood to discover persons who have traveled in lands of the Eastern Hemisphere. Find out the countries they have visited and the ones they know well. Make a list of these people's names for use later on in the year as you study about these countries.

2. Suggest to your student council that your school and a school located somewhere in the Eastern Hemisphere become partners. You could then think of things you might send to the other school to help them understand how you live. For example, you could send information about what you study and the kinds of jobs people in your community do. You might like to tell what you do during vacations. You might send them some pictures. They will probably share information about their school with you.

Your local Red Cross chapter might be able to help you find a school in the Eastern Hemisphere that is interested.

3. If you would like to find a "pen pal" with whom you could exchange letters, write to the International Friendship League, 40 Mt. Vernon Street, Boston, MA 02115.

4. Help prepare a bulletin board for the classroom called "This Week's News About the Eastern

Islands can be of many sizes. They serve many purposes. What do you think this island in Yugoslavia has been used for? Does it still serve that purpose?

Hemisphere." Plan ways of keeping the news up-to-date. You might want to have a committee that would be responsible for the bulletin board each week.

5. If an opaque projector is available, and with help from your teacher, project a map of the Eastern Hemisphere upon a large chart or piece of paper. Draw an outline map of the area by using black crayon or a felt pen. Label each continent and the oceans. As you study each region of the hemisphere, add information to the large outline map.

Importance of the Eastern Hemisphere

Some of you may still ask, "Why should we study about the Eastern Hemisphere? We don't live there." There are several reasons why the Eastern Hemisphere is important. Some of them are listed below:
● The first civilizations on the planet Earth probably started in the Eastern Hemisphere.
● People living in the Eastern Hemisphere probably were the first to practice farming. They were also the first to irrigate their fields.
● The alphabet we use was invented by people living in the Eastern Hemisphere.
● Ways of spinning thread and of weaving thread into cloth were first invented by people living in the Eastern Hemisphere.
● The first paper was made in the Eastern Hemisphere, and the first printing press was invented there, too.

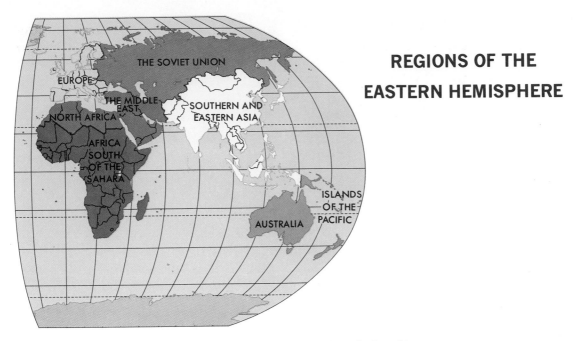

REGIONS OF THE EASTERN HEMISPHERE

Shown here are the six regions of the Eastern Hemisphere described in this book.

• The world's main religions all had their beginnings in the Eastern Hemisphere.

• Until the year 1800, almost all the important discoveries and inventions were made by people living in the Eastern Hemisphere.

Rapid Change

Another important reason for studying about the Eastern Hemisphere is that it is changing rapidly.

The Eastern Hemisphere has more young nations than any other part of the world. From about 1800 until about 1950, much of the Eastern Hemisphere was controlled by European nations. Since 1950 many new nations have been formed. Most of these young nations are in Africa and southern Asia.

One of the main problems young nations have is developing effective governments. Our own nation had this problem in its early history. The same is also true of many of the young nations in Africa and Asia. In many young nations, people belonging to the military have taken over the government. They rule by force. In some places the people have little to say about what the government does.

Another problem exists in Africa. The boundaries of many nations were set years ago by Europeans. The boundaries do not "fit" the people. Some members of a family or kinship group live on one side of a border. Others in the same group live on the other side. The border does not make sense to these people. They are loyal to each other rather than to the government.

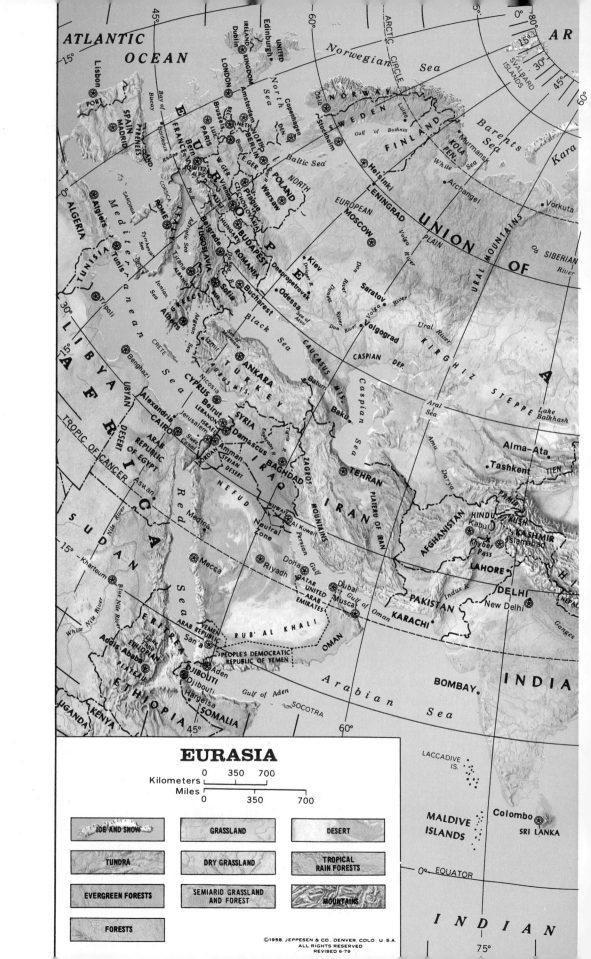

EURASIA

Kilometers 0 350 700

Miles 0 350 700

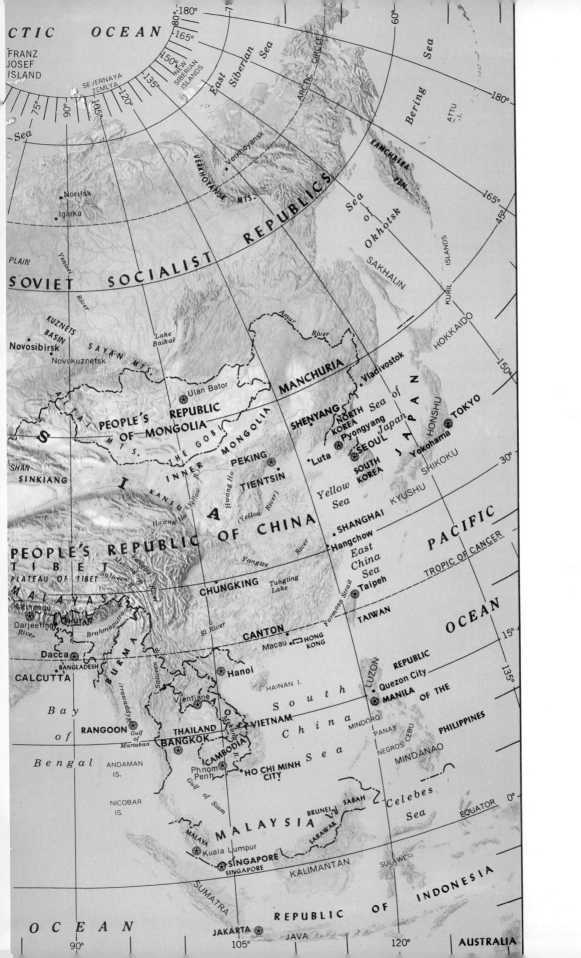

REMEMBER, THINK, DO

1. Where do you live? (In answering, use all the following: planet, hemisphere, country, region, state, county, city or nearest post office, street or rural route number.)
2. Which large continent is sometimes, as in this book, thought of as two continents?
3. On which continent do most people live? (See Appendix.)
4. Do as many people live in the Western Hemisphere as live in Asia?
5. Look at maps in this book, in atlases, or in encyclopedias and find answers to the following questions:

    a. How high is the world's highest mountain? What is its name and where is it located?

    b. How long is the world's longest river? What is its name and where is it located?

    c. How large is the world's largest desert? What is its name and where is it located? How many countries have land in this largest desert?
6. How many of the world's ten largest cities are in the Eastern Hemisphere? List them. (You may want to check several sources. If the figures you find do not agree, think about why.)

Rivers, such as this one in Kuei-Lin, China, are beneficial. They can also pose problems for people. What are some of the positive and negative aspects of a river?

Deserts are located on all continents of the world. The one pictured above is in Australia. Because of climate, the actions of people, and of animals, many of the deserts are growing in size.

Idea Conflicts

Other problems exist, too, in the Eastern Hemisphere. For many years, Arabs and Jews have been fighting near the eastern end of the Mediterranean (*mehd* ih ter *rain* ee uhn) Sea. Short wars have started several times in the past 30 years. The United Nations, the United States, and other countries have helped in stopping these wars.

Parts of Africa have been—and are—trouble spots, too. In some places a small number of white people control the government. They force a large number of black people to follow laws that do not apply to whites. Many persons think that these governments must change soon. If they do not, wars may break out.

Parts of the Eastern Hemisphere have for many years been areas for *idea conflicts*. Differences in what people believe that may cause wars or other uses of force by countries or groups are **idea conflicts**. Sometimes these idea conflicts have led to large wars. That happened in the 1940s when World War II was fought. It also happened in the 1950s and 1960s when the Vietnam War was fought. In the first case, Adolf Hitler planned to take over the world. He believed that the German people were superior to others and, thus, should rule over them. He built a huge army, navy, and air force. The Germans conquered much of Europe and part of North Africa before being defeated.

In the second case, the idea conflict was between communist and noncommunist nations. France, a noncommunist nation in Europe, had controlled Vietnam as a colony for many years. The people in the

Tropical rainforests are found in areas with a hot and humid climate. This rainforest is located along the equator in Western Equatorial Africa.

northern part of Vietnam started a war to gain control of their country. Communist nations, especially China and the Soviet Union, helped them by supplying guns and other weapons of war. The French were defeated, and they left Vietnam. Many Vietnamese did not like what happened when the Communists took over the government. They fled to the southern part of Vietnam and started another country, South Vietnam. The United States tried to help them defend themselves. After years of war and much destruction, the war finally ended. The Communists won and now control all of Vietnam.

The idea conflict goes on between Communist and non-Communist nations. Sometimes idea conflicts also occur between Communist na-

tions. The Union of Soviet Socialist Republics (U.S.S.R.), or the Soviet Union, is the nation with the largest land area on the planet Earth. China is a nation with the third largest land area and the most people. Both have governments controlled by Communists. But China and the Soviet Union have not been on friendly terms with each other for a number of years. Each nation is fearful of the other. Their armies are in place along their shared border in Asia. They seem, at times, to be almost ready to fight.

Many nations in the Eastern Hemisphere have not taken sides in these idea conflicts, however. These nations usually take positions different from those taken by the Soviet Union, China, and the United States of America. These countries call themselves nations of the **Third World**. While a few are large (as India), many of the Third World Nations are rather small in population and land area. Usually they have little military power. Sometimes, however, they do help the cause of world peace by urging the more powerful nations not to start wars.

Rapid Population Growth

Population growth throughout much of the Eastern Hemisphere is rapid. In Africa and in parts of Asia population growth is very rapid. At one time, because of high death rates, the population did not grow rapidly. In the past 20 years, however, the World Health Organization (WHO)—an agency of the United

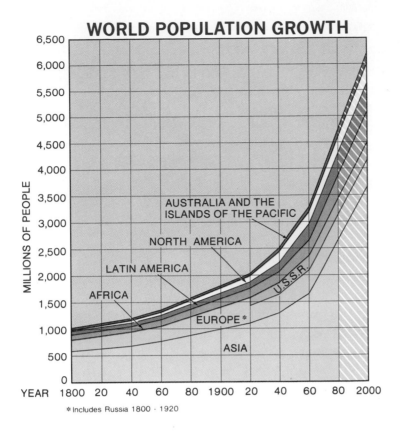

WORLD POPULATION GROWTH

MILLIONS OF PEOPLE

6,500
6,000
5,500
5,000
4,500
4,000
3,500
3,000
2,500
2,000
1,500
1,000
500
0

AUSTRALIA AND THE
ISLANDS OF THE PACIFIC

NORTH AMERICA

LATIN AMERICA

AFRICA

EUROPE *

U.S.S.R.

ASIA

YEAR 1800 20 40 60 80 1900 20 40 60 80 2000

*Includes Russia 1800 - 1920

This chart shows the population growth rate since 1800 for eight regions of the world. When did the world's population begin to grow rapidly? What region has had the greatest increase since 1800? Since 1960?

Nations—has taught the people how to control diseases. As a result, the death rate has decreased. People are living longer. In spite of wars and accidents, the world's total population is growing rapidly. Feeding all of the world's people has become a serious problem.

Farm scientists have learned much in recent years about how to increase food production. Twice as much corn or rice can be grown on farmland today as could be grown 20 years ago. Scientists probably will continue to learn more about how to raise more and better food. Most of the good farming areas on Earth already are used for growing food, however. Most of these are areas

where many people already live. Uncrowded parts of the world are generally cold or mountainous or dry. Whether or not food production can keep up with population growth worries many people today.

Never before in history have the Earth's resources been used as quickly as at present. Many persons are beginning to wonder if the world can support the people that may soon be living on it. If the resources are used wisely and if population growth slows down, good living for all people should be possible. If resources are wasted, the people who live 50 or 100 years from now may not have some needed resources.

How Many People Is Too Many?*

About April 1, 1976, the number of people on Earth reached 4 billion! How many is 4 billion? Well, as you know, it is 4,000 million! But how many is that? You can use time to get some idea of how many 4 billion is. You know that there are 60 seconds in a minute and 60 minutes in an hour. You also know that there are 24 hours in a day, 365¼ days in a year. So, in a year there are 31,557,600 seconds (60 × 60 × 24 × 365¼ = 31,557,600). In about 127 years there are about 4 billion seconds! That is a lot of seconds, and 4 billion is a lot of people!

Some people think that 4 billion people are too many for the resources of the planet Earth. Others think that 4 billion are enough. And some people think that the number of people should be permitted to grow without worry. "How many people are too many?" has become a serious question.

The planet Earth did not have 1 billion people until 1850. It had 2 billion in 1930—80 years later. The number reached 3 billion in 1961—31 years later. And it reached 4 billion in 1976—15 years later. If present rates of population growth continue, there will be 5 billion people in 1989. That will take only 13 years!

The question of "how many people are too many" is really a question of food and resources. Many people on Earth—especially in Africa and Asia—are always hungry. There simply isn't enough food for them. Many die each year of starvation.

The birth rate is slowing down in many parts of the world. At the present rate of growth in the United States, it will take 116 years for the population to double. In Africa, however, it will take only 27 years for the population to double. In southern Asia it will take only about 30 years for the population to double. There are already more than 1 billion people living in southern and southeast Asia! In east Asia, where another billion people live, the population will double in 43 years.

How many people are too many? No one knows for sure. But everyone knows that going to bed hungry is not fun. Have you ever had to try to go to sleep with your stomach growling because it is empty? Some youngsters do it every night. Too many people are too many—that's for sure!

---

*Basic data adapted, with permission, from *Intercom*, a publication of the Population Reference Bureau, 1337 Connecticut Avenue, N.W., Washington, D.C., 20036 (Vol. 4, No. 3, March 1976), and from the "1977 World Population Data Sheet" of the same organization. Interpretations are the authors'.

What kinds of problems do you think overcrowding creates for the people of the Eastern Hemisphere?

LEARN MORE ABOUT THE EASTERN HEMISPHERE

You should now be able to tell something about the Eastern Hemisphere. In addition to doing the activities at the end of this chapter, you may wish to do some things on your own. Here are some ideas for you:

1. Look in an encyclopedia for articles on *sphere* and *hemisphere*. Think of a way you can share what you learn with other members of the class.

2. Look at the Table of Contents for this book. Notice what is coming in a few weeks, and begin to do some reading about Europe. Librarians will be glad to help you find material.

3. Study a globe carefully—especially the part of it showing the Eastern Hemisphere. Study the general shape of the four continents.

Learn also the names of the main bodies of water around the continents.

4. Ask the school librarian to help you find filmstrips about the Eastern Hemisphere and about Earth-Sun relationships. Make a model of the Universe, or a drawing showing the planets revolving around the Sun. Label each planet, of course.

5. Begin asking around the neighborhood for people who have lived in Europe. Some of them might have been born there. Ask them what they remember and share it with the class.

Near the end of each chapter of this book, there are some review activities and some things for you to do. These activities will help you learn more about the Eastern Hemisphere.

—————— WHAT HAVE YOU LEARNED? ——————

Each chapter has some review questions under a heading called "What Have You Learned?" Usually, you will be able to answer these questions by using this book. In a few cases you may need to look in encyclopedias or almanacs. When the chapter is the last one of a unit, some of the questions will ask you to compare the countries.

1. How many continents are there? How many continents are there in the Eastern Hemisphere?

2. Why do some boundaries of countries on the continent of Africa not work well today?

3. What are idea conflicts? What are some of the more important idea conflicts occurring in the Eastern Hemisphere today?

4. Why does solving one problem, such as human disease, often cause another problem? Can you give some examples?

5. Answer the following questions:
   a. Which hemisphere has the most people?
   b. Which hemisphere has the country with the most land?
   c. Which hemisphere has the country with the most people?
   d. Which hemisphere has the Earth's highest mountain?
   e. In which hemisphere do you live?

———————— BE A GEOGRAPHER ————————

Each chapter also has a section called "Be A Geographer." Activities in this section usually require you to look at globes, maps, and charts. Sometimes you will need to make maps or charts. And, the activities often will ask you to use the information you get from these sources.

Study carefully the side of a globe showing the Eastern Hemisphere, or the second drawing of the globe on page 2, or the map on pages 542-543. Answer the following questions:

1. What sea separates Europe and Africa? What sea separates Africa and Asia? What ocean separates Africa and Australia? What ocean is north of Europe and Asia?

2. Which is the largest ocean in the Eastern Hemisphere? Is this the largest ocean on the planet Earth?

3. In the Eastern Hemisphere is there more land north or south of the equator?

4. Can you quickly point to each of the following bodies of water? (a) Baltic Sea, (b) South China Sea, (c) Mediterranean Sea, (d) Yellow Sea, (e) Black Sea, (f) Caspian Sea, (g) Red Sea, (h) Indian Ocean, (i) Sea of Japan.

─────── QUESTIONS TO THINK ABOUT ───────

In each chapter of this book some questions are given that may or may not have absolutely correct answers. Many of these ask you to use what you have learned in a different way. Some of the questions ask you to predict what might happen. For some questions, there will be no clear answer, because no one knows for sure. Some may ask you how to solve some of the world's toughest problems. Some ask for your opinion or for you to choose. The questions have been provided to help you think about our planet, its resources, the people who live on it, and the problems they face. We hope you will enjoy "Questions to Think About."

1. Population experts tell us that if present growth rates continue, the world will have twice as many people by the year 2014. What ideas do you have to double the available food supply?

2. What advantages and what disadvantages are there for a country that has rapid population growth?

3. How did early people probably learn to raise grains and other crops? In what areas of the Eastern Hemisphere do you think farming first took place?

4. If right now you could choose to live any place in the Eastern Hemisphere where would you choose? Why?

5. Which would you rather do? Why?
   a. climb the highest mountain
   b. cross the largest desert
   c. float down the longest river on a raft
   d. travel across the ice on foot to the North Pole

WHAT IS
—THE EASTERN HEMISPHERE?—

Hemisphere, as you will remember, means "half a sphere." A globe can be divided into hemispheres in several ways. The usual way of dividing a globe into hemispheres is to use the *equator* as the dividing line. The **equator** is the line drawn east and west around the globe halfway between the North and the South poles. It divides the globe into the Northern Hemisphere and the Southern Hemisphere.

Parallels, Meridians, and Hemispheres

The equator is a *parallel of latitude*. **Parallels of latitude** are circles that are drawn east and west around the globe. All these lines are drawn so that they are parallel to each other and to the equator. Only one parallel is a **great circle**—a line that divides the globe into two equal parts. Which parallel is a great circle?

Another way of dividing a globe into hemispheres is to use two *meridians of longitude*. **Meridians of longi-** **tude** are lines drawn north and south on a globe between the North Pole and the South Pole. Each meridian is *half* of a great circle. Find the meridians of longitude and the parallels of latitude on a globe.

There are many ways of dividing a globe into hemispheres using the meridians. An easy way is to use 0° Longitude and 180° Longitude as a great circle around the globe. Find these meridians on a globe. You will find that 0° Longitude passes near London, United Kingdom. In 1884 an international conference agreed to draw the meridian 0° where the Royal Greenwich (*grin* ij) Observatory was located. Because it is the meridian from which all other meridians are numbered, it is called the **prime meridian**. Halfway around the globe from the prime meridian you will find 180° Longitude. This meridian through much of its length serves another purpose. Do you know what that purpose is?

These two meridians can be used to divide the globe into hemi-

Large plains can be found on all continents. Some are inhabited by many people. Other plains, such as this one in Kenya, are still in their natural state and have few people living on them. Why do you think this is so?

spheres. It should be easy to figure out how the other meridians can be used for the same purpose. Each other meridian is numbered to tell in degrees how far east or west it is from the prime meridian. Therefore, they are called meridians of east or west longitude. Find the meridians 90° West Longitude and 90° East Longitude on your globe. Together they divide the Earth into hemispheres. Can you figure out what meridian of east longitude would form a great circle with 120° West Longitude?

The terms *Northern Hemisphere* and *Southern Hemisphere* clearly mean half a sphere and no more. The terms *Western Hemisphere* and *Eastern Hemisphere*, however, are used with a slightly different meaning. The term **Western Hemisphere** is used to describe the part of the planet Earth that contains North America and South America. Islands close to these continents are included in the Western Hemisphere. Look again at a globe. You can see that 20° West Longitude and 160° East Longitude might be used to show the half-sphere that is called "Western." However, that half-sphere also includes the islands of New Zealand and part of the continent of Asia. Therefore, the Western Hemisphere is usually used to mean less than half the planet.

The **Eastern Hemisphere**, then, is larger than half the world. It contains the continents of Europe and Asia (sometimes called *Eurasia*), Africa, and Australia. The islands near

What Is the Eastern Hemisphere?

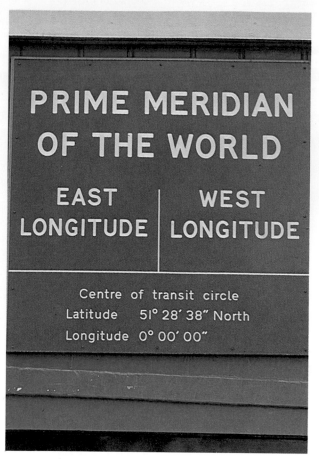

PRIME MERIDIAN
OF THE WORLD

EAST
LONGITUDE | WEST
LONGITUDE

Centre of transit circle
Latitude   51° 28' 38" North
Longitude 0° 00' 00"

Where do you think this marker is located? Why is it located there?

these continents are also part of the Eastern Hemisphere. Eurasia itself extends more than halfway around the world. It extends from about 10° West Longitude eastward to about 170° West Longitude. This distance is 20° more than half a sphere. The coast of western Africa, moreover, extends about 7° farther to the west.

Trying to divide the planet Earth into Eastern and Western Hemispheres for geographic study leaves out one continent. It is the one continent on which almost nothing grows. The only people living there are explorers and scientists. That continent, of course, is *Antarctica*—the land surrounding the South Pole. More than half of Antarctica is in the Eastern Hemisphere. But much of the best-known land in Antarctica is in the Western Hemisphere. For that reason, the study of Antarctica is placed in another book in this series, *The Western Hemisphere*.

REMEMBER, THINK, DO

To answer the following questions, use a globe. If a globe is not available, use the map on pages 542-543.
1. What meridian crosses the greatest amount of land in the Eastern Hemisphere?
2. What meridian crosses the least amount of land in the Eastern Hemisphere?
3. What continent in the Eastern Hemisphere has land farthest to the north?
4. Locate each of the following:
   a. Mount Everest
   b. Dead Sea
   c. The Sahara
   d. Soviet Union
   e. China
   f. Vietnam
   g. Nile River
   h. Mediterranean Sea

Look again at a globe or at the world map on pages 542-543. You will see that most of the land in the Eastern Hemisphere is north of the equator. Much of the land is in *low* and *middle latitudes*. Only a little of the land is in *high latitudes*.

Low latitudes are the areas located north and south of the equator between the Tropic of Cancer and the Tropic of Capricorn. Dashed or broken lines are used on maps and globes to show the Tropic of Cancer (at 23½° North Latitude) and the Tropic of Capricorn (at 23½° South Latitude). In other words, low latitudes extend from 0° Latitude (the equator) to about 23½° North and South Latitude. In low latitudes, the sun at noon is always high in the sky. Although some snow falls on high mountains, most places in low latitudes have warm weather every day of every year. As the map on page 540 shows, some areas in low latitudes are covered with tropical rain forests. In some areas, however, deserts are found.

Middle latitudes in the Northern Hemisphere are the areas between the Tropic of Cancer and the Arctic Circle. In the Southern Hemisphere, middle latitudes are between the Tropic of Capricorn and the Antarctic Circle. As with the Tropics, a dashed line is usually used to mark the Arctic and Antarctic Circles on globes and maps. Middle latitudes extend from about 23½° to about 66½° North and South Latitude. In middle latitudes, the sun is higher in

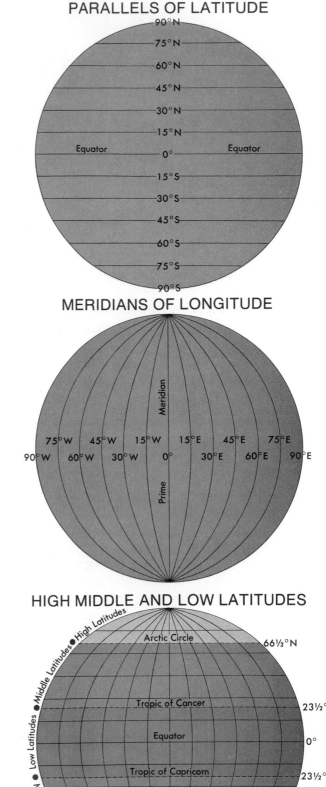

PARALLELS OF LATITUDE

MERIDIANS OF LONGITUDE

HIGH MIDDLE AND LOW LATITUDES

21

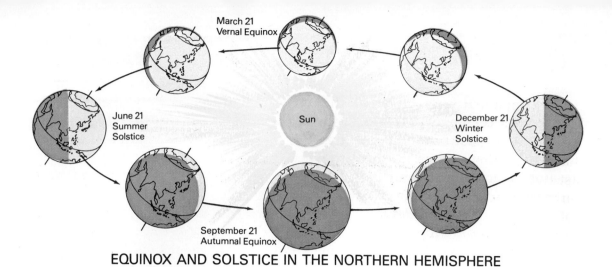

EQUINOX AND SOLSTICE IN THE NORTHERN HEMISPHERE

This chart shows the positions of the Earth in relation to the sun at different times of the year.

the sky during summer months than during the winter. Most places in middle latitudes have four seasons: summer, fall, winter, and spring.

High latitudes are the areas north of the Arctic Circle and south of the Antarctic Circle. High latitudes in the Northern Hemisphere extend from about 66½° North to the North Pole (90° North). In the Southern Hemisphere, high latitudes extend from about 66½° South to the South Pole (90° South).

Everywhere within the high latitudes, for at least one day each year, the sun is not seen above the *horizon*. The **horizon** is where the earth and the sky seem to meet. The nearer to the poles one goes, the more days there are when the sun does not rise above the horizon. At the poles, the sun is not seen for six months.

REMEMBER, THINK, DO

1. Using the map on pages 542-543, name one country in Africa and one country in Asia that is located completely in low latitudes. Be prepared to defend your choice.
2. Name one country in Europe and one in Asia with some land in high latitudes. Be prepared to defend your choice.
3. In the Eastern Hemisphere, does any country south of the equator have land in high latitudes?
4. Check a globe or map to see if the following countries have land in low, middle, and/or high latitudes. Write your answers on a separate sheet of paper. When you have finished, check your work with a friend.

   a. Soviet Union   d. Nigeria   g. Australia
   b. China   e. Sweden   h. France
   c. Israel   f. Indonesia   i. Japan

The **Vernal** (or Spring) **Equinox** (*ee kwih nawks*) occurs in the Northern Hemisphere about March 21. On that day sunlight and darkness each last twelve hours every place on the Earth except at the poles. At both poles, at the time of the Vernal Equinox, the sun can be seen on the horizon all day. For 24 hours the sun appears to follow a path around the skyline but does not rise or set.

The day after the Spring Equinox at the North Pole, the sun again appears to follow a path around the horizon. But now its path is slightly higher in the sky. It can be seen, unless it is a cloudy day, for the entire 24 hours. You can go about 1.6 kilometers (one mile) south from the North Pole and see the sun all day. The area of constant sunlight makes a small circle on the Earth. The North Pole is at the center of this circle. On each day that follows, the path of the sun gets a little higher in the sky. The circle of constant sunlight becomes a little larger. This continues until the **Summer Solstice** (*sawl* stihs) which is usually about June 22. On that day, the circle of 24-hour sunlight covers all the area north of the Arctic Circle.

On the following day this process turns around. Each day the path of the sun is a little lower in the sky. The circle of constant sunlight is smaller. This continues until the **Autumnal** (aw *tuhm* nahl) **Equinox**, which usually occurs in the Northern Hemisphere about September 22. On that day in the Arctic, the sun

is just above the horizon for 24 hours at the North Pole. As far as the sun is concerned, everything seems to be just as it was on the day of the Vernal Equinox. If you had stayed at the North Pole, you would have been in daylight every minute for six months.

Starting on the day after the Autumnal Equinox, the sun does not appear above the horizon at the North Pole for six months. On each day that follows the Autumnal Equinox, the circle without sunlight becomes larger. On or about December 22, the sun is not seen anywhere in the high latitudes north of the Arctic Circle. December 22 is usually the day of the **Winter Solstice**. (You will remember that at the time of the Summer Solstice sunlight covered this same area for 24 hours.)

On the following day, the circle without sunlight in the Arctic begins to get smaller. This continues until about March 21 when the sun again appears above the horizon at the North Pole. If you remained at the North Pole from about September 23 until March 20, you would not see the sun at all. The polar "night" lasts six months. Although the sun never appears above the horizon, it is close to it. The sun casts a faint glow for about a month before and after the equinoxes. During these times, the polar "night" is much like twilight.

At and near the South Pole the same thing happens. It happens, however, at opposite times of the year, March 22 to September 21. When the area within the Arctic Circle is without sunlight (about

23   *What Is the Eastern Hemisphere?*

December 22), sunlight covers all the area within the Antarctic Circle. While the North Pole is having its six-months' "day," the South Pole and the area within the Antarctic Circle is having its six-months' "night."

REMEMBER, THINK, DO

1. What parallel of latitude separates a globe into the Northern and Southern hemispheres?
2. What other lines can be used to divide a globe into hemispheres?
3. Why are the Western and Eastern hemispheres not equal in size as are the Northern and Southern hemispheres?
4. In which of the latitudes is the sun always high in the sky at noon?
5. In which of the latitudes is the sun always low in the sky?
6. When is the Vernal Equinox? What happens then?
7. When is the Autumnal Equinox? What happens then?
8. What happens north of the Arctic Circle at the Summer Solstice? What happens south of the Antarctic Circle on this same day (usually about June 22)?

LOCATING PLACES BY THEIR GRID

Use this map to practice locating places by their grid. Cities have been located at different points on this map. A dot for each city has been placed where lines for a parallel and a meridian cross. Geographers, in describing a place by its grid, give the parallel, or degree of latitude, first. The meridian, or degree of longitude, is given second. Find the degree of latitude and longitude for each of the five cities on the grid.

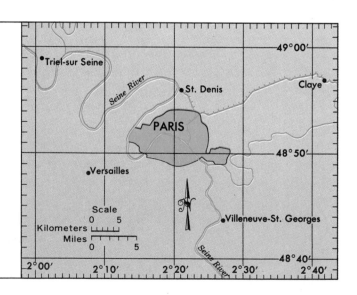

Places on the planet Earth can be located accurately by using parallels and meridians. Usually a place location is given in degrees. For example, find 40° North Latitude (40° N.), 20° East Longitude (20° E.) on the map on pages 8 and 9. If you do it correctly, you will be near the southern border of the country of Albania.

Any place on Earth, the distance between one degree of latitude and the next degree is about 111 kilometers (69 miles). Any place on the parallel 40° North Latitude, for example, is about 4440 kilometers (2,760 miles) north of the equator (111 kilometers × 40 = 4440 kilometers). Degrees of longitude cannot be used in this way, however, to estimate accurately distances on a globe. Study a globe or the map on pages 8 and 9 to figure out why this is true.

Locating places on the Earth can be made even more exact than is possible with only degrees. Degrees can be divided into *minutes*. Each **minute** is one-sixtieth (⅟₆₀) of a degree. A place can, therefore, be accurately located using degrees and minutes. The abbreviation for minutes is a single mark like this ('). Thus, the Tropic of Cancer is drawn at 23°27' N. Paris, France, is located at 48°52' N. and 2°20' E. (See map, page 24.) For even greater accuracy, each minute can be divided into seconds. We won't use seconds this year.

REMEMBER, THINK, DO

Work with a friend to do the following:

1. Look at the map of Eurasia on pages 8 and 9. Find each of the following cities:
   a. Paris, France
   b. Moscow, U.S.S.R.
   c. Jerusalem, Israel
   d. New Delhi, India
   e. Peking, China
   f. Tokyo, Japan
   g. Hanoi, Vietnam
   h. Jakarta, Indonesia

2. Locate as nearly as possible each of the above cities by latitude and longitude. Write your answers on a sheet of paper.

3. Check your figures by doing it again for each of the cities, using maps on pages 87, 260-261, 311, 445, 464, and 493. On your sheet of paper make a second column. Place your new figures in this column. How close are the figures? Which maps are easier to use? Why?

4. Finally, check your work by using an almanac or an atlas that gives locations of cities. Place these figures in a third column on your sheet of paper.

5. What have you learned about figuring locations of cities? What do you think geographers do when they want to know the location of a city?

Time Zones

We have learned that meridians can be used in locating places on Earth. Meridians can also be used in figuring what time it is at different places. The Earth rotates once, or 360°, every 24 hours. It therefore rotates 15° in one hour (360° ÷ 24 = 15°). On most globes a meridian is drawn every 15° from the prime meridian. **Time zones** extend about 7½° east and 7½° west of these meridians. On globes with meridians drawn every 15°, the time along any meridian is one hour different from the time along the meridians on each side.

In some areas, time-zone boundaries have been changed because of the locations of state or country boundaries. A map showing time zones is on page 27.

The International Date Line

The meridian of longitude at 180° is used along much of its length as the *International Date Line*. Midnight at the **International Date Line** marks the end of one calendar day and the beginning of a new one. As the Earth rotates from west to east, the new calendar day begins one hour later at 165° East Longitude. It begins two hours later at 150° East Longitude. Twelve hours later, that new calendar day starts along the prime meridian. And 24 hours later, at the International Date Line, another new calendar day begins.

The International Date Line does not follow the meridian at 180° throughout its entire length. North of the equator, the line is bent eastward so that all of Asia has the same day. The line also is bent westward so that all of North America has the same day. South of the equator the International Date Line is again bent eastward so that islands east of New Zealand have the same day.

Directions on the Earth

Places on the Earth can also be located by using compass directions from a known place. One can say that Maria Garcia's house is two blocks north and one block west of the school. Or, one can say that San Francisco is about 3700 kilometers (2,300 miles) west of Washington, D.C. To locate places fairly accurately by compass direction, 16 different directions are used. These directions and their abbreviations are shown in the picture on this page. Notice that northwest is halfway between north and west. Notice also that west-northwest is halfway between west and northwest. If you do not know the 16 compass directions you should learn them. You will be using them as you read this book.

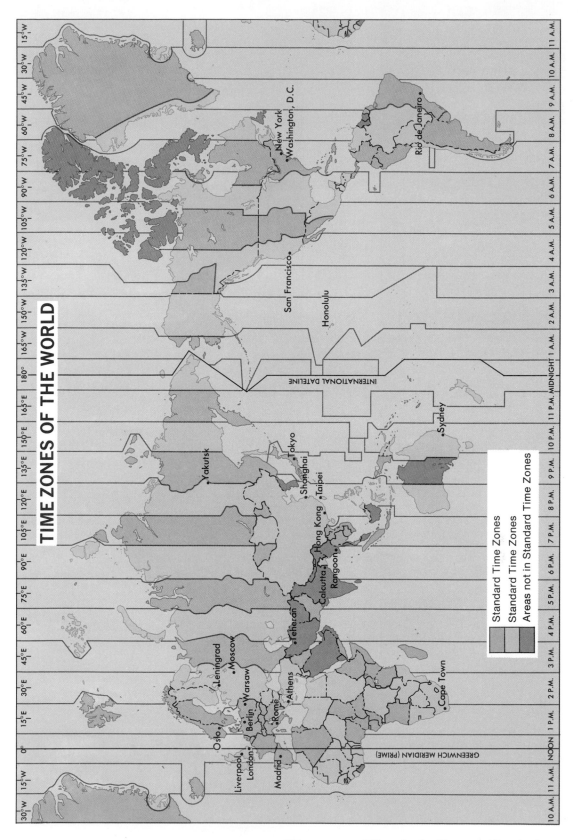

TIME ZONES OF THE WORLD

Standard Time Zones
Standard Time Zones
Areas not in Standard Time Zones

GREENWICH MERIDIAN (PRIME)

INTERNATIONAL DATELINE

10 A.M. | 11 A.M. | NOON | 1 P.M. | 2 P.M. | 3 P.M. | 4 P.M. | 5 P.M. | 6 P.M. | 7 P.M. | 8 P.M. | 9 P.M. | 10 P.M. | 11 P.M. | MIDNIGHT | 1 A.M. | 2 A.M. | 3 A.M. | 4 A.M. | 5 A.M. | 6 A.M. | 7 A.M. | 8 A.M. | 9 A.M. | 10 A.M. | 11 A.M.

What Is the Eastern Hemisphere?

Ice and Snow

Mixed Forest

Desert

Savanna

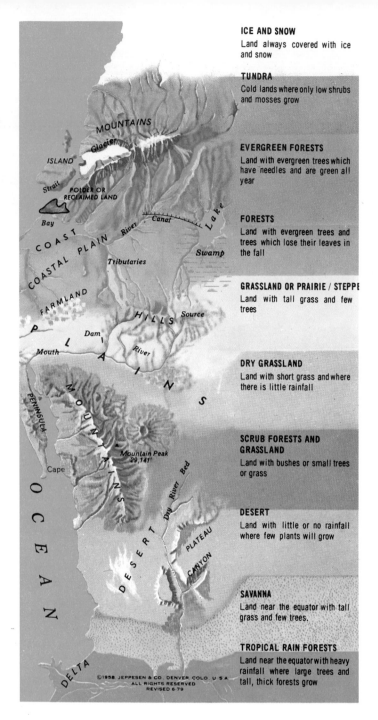

ICE AND SNOW
Land always covered with ice and snow

TUNDRA
Cold lands where only low shrubs and mosses grow

EVERGREEN FORESTS
Land with evergreen trees which have needles and are green all year

FORESTS
Land with evergreen trees and trees which lose their leaves in the fall

GRASSLAND OR PRAIRIE / STEPPE
Land with tall grass and few trees

DRY GRASSLAND
Land with short grass and where there is little rainfall

SCRUB FORESTS AND GRASSLAND
Land with bushes or small trees or grass

DESERT
Land with little or no rainfall where few plants will grow

SAVANNA
Land near the equator with tall grass and few trees.

TROPICAL RAIN FORESTS
Land near the equator with heavy rainfall where large trees and tall, thick forests grow

©1958 JEPPESEN & CO. DENVER COLO. U S A
ALL RIGHTS RESERVED
REVISED 6-79

LEGEND

LONDON .... over 2,000,000	All other cities .... •
Kinshasa .... 500,000 to 2,000,000	Railroad .... ——|——|——
Patna........................... under 500,000	Canal .... —⊥—⊥—⊥—
National capitals .... ⊛	Undefined boundary .... — — — — —
State or colony capitals .... ⊛	

Those who make maps have a hard problem. They try to show on a flat surface information about a round world. That is not easy—especially if the map is showing a large area like the Eastern Hemisphere. Maps can be made to show many kinds of information. Some maps are made to show mainly what grows on the land during summer months. Some maps show the kinds of soil at different places. Some maps show the height of the land, including mountains. Some maps show **prevailing** (or usual) wind direction. Some maps show average temperatures for a period of time such as during the summer months. Some maps show average rainfall for a long period of time. Some maps show what the weather will proba-bly be like tomorrow. You may have seen these maps on television news programs. Maps can show where people live and about how many people live there. They can show what languages are spoken. And they can help you figure out time differences any place on the planet. Maps can show many, many kinds of information. They are very useful.

Until you look at a *map legend*, however, you cannot tell for sure what any map shows. A **map legend** tells what symbols or marks or colors the map maker used and what they stand for. Look at pages 28 and 44. These pages show legends that can be used with many of the maps in this book. Turn to these pages often as you look at maps about the Eastern Hemisphere. They will help you read the colors and the symbols on the maps. They will help you get meaning from the maps.

REMEMBER, THINK, DO

Use the Time Zones Map on page 27 or a globe to answer the following:
1. If it is 12 noon in London, United Kingdom, what time is it in Rome, Italy?
2. If it is 12 noon in London, United Kingdom, what time is it in Washington, D.C.?
3. If it is 8:00 P.M. in Hong Kong on Tuesday, what day is it in San Francisco, California?
4. If it is 10:00 A.M., Tuesday, in Sydney, Australia, what time and what day is it in Rio de Janeiro, Brazil?
5. If it is 6:00 A.M., Saturday, in Moscow, U.S.S.R., what time and day is it in Washington, D.C.?
6. Make up several other time problems and give them to a friend. Be sure you know the right answer so that you can check your friend's work.

LEARN MORE ABOUT
THE EASTERN HEMISPHERE

You should now be able to tell more about the Eastern Hemisphere and about Earth-Sun relations. Remember, the Eastern Hemisphere is larger than half the planet Earth. It has land in low, middle, and high latitudes. And the Eastern Hemisphere has land in 14 time zones.

You may wish to do some things on your own to learn more about the hemispheres. Or you may want to know more about the grid system (see map and caption, page 24) that is used by geographers. Here are some ideas that might help:

1. In your school or public library, locate a volume of an encyclopedia that treats the subject of MAPS. Read the article. If you have trouble reading or understanding some parts of the article, ask a parent or a teacher for help.

2. Ask the school librarian for help in finding books about—

   a. countries in the Eastern Hemisphere
   b. globes and maps
   c. people in the Eastern Hemisphere

3. Look in an atlas, such as *Goode's World Atlas,* and see how many different kinds of information can be mapped. Note the differences among areas of the Eastern Hemisphere. Especially note the map showing population density (*i.e.,* how many people live where). What problems are created when people are not evenly distributed across the land? What are the advantages?

As you have learned already, there are many countries in the Eastern Hemisphere. Not all of them can be presented in this book. Choices had to be made. The authors have chosen to give you detailed information about a few countries in each major region. Then they have written a little about other countries in the region. The authors hope that you will want to learn some more about the other countries. There are several ways that this can be done:

1. You can study on your own. Some of you may wish to do a special report on the other countries for extra credit.

2. Small groups can be formed to learn about the other countries. Reports can be given to the whole class.

3. Larger groups can be formed to learn about *some* of the other countries. Choices will need to be made about which countries to study. One good way to choose would be to find out where the ancestors of class members came from. Do special large-group studies of those countries from which the ancestors of several students came. Have several small groups and individuals study the other countries.

A Study Guide

To help you think about and begin your study of other countries, this Study Guide has been prepared. Use these questions to help you learn more about the Eastern Hemisphere.

Learn about—

1. The location and physical features of the country
   a. Where is it located? Is its location unusual? Why?
   b. What are the major physical features of the land?
   c. Are there any important bodies of water in this land?
   d. What type of climate and what seasons does it have?
2. The history of the people of this country
   a. Who are some of the important people, past and present?
   b. What are some of the important past events?
   c. What are some of the important achievements of the people?
   d. How did this nation become independent?
3. How the people live on the land
   a. In what part of the country do most of the people live?
   b. Why do the people live in these areas?
   c. How do the people earn a living?
   d. How much of the land is farmed?
   e. What farming methods are used by the farmers?
   f. What are the major crops grown?
   g. Are the farmers able to grow enough food for all the people?
4. The mineral resources of the country
   a. What are the known mineral resources? Where are they located?
   b. Which minerals are currently being mined?
   c. Are some of the country's minerals not being mined? Why?
   d. What is the country doing to help conserve its natural resources?
5. The industrial development of the country
   a. Where are the industrial centers located?
   b. What kinds of manufacturing are done?
   c. What types of manufactured items are imported? Exported?
   d. What natural resources are imported? Exported?
   e. What types of pollution problems exist?
   f. What is the government doing to stop pollution?
6. The transportation system of the country
   a. How do people and goods move within the country?
   b. How do people and goods move from the country to other countries?
   c. Does the country have any unusual transportation problems?
7. The daily life of the people
   a. What health facilities are available to the people?
   b. What type of education is available to the people?
   c. How do the people entertain themselves?
   d. What are some of the special holidays and customs?
   e. How are the people governed?

UNIT I REVIEW

WHAT HAVE YOU LEARNED?

1. Why is the Eastern Hemisphere larger than half a sphere?

2. Where are low latitudes, middle latitudes, and high latitudes on Earth?

3. How long is the sun not seen at the Poles?

4. Where on Earth is the sun always high in the sky at noon?

5. What ways can be used to locate places on Earth accurately?

6. Why is the prime meridian of special importance in locating places on Earth?

7. What parallel of latitude serves a purpose similar to the one served by the prime meridian?

8. What happens every midnight at the International Date Line?

9. What were some early accomplishments or inventions of people living in the Eastern Hemisphere?

BE A GEOGRAPHER

With a friend or two look carefully at the Eastern Hemisphere on a globe. Be prepared to answer the following questions:

1. What country has the Baltic Sea to the west and the Black Sea, Caspian Sea, and the Aral Sea near its southern border? What ocean is north of this country? What seas are east of this country?

2. What country has land the farthest north on the mainland (not an island)?

3. Find three of the world's larger islands: New Guinea, Borneo, and Madagascar. Be prepared to show where each of them is located.

4. Which of the four continents in the Eastern Hemisphere appears to have the most land? The least land?

5. What is the largest country that you can see in the Eastern Hemisphere?

6. What two continents of the Eastern Hemisphere have all their land north of the equator? One continent in the Eastern Hemisphere has land both north and south of the equator. What is it? What inhabited continent in the Eastern Hemisphere has all of its land south of the equator?

7. Could you sail by water from London, United Kingdom, to Sydney, Australia? How? Plan several routes.

8. Make up a question about the Eastern Hemisphere that can be answered by looking at a globe. Be prepared to write it on the board or a chart for other groups to answer.

—————— QUESTIONS TO THINK ABOUT ——————

1. Why, if the sun shines for six months, is there a lot of ice near the poles?

2. What do you think would happen if suddenly, or over a period of time, the axis of the Earth changed so that it was not tipped 23½°?

3. The Royal Observatory is no longer located in Greenwich. What difference did it make on globes when it moved? What difference did it make in figuring time?

4. What do you think it would be like not to see the sun at all for several months? Would you like to live in such a place? Why do you suppose people do live north of the Arctic Circle?

5. How would you describe the approximate location of your house?

6. What direction is up? Down? North? South? East? West? Think up good definitions for each of the six main directions.

Open-air markets, such as this one in Otley, United Kingdom, are common sights throughout Europe.

Unit II

EUROPE

Unit Introduction

Chapter 3
WHY STUDY THE CONTINENT OF EUROPE?

Unit Preview

LOOK

- Could this photo have been taken in the state where you live? Why or why not?

- Make a list of things shown in the photo that seem unusual to you. (There may not be any, of course.)

- Make another list of things shown that you would expect to find in the largest city in your state.

GUESS

- Where was the photo taken? Why did you answer as you did?

- What season of the year was it when the photo was taken?

- How many people probably live in this metropolitan area?

DO YOU KNOW?

- Did any of your ancestors come from Europe?

- Are cities like London and Paris farther north or farther south than where you live?

- Was Columbus the first European to reach the "New World"?

- Why do countries in Europe, even though some are small, seem to have so much influence in world affairs?

Unit Introduction

Europe

Europe is part of the largest land mass on Earth. This large land mass is called **Eurasia**. It is made up of both Europe and Asia. Find it on the globe. A map of Eurasia is given on pages 8 and 9. As you already know, some geographers do not think of Europe as a continent. Whether Europe is—or is not—a separate continent is not really very important. What is important is where it is and what it is. In this book, figures about Europe are usually given as if Europe is a separate continent.

On the map on pages 8 and 9 find the Union of Soviet Socialist Republics. It is the very large country that extends across most of the top of the map. The U.S.S.R. extends from the Baltic (*bawl* tihk) Sea on the west to the Bering (*beer* ing) Sea on the east. It extends from about 20° E. to about 170° W. Most geographers do not include the U.S.S.R. when giving information about Europe. However, the part of the U.S.S.R. west of the Ural (*yur* uhl) Mountains is considered as part of Europe. You will find the Ural Mountains on the map on pages 8 and 9 at about 60° N. 60° E.

In this book, Europe includes the large *peninsula* (puh *nihn* suh luh) west of the U.S.S.R. A **peninsula** is land almost surrounded by water. Also included are the islands near this peninsula.

Europe is one of the most industrialized regions of the world. However, many old customs have been kept. Cheese merchants in The Netherlands still wear the traditional clothing.

Things You Might Like To Do As You Study About Europe

1. Prepare an outline map of Europe on a large sheet of paper. If one is available, a projector can be used in preparing such a map. Then, as you learn about Europe, place on it:

    a. names of countries

    b. names of seas and rivers

    c. names of capital cities and other important cities

    d. shading to show lowland and highland areas

2. Form a group and prepare a report on some topic of interest, such as one of the following:

    a. The Industrial Revolution

    b. Queen Elizabeth II and the Royal Family

    c. Living North of the Arctic Circle

    d. Famous European Explorers

    e. The Languages of Europe

    f. The North Sea: Extraordinary Oilfield

3. Collect and post on a bulletin board important news stories about Europe.

4. Interview several people who were born in Europe. Find out from them what they remember about their childhoods. Ask if they have returned to Europe recently. Find out from them if any changes have taken place in their homelands.

5. Find out all you can about NATO—the North Atlantic Treaty Organization. Do you believe such organizations are worthwhile? Why or why not?

❧ 3 ❧
Chapter

WHY STUDY THE CONTINENT ——————OF EUROPE?——————

Europe and its people are of interest to us for several reasons. Almost 500 years ago, Europeans "discovered" North and South America. Of course, people already lived in the Americas when the Europeans first arrived. The Europeans were looking for a shorter way to India. When they arrived they thought they had arrived in India. So they called the people they found here "Indians." In time many people came from Europe to North and South America. They settled on the land. They began to use the continents' natural resources.

Most of the people living today in North America have ancestors who came from Europe. Some of us, of course, have ancestors who were already here when the first Europeans came. Some of us have ancestors from Africa, Asia, South America, or Australia. Some of us have ancestors who came from islands in the Caribbean Sea. And some of these people had ancestors who had come from Europe. Where did your ancestors come from? Ask your parents or grandparents.

By 1900 Europeans had spread their languages and their learning over much of the world. They had also spread their ideas of government and their ways of living. The language used in this book came from Europe. Most of our ideas of government can be traced back to Europe as can many of our ways of living. Of course the reverse is also true. American ways of doing things have changed some ways of living in Europe, too. This is especially true today.

The Population of Europe

For its size Europe has more people living on it than any other continent. Asia has many more people, but Asia has much more land. Europe (including the U.S.S.R.) has about 65 people for each square kilometer (168 per square mile) of land. Asia (including the U.S.S.R.) has about 57 people per km² (148 per sq. mi.). For the whole world, the average is about 29 people per km² (75 per sq. mi.). The United States has about 24 persons per km² (62 per sq. mi.).

The bagpipe is played by people from many parts of Europe and Asia. It has been used for hundreds of years.

Canada, however, has only about 2 people per km² (5 per sq. mi.). The graph below shows how many people per square kilometer live on each continent.

Perhaps you are wondering why so many people live in Europe. There is no simple answer to this question. Part of the answer may be the location of Europe. Part may be the climates. And part may be the natural resources. People tend to live where they can be comfortable and make a good living.

The Location of Europe

One reason why Europe has been so important in the world is its location. Europe is close to the large continents of Africa and Asia. In addition, Europe is surrounded on

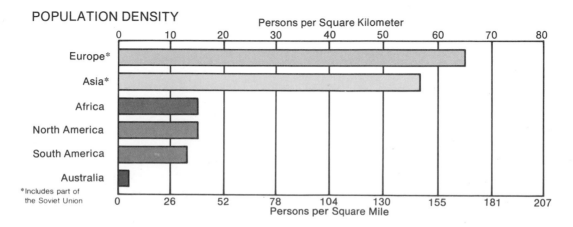

POPULATION DENSITY

Europe

three sides by the sea. Oceangoing ships can sail to within about 500 kilometers (310 miles) of any spot in Europe. The area also has many rivers. And people have built canals to connect many of these rivers. Partly as a result of its location, Europe has become a great center for world trade.

Most of Europe is located at about the same latitudes as eastern Canada. Use a globe to check and make sure that this statement is true. Most parts of eastern Canada and southern Alaska have very cold winters. Not much of the land can be used for farming during summer months. What is Europe's climate and land like at these same latitudes? Can much of Europe's land be used for farming?

The Climates of Europe

Although Europe is a rather small area, it has several different climates. Much of Northwestern Europe has what is called a *mid-latitude marine* (muh *reen*) climate. The term **mid-latitude** is used because it is a kind of climate that occurs in the middle latitudes. **Marine** means having to do with the ocean or sea. A marine climate is a **humid** climate—one that is moist or damp. A marine climate is usually found near the coast on the western sides of continents. Such climates occur where warm ocean currents flow close to the shores.

Northwestern Europe has a mid-latitude marine climate because of the **North Atlantic Current**. This is

The land of Europe is crowded in many areas. There is little room to raise cattle. As a result there are more goats and sheep raised in Europe than cattle.

an ocean current that carries warm water across the North Atlantic Ocean from the Gulf of Mexico. The North Atlantic Current flows northeast past the United Kingdom. It flows along the western coast of Norway. Winds in this area—the prevailing winds—blow rather steadily from the west. They blow over the waters that are warmed by the North Atlantic Current. Thus, these winds bring warm, moist air to much of Northwestern Europe.

Farther eastward in Europe, the ocean winds have less effect upon the temperatures. The winters in Central Europe, therefore, are colder than winters near the coast. As one goes farther northward, winters are also colder. The summers in Central Europe are hotter, too. Central Europe has a climate much like that in the eastern part of the North Central States in the United States. This type of climate is called **humid continental**. Usually enough rain falls so that crops can be grown. Once in a while, however, there are long dry spells called **droughts** (drouts).

Southern Europe has a *Mediterranean* climate. In a **Mediterranean** climate, less rain falls during the year than in the other climates mentioned so far. Most of the rain comes during the fall and winter months. Summers are long, hot, and dry. Crops have to be **irrigated**, or watered. Winters are mild, with some snow falling in the high mountains. Most days are sunny and bright. Southern California in the United States has a Mediterranean climate.

Parts of Central and Southern

REGIONS OF EUROPE

The continent of Europe in this book has been divided into three major regions. Each region will be studied in a separate unit.

Europe have a **highland** or **mountain** climate. This climate depends largely upon how high the land is above sea level. High in the Alps, the climate is cold most of the year. Snow and ice cover the highest mountains during both winter and summer. On many of the lower slopes of the mountains, the climate is much milder. Find the Alps on the map on page 39.

Parts of Northern Europe have a *subarctic* climate. In a **subarctic** climate, like that found in northern Norway, Sweden (*swee* d'n), and Finland, winters are long and cold. Summers are cool. Only a few hardy crops are grown. Most of the land, as in large areas of Canada, is covered by evergreen *coniferous* forests. **Coniferous** trees, such as pine or spruce trees, are ones that bear cones.

Europe

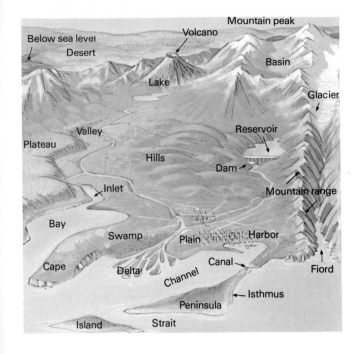

North of the areas having a subarctic climate, the land is frozen most of the year. Only mosses, coarse grass, and a few dwarfed bushes and trees can grow. The climate of these areas is called **tundra**. Where in the United States is that kind of climate found?

The Land of Europe

Europe has many different kinds of landforms. *Mountains*, *plateaus*, and *plains* are among the kinds of landforms found in Europe. Do you know what all those words mean? A **mountain** is a high, often rugged, and usually steep land mass. A **plateau** is high, rather flat land, which may have one edge that drops quite steeply to lower land. A **plain** is level land usually at low altitudes.

The mountains in Europe were created at three different times. The oldest mountains are located in the United Kingdom, Ireland, and Norway. For millions of years these old mountains have been worn down by ice, water, and wind. This process is called **erosion** (ee *roh* zhun). During the Ice Ages these mountains were scraped and rounded by *glaciers*. **Glaciers** are large rivers of ice that move slowly over the land. Now many of the old mountains are hills.

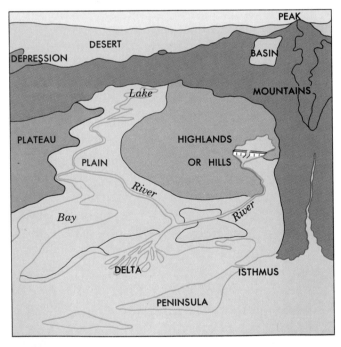

Landforms

	Below sea level
	Basins and plains
	Hills
	Plateaus
	Mountains

The picture at the top of this page shows many of the landforms that will be discussed in this book. The map to the left is a Landform map of the picture above. The map is like the ones that are used throughout the book. Study the picture and the map so that you will be able to read the Landform maps that are in the book.

The youngest mountains in Europe are the Alps and the Pyrenees (*pihr* uh neez). Find these mountains on the map on pages 200-201. These mountains are high and rocky. They have not been worn down by erosion as much as the other mountains.

The mountains, especially the youngest ones, divide Europe into *natural regions*. **Natural regions** are areas where the land, climate, and resources are much alike. The mountains have also been boundaries for countries for a long time. The Pyrenees Mountains, for instance, divide Spain and France. The Alps separate Italy from France, Switzerland, Austria (*aws* trih uh), and Yugoslavia (yoo go *slahv* ee uh). Find all these countries on the map on page 39.

In earlier days the mountains in Europe made travel between peoples difficult. Today, it is fairly easy to go by automobile from one country to another through mountain **passes**, low places in mountain ranges. The Pyrenees Mountains, however, do not have good passes. Roads have been built around them near both coasts. Of course airplanes fly easily over any of the mountains. Railroad tunnels have been cut through many of the highest ranges. The longest one in Europe is the Simplon (*sihm* plahn) Tunnel. Cut under the Alps between Switzerland and Italy, this tunnel is 19.8 kilometers (12.3 miles) long.

In many ways, the most important part of Europe is the *North European Plain*. This plain reaches from south-

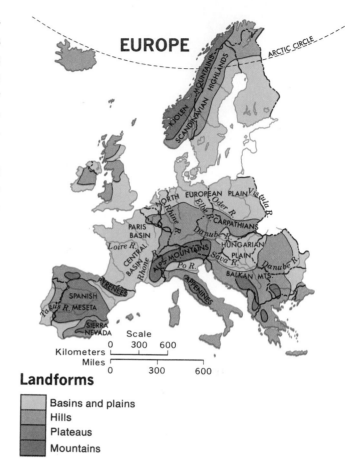

EUROPE

Landforms

	Basins and plains
	Hills
	Plateaus
	Mountains

western France northeastward into the U.S.S.R. See the map on pages 8 and 9. This plain includes much of the land of France, Belgium (*bell* jum), The Netherlands, West Germany, Denmark, East Germany, and Poland. Parts of England and Sweden are also a part of this plain. Find all these countries on the map on page 39. Most of the land in this plain is gently rolling, not flat. The plain provides much of Europe's good farmland. Many of the largest cities of Europe are located on this plain. Many of these cities are seaports or river ports, and serve as main centers for trade and transportation.

45

On January 19, 1979, the European Parliament held its first meeting in the new Palace of Europe located in Strasbourg, France.

A United States of Europe?

In 1957 six nations of Europe agreed to join what they called the *European Economic Community*. Later, they agreed to other ways of working together and formed the *European Community*. Sometimes the European Community is referred to as the *European Common Market*. The six nations that started the European Community were Belgium, France, Italy, Luxembourg (*luhk* sehm burg), The Netherlands, and West Germany. Find them all on the map on page 39. In 1973 three other nations joined the European Community. They were Denmark, Ireland, and the United Kingdom. Greece joined in 1979.

Together these nations are working toward a single economic system. They have stopped all **tariffs**, or taxes on goods, among members of the Community. They have fixed tariffs on all goods imported from nonmembers. They hope soon to have a common system of money. At present each nation has its own money system. The British, for instance, use the pound as the basis for their system. The French use the franc; the West Germans, the mark; and the Italians, the lira.

The Community has been very successful! Together these nations import and export more than any country in the world, including the United States of America. Together they produce more steel and more automobiles than the United States of America. They are trying to work together like a United States of Europe.

To do their work, the European Community has three branches of government. A Commission of 13 members and a Council of Ministers together serve as the *executive* branch. The Commission proposes needed laws. The Council of Ministers accepts or rejects the proposals. Before 1979, the members of the European Parliament were appointed by the parliaments of the member nations. Now, however, members are elected by the people in each of the nations. It serves as the *legislative* branch—debating proposals for action. It can, by a two-thirds (⅔) vote of its members, remove the members of the Commission. The nine-member Court of Justice serves as the *judicial* branch. It is the Supreme Court for the European Community. As in our country, decisions of the Court of Justice are final.

Will the European Community become a United States of Europe? In many ways the Community is already acting like a group of united states. Each nation has kept its own language and customs. But each has also given up some of its power for the common good. This is like the states in the United States of America and the provinces in Canada. The members of the Community have not yet worked out a common money system, however.

A Look Ahead

In this book Europe is divided into three main sections: Northwestern Europe, Central Europe, and Southern Europe. A unit is devoted to each of these sections. Unit III, on Northwestern Europe, includes three chapters that highlight the United Kingdom, France, and Sweden. Unit IV, on Central Europe, has four chapters that give considerable detail about West Germany, Switzerland, Poland, and Romania. Unit V, on Southern Europe, also has four chapters. They provide information about Spain, Italy, Yugoslavia, and Greece. At the beginning of each unit on Europe, there is a brief introduction to the region. At the end of the final chapter of each unit, brief descriptions are given about other countries in the region. Each chapter and unit ends with questions to help you review what you have learned, activities and questions to help build map and globe reading skills, and "thought" questions to help you to think in new, different ways about the land and people of the region.

UNIT II REVIEW

———— WHAT HAVE YOU LEARNED? ————

1. Is Europe a continent? Why or why not?

2. Which continent has the most people? Which continent has the most people for its size?

3. Why did Europeans call the people they found in North and South America "Indians"?

4. Why do people live where they do on Earth?

5. What surrounds Europe on three sides? Why is that important to the people living in Europe?

6. What is a natural region? In what natural region do you live?

———— BE A GEOGRAPHER ————

1. Ask five people (from different families) near where you live where their ancestors came from. Make a class chart that shows what you learned.

2. Look in a daily newspaper for weather information. Many papers provide temperatures for cities around the world. Compare the high and low temperatures where you live with the temperatures of London, Paris, Stockholm, and Rome.

3. What kind of climate do you have where you live?

4. What ocean is north of Europe? What ocean is west of Europe? What large sea is south of Europe? Do you live near an ocean or sea? If you answer yes, what difference does that make in the way you live? If you answer no, what differences might there be in the way you live if you did live near a sea or ocean?

5. Find a map in your classroom or the library that shows where the North Atlantic Current flows. Look at other currents in the oceans of the world that are shown on maps. What marks are used to show ocean currents?

QUESTIONS TO THINK ABOUT

1. If you had your choice, what kind of climate would you prefer to have where you live? Why?

2. What differences might it make in your living habits if you lived where the sun is not seen for two or three months in the winter? If you lived where the sun is above the horizon for only two or three months in summer?

3. Does the state where you live have more or fewer people per km² (or per sq. mi.) than Europe? One nation in Europe, The Netherlands, has about 342 people per km² (886 per sq. mi.). Does the state where you live have more or fewer people for its area? Why are some areas more crowded with people than others?

4. Do you live in an area that is mountainous, or on a plateau, or on a plain? In which of these three kinds of land do you think most people on Earth live? Why do you think most people live there?

5. Have you ever seen a glacier? What did it look like? Did it appear to be moving?

6. Are there any signs of erosion where you live? Is erosion good or bad? How can you tell?

Although not as common as they used to be, reindeer caravans can still be seen in Lapland hauling supplies for the people who live in this region.

Unit Review

The people of Europe enjoy and use flowers on many occasions. Flower markets are a common sight throughout this region.

Unit

NORTHWESTERN EUROPE

Unit Preview

Pictures on this page were taken in London, United Kingdom; Paris, France; and Stockholm, Sweden. Try to figure out where each picture was taken.

- How do the nations of Northwestern Europe compare in size to states in the United States?

- Why are many European students able to speak three—or even four—different languages?

- Why is the climate of the British Isles so mild for a nation so far north?

- Why does France attract so many tourists each year?

- Why is Norway often called "The Land of the Midnight Sun"?

Unit Introduction

Northwestern Europe

Northwestern Europe, in this book, includes the countries in northern and western Europe. These are the European countries whose climates are most influenced by the North Atlantic Current. Three of these countries are described in some detail: the United Kingdom, France, and Sweden. You may wish to learn more about the other countries in Northwestern Europe. They are Belgium, Denmark, Finland, Ireland, Luxembourg, The Netherlands, and Norway. Iceland, an island-nation in the North Atlantic Ocean, once belonged to Denmark. It is usually considered to be a part of Northwestern Europe, also.

For most Americans, one of the most striking things about Northwestern Europe is the small size of the countries. France, the largest country, has much less land than the state of Texas. Luxembourg, the smallest country in Northwestern Europe, is smaller than Rhode Island!

Another great difference between our country and Northwestern Europe is language. Many languages are spoken in the United States. But English is the main language, and almost everyone uses it. Of course, in the United States (or Canada), English may be the second language for many people.

Northwestern Europe contains the world's most densely populated country, The Netherlands. Rotterdam is one of its large cities.

In Northwestern Europe, though, many different languages are used. Each country usually has its own language. Because of that, one of three languages is used as a second language by many people. English, French, and German are taught in almost all the schools. If you grow up in Sweden, for instance, you first learn Swedish. Then, beginning in the fourth grade, you would probably learn English. In the seventh grade you would learn either French or German. By the time you finished high school, you could communicate in any of the three languages. And you might be able to speak all four! Of course, you would probably be best in Swedish. That is only natural.

Things You Might Like To Do As You Study
About Northwestern Europe

Many learning activities are given later in this chapter to help you learn more about Northwestern Europe. You may wish to do some of the following as you begin your study:

1. Make a large outline map of Northwestern Europe. Place on it only the names of each country and the location and name of its capital.

Add important information as you continue your study.

2. Separate into committees to do the following projects.

a. One committee might choose to learn about the climates in the three countries presented in detail in this book. Other members of the committee might like to find out about one or two other countries, like Iceland or The Netherlands.

b. Another group might like to learn about the land and how it is used.

c. A third committee might like to study industries and products.

d. A fourth group might like to learn about the peoples and their customs, including special holidays.

e. A fifth group might wish to study transportation.

In each case, the group should be prepared to share with the whole class what its members have learned. Each person in each group should use the library and people living in the neighborhood to learn more than is included in this book.

3. Keep an up-to-date bulletin board about the Eastern Hemisphere. Try to get articles mainly about Northwestern Europe.

4. Invite a person who was born in Northwestern Europe to come visit the class and tell about his or her childhood. Almost every community has someone who would be glad to share memories.

5. Look in magazines for ads that try to get Americans to visit Northwestern European countries. Make a display of these ads. Think about which country seems to be the most interesting to you. Plan a trip to that country. Airlines and travel agencies may have some materials you could use to plan your trip.

6. Perhaps you already know how to speak French or Swedish. If not, learn how to say "Good Morning," "Goodbye," "Thank you," and some other common phrases in these languages.

Northwestern Europe has much open land and numerous farms. This farm is located in northern England.

Chapter

——THE UNITED KINGDOM——

Look at the map on page 39 and locate France. The map shows two large islands and many smaller islands northwest of France. Together these are usually called the *British Isles*. The two large islands are Great Britain and Ireland. At one time, all of Ireland was ruled by Great Britain. Now, most of the island has become the independent Republic of Ireland. Northern Ireland, however, is united with Great Britain. The official name of this nation is the *United Kingdom of Great Britain and Northern Ireland*. The name is so long that the country is usually called the United Kingdom or, sometimes, Great Britain.

The island of Great Britain is divided into three parts: England, Scotland, and Wales. People from these areas are known as the English, the Scottish or Scots, and the Welsh. Grouped together, they are known as the British. The present form of the government of Great Britain was first established in England. Scotland, Wales, and Ireland at different times in history were brought under the control of En-

gland. To this day, however, the people in Wales, Scotland, and Northern Ireland think of themselves as different from the English. At times some Scots and Welsh have talked about becoming separate nations. It is hard to predict what will happen in this matter. Probably, some way of giving Wales and Scotland more independence within the United Kingdom will be found. Both areas may, in time, have their own **parliaments**, or lawmaking bodies.

The United Kingdom is the European country whose history is most closely related to the history of the United States. The 13 colonies along the North American Atlantic coast, that became the United States of America, were British colonies. They were first settled mainly by people from England and were also ruled by England. In time the American colonists began to object to the way England ruled. The colonists united to fight against England. Led by George Washington, they fought the War for Independence, or the American Revolution. The colonists won the war. They became self-governing and formed the United States of America.

The United States of America and the United Kingdom have had friendly relations for more than 150 years. Today, no European people have closer ties with the people of the United States.

The People and History of the United Kingdom

The ancestors of the people of the United Kingdom came from many parts of the mainland of Europe. The first known invaders of the islands was a tribe called the *Celts* (selts). The Celts were fierce fighters. They successfully invaded England and ruled it from about 600 B.C. until about the year 55 B.C. Very little is known about the Celts. We know, however, that they lived in different tribes and worshipped gods of nature.

In 55 B.C., people from the Roman Empire invaded the British Isles. The Romans had fine armies. They had already conquered large parts of southern and western Europe and parts of central Europe. The Romans never completely conquered the Celts, however. They finally had to build a wall across northern Great Britain to protect themselves against the Celts.

The Romans were in the British Isles for nearly 400 years. When they left, tribes from north-central Europe invaded Britain. The *Angles* and the *Saxons* were the most important. These people became the main ancestors of the British people of

This cottage was the home of William Brewster in Scrooby, United Kingdom. From here he led a group of people who eventually founded a colony at Plymouth, Massachusetts, in 1620. By what name are these people called today?

today. Together they developed a language which they called *Englisc*. This language became a main part of our English language. In time, the country became known as *Angle-land*, a name very close to today's spelling of England.

The next invaders of the British Isles came from France, but they were really Danes. The Danes were daring sailors from northern Europe. They are also known as the *Vikings*, or the *Norsemen*. Perhaps you have read some stories about their raiding expeditions along the coast of Europe and the British Isles. Some of the Norsemen (Danes) invaded, conquered, and settled the mainland of northwestern France. In time these people became known as *Normans* and they spoke the French language. A group of Normans invaded England in A.D. 1066. The Normans defeated the Angles and Saxons in battle. They then set up a strong central government. In time the French language of the Normans blended with the Englisc spoken by the Angles and Saxons to form our English language. The Norman conquest of England was the last successful landing on Great Britain by an invading army. It was not until 1707, however, that England, Wales, and Scotland became a single nation.

The British Empire

Many British people also became daring sailors. They explored many lands not previously seen by Europeans. They founded colonies in different parts of the world. By 1900 the United Kingdom controlled about one-fourth (¼) of the land and some 500 million of the Earth's people. At that time, the British Empire was the world's greatest *empire*. An **empire** is a group of nations or states controlled by or united under one nation or ruler. To carry on trade with and have control over their large empire, the British built a large navy.

The British ruled their colonies for their own economic gain. They did some things quite well. For instance, they built railroads, developed large farms, and started some industries. They also started schools to provide some education for some of the people. But through the years, they received more from the colonies than they gave to them. Thus, the United Kingdom became a very rich and very powerful nation.

Most British colonies now have become independent. Former colonies now sell their goods to other nations as well as to Great Britain. They also buy from other countries. Almost everything the people in the United Kingdom buy costs much more than it did in the past. It also costs more to make things because the materials used to manufacture goods—the **raw materials**—cost more. The goods the British have to sell must now compete with goods from other countries. This is one reason why times have been a bit hard for the British people recently.

The Commonwealth of Nations

Through the years, many people from Britain moved to the British colonies. They took British customs and the English language with them and these spread to the people living in the different colonies. As the colonies grew in importance, many of them became independent countries. Most of these newly independent countries wanted to keep a relationship with the United Kingdom. So an organization was started called the *British Commonwealth of Nations*.

The organization was created so that all former colonies of the United Kingdom could participate as equal partners if they wished to. The members have worked together on many problems. They have made trade agreements with each other and with the United Kingdom. They have helped each other in education, health, welfare, and scientific research.

When India and Pakistan became independent nations in 1949, they wished to remain members of the British Commonwealth of Nations. They objected to the word "British" in its title, however, so "British" was dropped.

In many ways the Commonwealth of Nations is like a large family. Meetings are held every once in a while much like a family reunion. The heads of all the countries that were former colonies and the heads of the present colonies may attend if they wish to. Not everyone agrees on what should be done about the problems they discuss. And members sometimes go back home and do not do what the Commonwealth leaders agreed to do. But, as in a large family, the Commonwealth members like each other and try to work together even when they do not agree.

As of 1980 the countries indicated on this map were members of the Commonwealth of Nations.

COMMONWEALTH OF NATIONS

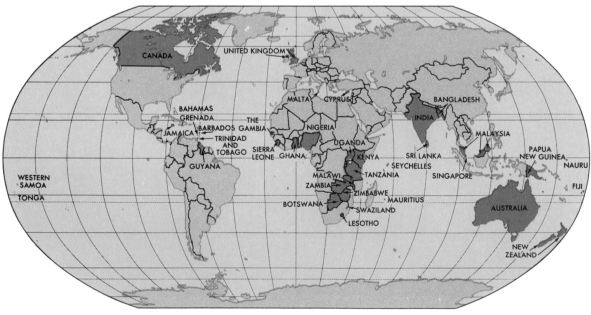

The United Kingdom

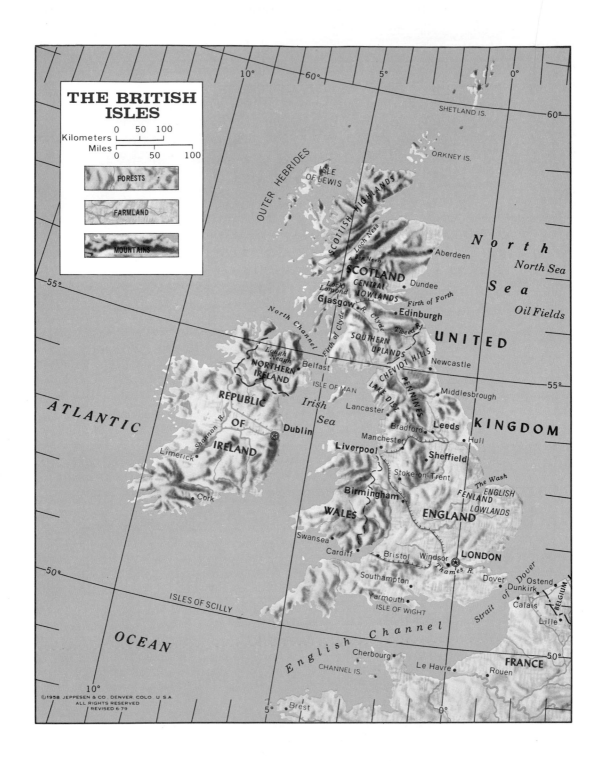

THE BRITISH ISLES

Kilometers 0 50 100

Miles 0 50 100

FORESTS

FARMLAND

MOUNTAINS

©1958 JEPPESEN & CO. DENVER COLO. U.S.A.
ALL RIGHTS RESERVED
REVISED 6-79

SHETLAND IS.

ORKNEY IS.

OUTER HEBRIDES

ISLE OF LEWIS

SCOTTISH HIGHLANDS

Loch Ness

Ben Nevis

Aberdeen

SCOTLAND

CENTRAL LOWLANDS

Loch Lomond

Dundee

Glasgow

Firth of Forth

R. Clyde

Edinburgh

Firth of Clyde

Tweed R.

SOUTHERN UPLANDS

CHEVIOT HILLS

Newcastle

North Channel

Lough Neagh

NORTHERN IRELAND

Belfast

ISLE OF MAN

LAKE DIST.

PENNINES

Middlesbrough

REPUBLIC

Shannon R.

OF

IRELAND

Dublin

Irish Sea

Lancaster

Bradford

Leeds

Hull

Limerick

Manchester

Liverpool

Sheffield

Cork

Stoke-on-Trent

Birmingham

FENLAND

The Wash

ENGLISH LOWLANDS

WALES

ENGLAND

Swansea

Cardiff

Bristol

Windsor

LONDON

Southampton

Thames R.

Dover

Dover

Ostend

ISLES OF SCILLY

Yarmouth

ISLE OF WIGHT

Strait

of

Dunkirk

Calais

Lille

OCEAN

English

Channel

Cherbourg

CHANNEL IS.

Le Havre

FRANCE

Rouen

Brest

North

North Sea

Sea

Oil Fields

UNITED

KINGDOM

ATLANTIC

North Sea

BELGIUM

A Favorable Location

One reason why the United Kingdom became such a great and powerful nation is its location. East of the British Isles is the North Sea. Look at the map on page 60. You will see that the North Sea touches a number of countries in Europe. West and south of the islands is the Atlantic Ocean. The Atlantic provides sea lanes to the Mediterranean Sea and to North America, South America, and Africa. British ships can sail through the Suez Canal from the Mediterranean. They can move through the Panama Canal from the Atlantic Ocean. These canals make it possible for ships to sail to the Indian and Pacific oceans. Of course the ships could sail to these oceans by going other ways. Going around the continents of Africa or South America is much longer, however.

The United Kingdom is very close to the mainland of Europe. Only a narrow body of water, the English Channel, separates it from the mainland. The water around the United Kingdom makes it easy for the British to trade with their European neighbors and the rest of the world. At the same time, it has helped to protect them during wars.

REMEMBER, THINK, DO

1. Do you think that Scotland and Wales should be permitted to become independent nations? How should such questions be decided?
2. Why do you suppose the United States of America is not a member of the Commonwealth of Nations? Do you think it should be?
3. Compare the latitude of the following pairs of cities:
   a. London, United Kingdom, and Washington, D.C.
   b. Glasgow, United Kingdom, and Anchorage, Alaska
   c. Belfast, United Kingdom, and Minneapolis, Minnesota
4. Is any part of the 48 touching states as far north as the southern tip of England?
5. Using a globe, measure with a string, a flexible ruler, or the side of a sheet of paper, the shortest distance from New York City to London. Record the number of centimeters (or inches). Then measure the distance from New York City to London staying south of 50° N. until you get to the English Channel. Record that measurement too. Which is shorter? Why?
6. What direction is it from London to:
   a. Belfast?   d. Dover?
   b. Southampton?   e. Hull?
   c. Edinburgh?   f. Birmingham?

The Climate of the United Kingdom

The climate of the British Isles is mild for land so far north. You have already learned why that is so. The winters in the United Kingdom seem cold, but usually little snow falls. The harbors are open all year. The rivers that are used as waterways seldom freeze over. The Great Lakes, which are farther south in North America, usually are frozen and closed to shipping for several months each winter. Summers in Great Britain are usually very pleasant. Rarely does the temperature get so high that people are too hot to work comfortably.

To see the differences in climates between New York City and London study the chart that is shown below. You will remember that London has a mid-latitude marine climate. Climates are very similar at the same latitudes on the western coasts of North America and Europe. To show this, the same information for Seattle, Washington, is given in the table. Seattle is not quite as far north as London. Note how much the summer and winter temperatures are alike in the two places.

The British Isles have many cloudy and rainy days. London once had 72 rainy days in a row! Seattle, Washington, however, usually has more rainfall each year. London gets only about 60 centimeters (24 inches) of rainfall. Seattle gets about 80 centimeters (32 inches).

As the rainfall map on page 167 shows, enough rain for farming falls throughout the United Kingdom. The rainfall, moreover, is spread throughout the year. About half of it falls during the growing season— April to September. The rainfall map shows that the western parts of the islands get more rainfall than the land farther east. Compare the rainfall map and the landform map on page 45. You will see that the areas of heavy rainfall are mostly highland areas. As the moisture-bearing winds blow over the land, the air is forced upward by the land. As the air moves upward it expands and cools. As the air cools, the water vapor in it forms clouds. When these cool a bit more, raindrops fall to the earth.

The westerly winds lose some of their effect as they blow over the land. They bring less rainfall to the eastern parts of the islands. There, winds from the mainland of Europe often blow. Therefore, temperatures change more in eastern England and Scotland.

City	Latitude	Average Temperature	
		January	July
New York	41°N.	0.6°C (33°F)	24°C (75°F)
London	52°N.	5°C (41°F)	17°C (63°F)
Seattle	48°N.	5°C (41°F)	19°C (66°F)

The Firth of Forth is one of Scotland's largest inlets from the North Sea. Along its shores is the city of Edinburgh as well as many small farms. This ruined abbey, which stands along the shore, has become a popular site for picnics.

For many years London was famous for its fogs. Heavy, sooty fog would cover the city for a week or more at a time. During very bad fogs ships, trains, buses, and automobiles had to creep along. Planes could not land. Finally, the city government decided to do something about it. They decided to control pollution. The government ruled that coal had to be treated so that it would make less smoke. They placed pollution controls on all factories. As a result, London has not had a really bad fog since the early 1960s. Londoners now enjoy much more sunshine and breathe much cleaner air. It is a lesson many other cities could learn.

The Land of the United Kingdom

Several kinds of land are found in the United Kingdom. There are, however, no high mountains and no areas below sea level. The coastline of the islands, as you can see on the map, is very irregular. Many inlets reach far back into the land. There are also many peninsulas. No place on the islands is more than 115 kilometers (70 miles) from the sea. How far from the sea is your home?

Near the two main islands of the United Kingdom—Great Britain and Ireland—are many other smaller islands. Between them is the Isle of Man. To the west of Scotland is a

The United Kingdom

group of islands called the Outer Hebrides (*hehb* ruh deez). Just north of Scotland are the Orkney Islands. Even farther north are the Shetland Islands. In the English Channel, just south of Great Britain, is the Isle of Wight. The Channel Islands, near France, are also part of the United Kingdom.

Most of the land of the United Kingdom is either highland or lowland. The map on page 60 shows that the highland areas have two kinds of land: mountains and *uplands*. **Uplands** are areas of hills that are not quite as high as mountains.

Scotland

The Scottish Highlands. As the map on page 45 shows, most of Scotland is highland. The land farthest north is known as the *Scottish Highlands*. The Highlands are made of old, hard rock. For millions of years ice, wind, frost, and water have worked together to wear down the rock. The tops of most of the mountains have been rounded. Deep valleys have been cut. However, there are still steep cliffs and rough, rocky places in the mountains.

During the Ice Ages sheets of ice covered much of Great Britain. The island was covered about as far south as the Thames (tehmz) River on which London is built. Great rivers of ice moved down the valleys along the west coast of Scotland. These glaciers cut deep valleys as they moved toward the sea. After the ice melted the sea flowed back into these valleys, forming *fiords*. A **fiord** is a narrow inlet of the sea between steep cliffs. The map on page 60 shows several long narrow inlets that are fiords. The Scottish name for an inlet from the sea, however, is *firth*. You will find several firths on this map.

Few people live in the Scottish Highlands. Only a small part of the land there is good for growing crops. Many areas where the soil is too thin or the land is too steep for farming are used for grazing animals. Most people in this area live along the coast. Many of them are fishers.

The Central Lowlands of Scotland. Only about one-fifth (⅕) of Scotland is lowland. Most of the Scots, however, live in this part of the country. As you can see on the map on page 45, the lowland plain extends across the island. The Firth of Forth is on the North Sea side of the Central Lowlands. The Firth of Clyde is at the western end of the lowlands. Glasgow (*glas* koh) and Edinburgh (*ehd* 'n buhr oh), the two largest cities, are located in these lowlands. Glasgow is on the River Clyde which flows into the Firth of Clyde. The city is noted for its factories and shipyards. Edinburgh, the beautiful capital city of Scotland, is on the Firth of Forth. A canal connects the two firths and the two cities. The discovery of coal and iron ore made possible large iron and steel mills in and near Glasgow. Other important industries include the making of **textiles** (woven cloth or fabrics).

The Central Lowlands area is good farmland. Farming is especially important in the eastern half of the lowlands area. There is less rainfall in that area. Many farmers in the west raise cattle and sheep. Pastures grow well because of the heavy rainfall there.

The farmers of the United Kingdom are collecting seaweed that has been left by the out-going tide. The farmers will spread it on their fields before plowing.

Seaweed in Space*

When the American astronauts landed on the moon a little bit of Scotland was along! And that "wee bit" (as the Scots would say) had come from the sea.

Have you ever visited a seashore? If so, you probably have seen seaweed. It is a plant that grows near the shores of oceans. Some of it floats on the water and is washed ashore. There is a lot of seaweed in the world. Not much use is made of it except in a little village on Scotland's Isle of Lewis in the Outer Hebrides. (See map, page 60.)

There workers walk into the sea at low tide every other day. They stretch a rope around the section of seaweed that they

*Adapted with permission from an article in the *Christian Science Monitor*, October 20, 1971.

The United Kingdom

plan to cut. They use a small curved knife called a hand sickle to cut the weed. Then they pull the rope tightly around the cut seaweed. Some workers let the bundles of seaweed drift toward the shore as the tide comes in. Then they load it onto trucks. Other workers use cranes to lift the wet seaweed onto small ships.

At a small factory nearby, the seaweed is placed on a conveyor belt. It slowly moves through furnaces and is dried. After about three hours in the furnaces, the weed is brittle. It is then crushed into something looking like coarse tea. Then it is placed in bags and is shipped to a factory on the Scottish mainland. There a product called *alginate* (*al* jih nate) is taken from the crushed, dried seaweed. **Alginate** is a powder that is used in more than 500 different products. One of them is fireproof paper. And that is what the astronauts took with them into outer space!

You may have eaten some alginate and not have known it. If you like chocolate pudding, you probably have eaten a small amount of it. The alginate from the seaweed from the small village in Scotland is used in making products in more than 100 different countries!

The Scottish Southern Uplands. The Southern Uplands are south of the Central Lowlands. See the map on page 60. The Uplands have broader valleys, and the hills are not as high as those in the Highlands. Some farming is done, but the Uplands are most noted for sheep raising. The sheep, of course, furnish wool for the woolen mills. Many of these mills are located along the Tweed River. The Cheviot (*chehv* ih uht) Hills, where many sheep graze, form the boundary between Scotland and England. You may have heard of woolen goods called *Harris tweed* and *cheviot*. You may also have seen some of the famous Scottish *plaids* (plads). Each Scottish family has its own plaid. They are woven in this region. Many dairy cattle are also raised in this area, especially in the west where there is more rainfall.

England

The Highlands of England. There are two highland areas in northern England: The *Pennines* (*pehn* nyns) and the *Lake District*. The Pennine Chain of highlands has some hills about 900 meters (3,000 feet) high. The wind-swept tops of these highlands are treeless and are called **moors**. The soil on the moors is generally too thin for farming except in the lower valleys. Thousands of sheep roam over the slopes and eat the short grass. Coal fields are found on both sides of the Pennines. Many manufacturing cities have therefore grown up near them.

The Lake District in northwestern England is famous for its beautiful scenery. It is one of the most popular places in Great Britain for a summer vacation. Many waterfalls and long narrow lakes are found in this region. Sheep and cattle are raised in the Lake District, but few crops are grown. This area has heavy rainfall. Much of the land, however, is too steep for cultivation.

The English Lowlands. Most of England south of the Pennines is a plain. It is crossed in places by gently rolling hills. You can easily locate some of these hills on the map on page 60. Like most of the higher lands in England, these hilly regions are used for grazing sheep. Most of the English lowlands area, however, is good cropland. It is really part of the North European Plain.

On the map on page 60 you will see a broad inlet on the North Sea coast named "The Wash." Inland from The Wash is an area called "Fenland." *Fen* means "marsh" or "swamp." In order to make this land into good farmland, the British built *dikes* and drained the land. **Dikes** are banks of earth thrown up to hold back water. Do you know of any other countries where dikes have been built? Fenland is now excellent farmland. It is Britain's best wheat-growing region.

The lowlands area, generally, has many people. Study the population map on page 100. Villages and towns in this area are only a few kilometers (or miles) apart. London, one of the

Before the development of the railroads, canals were an important part of the economic life of the United Kingdom. Today most of these canals are used for recreation. However, some of the larger and more important canals are still used for commercial shipping.

world's largest cities, is in this region. Although this is the best farmland in the United Kingdom, it also has much manufacturing.

The lowlands region is crisscrossed with highways, railroads, and canals. These are used to move raw materials from seaports or farms to factories. The same transportation system is used to move manufactured goods to seaports for export. Look again at the map on page 60. Notice how London is connected with Bristol, Birmingham, Liverpool, Manchester, and Leeds by canals. These are the largest cities in England. Only one of the largest cities, Sheffield, is not linked to this canal system.

Wales

Wales is a mountainous peninsula with deep valleys, waterfalls, and lakes. It is much like the land in the Scottish Highlands. The mountains are not quite as high, however. The interior of Wales has few people. Much of the land there is of little value except for grazing sheep. Most of the people live near the sea in the lower part of valleys. These areas have some land that is used for farming.

The southern part of Wales has more people than the northern part because of coal. This area has become a main mining and industrial region. Mining coal and using it to produce tin, copper, and steel are big industries.

Northern Ireland

As the map on page 45 shows, Northern Ireland is mainly upland and highland. Near the center of Northern Ireland is a lowland area around Lough (lock) Neagh (nay). *Lough* is the Celtic (Gaelic) word for "lake." Lough Neagh is the largest lake in the British Isles. The land around the lake is swampy and made up of decaying plants. Such areas are sometimes called **bogs**. Little can be grown in this lowland area. Much of the higher land is good for grazing animals and for raising flax. The textile called *linen* is made from flax.

When Will Peace Come to Northern Ireland?

For many years Northern Ireland has had much violence. Most of it has been caused by conflicts between Protestants and Catholics. Many Protestants in Northern Ireland want the area to remain a part of the United Kingdom. Many of the Catholics want Northern Ireland to become a part of Ireland. A small group of people on both sides of the argument have used force to try to get their way. Bombs are set off in Catholic neighborhoods by the most determined Protestants. Bombs are set off in Protestant neighborhoods by the most determined

Catholics. Some of the latter people call themselves the Irish Republican Army (IRA).

As you know, bombs and other explosives can kill innocent people. Often, children are killed when bombs explode in stores or on streets. Most of the people in Northern Ireland would like the violence to stop. But it goes on, year after year.

Two Irish women, one Protestant and one Catholic, decided in 1976 to try to do something about the problem. They held many meetings throughout Northern Ireland, asking the people to stop the killing. Betty Williams and Mairead Corrigan won a Nobel Peace Prize, one of the world's most precious awards, for their work. Unfortunately, however, the killing went on.

The United Kingdom sent more and more soldiers into Northern Ireland to try to prevent violence. Many of them were killed by the determined people on both sides. As this book is written, the struggle goes on. One keeps asking, "When will peace come to Northern Ireland?"

Trying to end the violence in Northern Ireland, Betty Williams (left) and Mairead Corrigan (right) organized peaceful marches and demonstrations. On August 21, 1976, carrying telegrams of support from people all over the world, they led 20,000 people through the city of Belfast.

1. Why is the climate in Britain as mild as it is?
2. Keep a record for a month of temperatures and rainfall where you live. Make charts to show the data. How does your climate compare with London's?
3. What are uplands? Are there any uplands near where you live?
4. What is the difference between a fiord and a firth?
5. Have you ever seen a bog? Where?
6. In what part of the United Kingdom would you prefer to live? Why?

The Government of the United Kingdom

The United Kingdom has a government called a *constitutional monarchy*. A **monarchy** is a country that has a queen or a king. Do you know the name of the monarch of the United Kingdom? A **constitutional monarchy** is one in which the monarch must follow a *constitution*. A **constitution** is a set of rules (either written or unwritten) drawn up for the government of a country.

The United Kingdom's lawmaking body—its **Parliament**—is somewhat like the Congress in the United States. The Parliament has two houses, the *House of Commons* and the *House of Lords*. Members of the **House of Commons** are elected directly by the people as are all members of Congress in our country. The House of Commons is the most powerful house in Parliament. The head of the British government is called the *prime minister*. The prime minister is chosen by the political party that has the most members in the House of Commons. The prime minister must carry out the laws passed by the House of Commons.

Elections of members to the House of Commons must be held at least every five years. Sometimes they are held sooner. If the Commons votes against a proposal of the prime minister, the prime minister may decide to resign or leave office. The prime minister then asks the monarch to dissolve Parliament. If that is done, an election for a new Parliament is held. The monarch might ask the prime minister to think again about resigning. Sometimes after a few days, and more discussion, the prime minister decides not to resign after all.

Members of the **House of Lords** are not elected by the people. The members are called *peers*. **Peers** are people who hold either inherited titles, such as duke or duchess, or who were made nobles by the monarch. We do not have a similar house in our government. It is an honor to be a member of the House of Lords. Although about 1,000 people are members, only about 200 peers actually take part. The House of Lords has much less power in the British government than the House of Commons. It cannot pass laws and cannot really keep the House of Commons from doing so. It can delay passage of non-money bills.

Queen Elizabeth II and her husband, Prince Philip, visited the United States in 1976. They came to help celebrate the nation's 200th birthday. Do you think it is strange that the Queen of England would help the people of the United States celebrate this event?

Silver Jubilee for Queen Elizabeth II

In June, 1977, Queen Elizabeth II lit a bonfire on Snow Hill near Windsor Castle. People on other nearby hills saw the flame. They lit their bonfires. Others saw those fires. Within about an hour, bonfires were burning as far north as the Shetland Islands. They were burning as far southwest as the Isles of Scilly. (See map, page 60). Throughout Great Britain bonfires burned the joyous news. The Silver Jubilee had started!

Elizabeth II had been monarch of the United Kingdom of Great Britain and Northern Ireland for 25 years! Now a special week to honor her was starting. Elizabeth II made a trip in a small boat up the Thames River in London. Stops were made several places to greet the people. Whistles blew. That night fireworks lit up the city of London. During the week picnics were held where whole cows were roasted outdoors. There were street fairs in small villages. Special church services were held.

The royal family, of course, helped the Queen and the people celebrate. At Elizabeth's side, almost always, was her husband, Prince Philip. Prince Charles, Queen Elizabeth's son and next in line to the throne, also took part. So did his brothers and sister. Sometimes Elizabeth's mother was there, too.

Queen Elizabeth II is not only the monarch of the United Kingdom, but also monarch for many other countries. See whether you can make a list of all of them. A good place to look is in an encyclopedia under "Commonwealth of Nations." Also see whether you can find all the Commonwealth countries on a globe. Elizabeth is monarch for all British colonies, too.

What does the British monarch do? As Queen, Elizabeth II reigns rather than rules. As monarch she has much influence, but does not have any real power. A British monarch has three important political rights: to be consulted, to encourage, and to warn. The prime minister meets at least once each week with Queen Elizabeth. They discuss problems existing throughout the world. Suggestions are made about what might be done. Queen Elizabeth II has been doing that for 25 years.

More than anything else, though, the monarch unites the British people. Queen Elizabeth II is a gracious, kind person. She is the mother of one daughter, Princess Anne, and three sons: Princes Charles, Andrew, and Edward. The monarch is a symbol for the people of Britain.

Agriculture of the United Kingdom

For many years farming was the most important occupation in the United Kingdom. Before 1800 the country could export food to other countries. Then the population increased. As factories were built, cities began to grow. Now Britain has to import about half of the food it needs. Today, only about 5 out of every 100 workers in Britain are farmers. And the farms they have to work are rather small.

For the size of the country, there are a great many people in Great Britain and Northern Ireland. In the United Kingdom there are about 231 people per km² (598 per sq. mi.). In England alone, there are about 359 people per km² (930 per sq. mi.). Parts of the United States are crowded, of course. But we have on the average only about 24 people per km² (62 per sq. mi.).

Great Britain does not have enough land to grow food for all its people. Less than one-third (⅓) of the land can be used for growing crops. Farmers in England do produce a great deal of food on the available land. They use fertilizers, farm machinery, and other scientific methods.

Important crops in the United Kingdom include hay, turnips, potatoes, sugar beets, and many other kinds of vegetables. Fruits, such as apples, plums, and pears, are grown in the drier valleys and lowlands. Berries do well in the damp climate along the west coast. But corn does not grow well there.

Raising Sheep and Cattle

The United Kingdom has a very large number of domestic animals. Cattle and sheep are the most numerous. This situation is a bit unusual for a country with so many people and so little land. There are several reasons why this is so.

Grass grows well in Great Britain. The heavy rainfall helps. You will remember that Great Britain usually has mild winters. Mild winters make it possible for cattle and sheep to graze in the fields most of the year.

The British people like to eat a great deal of meat, especially beef and **mutton** (meat from sheep). British farmers cannot raise as many meat animals as are needed, however. Great amounts of meat and dairy products are imported from Ireland, New Zealand, and Australia.

For centuries wool was the most important product of the United Kingdom. Today raising sheep is still an important activity throughout the British Isles.

Drilling for oil in the North Sea is hard and sometimes dangerous work. Would you like to do this kind of work?

Energy for Industry

Every nation with many factories has to have power to run the machines. The United Kingdom became an early center for manufacturing because it had a huge supply of coal. The energy from coal was used to run the machines. There were few streams that could be used to produce electric power. And until recently Great Britain had almost no supply of natural gas and oil.

Coal. For the past 200 years coal has been the most important power source in the United Kingdom. It will probably continue to be the main source of power for manufacturing in the years ahead. Coal was not very important as a power source until the steam engine was invented. James Watt, a Scottish engineer, developed the first workable, coal-burning steam engine. Soon after that the United Kingdom became a leader in the *Industrial Revolution*. The **Industrial Revolution** came about when factories began using the power of coal-fueled engines to drive machines. The United Kingdom still is an important producer of many manufactured products. Most of them are made in factories using electric power that is made from coal.

At one time the British exported coal in exchange for other needed raw materials. As the population grew and manufacturing industries increased, more coal was needed at home. Now the British have little coal to export. Only four other countries in the world, however, mine more coal than the United Kingdom. Much of the coal that is easiest to mine has already been used. Mines, therefore, have been sunk deeper and deeper into the earth. Britain has enough coal for a long time to come, however.

The coal mines are very different places to work than they used to be. At one time almost all the work in coal mines was done by hand. Now a machine called a **power shearer** does much of the hardest work. It works on a layer of coal much like an electric razor does. With the power shearer, the mines can produce about four times as much coal per working day as they used to. The coal is usually burned right next to the mine to make electricity. Thus, transportation costs have been cut.

The United Kingdom has become concerned about being so dependent upon coal for power. And the cost of oil, another power source, has been rising on world markets. So the government began to build atomic power plants. Several are now operating. Today power from atomic energy costs more to produce than power from coal. The greatest disadvantage of atomic power is that wastes from power plants are dangerous and hard to get rid of. The advantage of such plants is that a small amount of fuel is needed to run a power plant for a long time.

Oil from the North Sea. Until a few years ago the United Kingdom did not have a good supply of natural gas or petroleum. Yet, millions of people in Great Britain drove automobiles. So the country had to import great amounts of oil. The oil bill for the country was huge.

But good news arrived in 1968. Oil had been found under the North Sea. A few years before, natural gas had also been found there. Some people thought that almost overnight Britain would be rich again. But developing an oil field underwater is a slow and costly process. Huge platforms had to be built. Drilling had to be done. And the workers had to try to prevent oil spills that would kill fish and birds and pollute the shores.

Years went by. During that time of waiting, the price of oil on world markets rose rapidly. The United Kingdom seemed to be almost out of money. Yet the people, like those in the United States, kept driving their cars. More and more oil had to be imported. Would the oil start flowing soon enough to prevent the nation from going broke?

On November 3, 1975, Queen Elizabeth announced that the flow of oil from a large British oil field in the North Sea had begun. It is located at about 58° N., near the middle of the North Sea. By 1977 several new fields had been found farther north. These newer fields are northeast of the Shetland Islands. By 1980 the United Kingdom expects to be producing enough oil for its own needs. It also hopes to be making lots of money by selling oil to other countries. Some experts think that the oil fields under the North Sea may be the world's second largest. Only the oil fields in Saudi Arabia may be larger.

Manufacturing in the United Kingdom

The United Kingdom is a world leader in manufacturing. Factories are located in all major cities. Most of them not only make products needed in Britain, but also make products needed in the European Community and in Commonwealth countries.

Iron and Steel. Many great changes in manufacturing came about because of steel. The invention of power-driven machinery made of iron and steel led to the Industrial Revolution. Products that once were made in the homes by hand were then made in factories by these machines. Most of the early ideas and inventions of the Industrial Revolution came from Great Britain. During most of the 19th century the United Kingdom led all the world in steel production. Today, however, several other countries make more steel.

The British had several advantages that helped them in producing iron and steel. First, they found iron ore close to their coal fields. Limestone was found nearby. Limestone is used in melting ore to get the needed metal, a process called **smelting**. Second, the iron ore, coal, and limestone were all found close to the sea. The main iron and steel producing areas in the United Kingdom are still near the sea. It is, therefore, easy to ship the heavy iron and steel products overseas. Location also made it easy for British companies to start importing higher grades of iron ore. Today about one-half (½) of the iron ore used in Great Britain is imported from other countries. Third, the British had a great fleet of ships to use for world trade. The ships would take the manufactured products to other countries. They could bring needed raw materials from other lands to factories in Great Britain.

The main centers for the steel industry are Glasgow, Sheffield, Newcastle-Middlesbrough, Manchester, Birmingham, and Swansea-

Cardiff. Find these places on the map on page 60. Sheffield is noted for its fine knives and scissors. Birmingham makes a large variety of tools and other hardware. Glasgow and Newcastle-Middlesbrough and Swansea-Cardiff make a great deal of **structural steel**. That is the kind of steel used for railways, ships, buildings, bridges, and other construction. Machinery for textile manufacturing is made in Manchester.

Textile Industries. The United Kingdom was the leading nation in the manufacture and export of cotton and woolen goods for more than 100 years. A third textile, linen, was also of importance. Northern Ireland, especially, is noted for its fine linen.

In early times, most woolen cloth was woven by hand. After the invention of steam-driven textile machinery, the woolen textile industry grew very rapidly. Woolen manufacture became centered east of the Pennine Chain. Find Leeds and Bradford on the map on page 60. These are the two leading centers for wool manufacture in this region.

Although British woolens are known throughout the world, the British are even better known for their cotton goods. The amazing fact is that no cotton is grown in the United Kingdom. From early times, cotton has been brought by ships to the Manchester region. It is the main center for cotton manufacturing. For many years, the cotton bales had to be unloaded at Liverpool and shipped to Manchester. Then a canal was dug so that the ships could take the cotton bales right to Manchester.

Iron and steel factories, such as "Coketown" in Henley, Staffordshire, contributed greatly to the United Kingdom's development as a powerful industrialized nation. Workers often lived close to their job in housing built and owned by the factory.

The Manchester region became the center for cotton manufacturing for several reasons. The damp climate there provides the humidity that is needed in making fine cotton goods. The streams that flow down from the Pennines furnished water power to run mills in the early days. Coal has since furnished most of the needed power. The sea was nearby, making it easy to move both the raw material and the finished products.

Many other countries now have machines to make cotton and wool cloth. In addition, new **synthetic** (or machine-made) fibers, such as nylon, rayon, and polyester have been developed. British textile industries, therefore, began to make less money. Many changes, therefore, had to take place in the British textile industries. Factories with new machinery to make these new fibers have been built.

London, the United Kingdom's Large Capital

The great city of London began as a trading center on the Thames River. Today it is one of the largest cities in the world. Five cities that probably are larger than London are Shanghai, China; Tokyo, Japan; Mexico City, Mexico; Peking, China; and New York, New York. It is very difficult to tell for sure which city is the largest at any time. All of them have many smaller cities around the central city.

London has all the problems of the world's largest cities. Too many people are on the sidewalks of downtown London at rush hours. Too many automobiles, buses, and

Many people from countries who were once part of the British Empire have been allowed to migrate to the United Kingdom. Some, such as these Pakistanis, work in the textile industry.

trucks try to use the streets. Too many people try to get to office buildings weekday mornings. Too many try to get home about the same time weekday evenings. Airports are too busy. And the air above the city is too polluted, even though it has greatly improved.

London is one of the busiest port cities in the world. Ships from all over the world bring food, raw materials, and manufactured products to the docks along the Thames. Products from the factories in London and other cities are loaded onto ships. Many of the goods are taken to other countries of the European Community. Many shiploads of goods are also sent to members of the Commonwealth.

The best-known buildings in London are government buildings. The Houses of Parliament are especially well known. Buckingham Palace is the London home of the British monarch. The royal family owns several other homes in Great Britain, too. Perhaps the best known is Windsor Castle. It is about 35 kilometers (21 miles) west of London. Many tourists visit Windsor Castle every year. Queen Elizabeth II lives there most of the time.

Westminster Abbey is the large cathedral where the monarchs of the

The Port of London is one of the world's busiest ports. In 1979 nearly 50 million metric tons of cargo were handled there.

United Kingdom are crowned. In addition, many of Britain's most famous people are buried in Westminster Abbey. During the Second World War Westminster Abbey was damaged by bombs. After the war the British rebuilt the damaged part of the cathedral. They built a special room as a memorial to Americans who lost their lives defending Britain during the war.

Imagine Changing the Money System!

Suppose when you got up next Monday morning you would have a new money system to use! That happened to everyone living in Great Britain in February, 1971. All the banks in the United Kingdom closed on a Wednesday afternoon. They didn't open again until the next Monday morning. On that day, the

The United Kingdom

Everybody in the United Kingdom had to learn to use a new monetary system. Students used giant play money to help them understand the changes. Can you see any differences between the money in the picture and the money that you use?

British started using a new money system based on **decimals**, or units of ten. All bank accounts had to be changed between Wednesday and Monday. And there were about 25 million bank accounts! For us, because we are used to a decimal system of money, the change would seem easy. If you had lived all your lives with another system, though, the change probably would have been confusing.

Until 1971 the British pound was divided into 20 shillings. Each shilling was worth one-twentieth of a pound. And each shilling was divided into 12 pence. Imagine doing an arithmetic problem with money using that system! Now the British pound has 100 pence—just as the American dollar has 100 pennies. The British have one more coin than we have, however. They have coins for one-half pence, 1 pence, 2 pence, 5 pence, 10 pence, and 50 pence. The British do not have a coin like our quarter. We probably should not have it either, because it does not fit into the decimal system.

Think of all the changes that had to be made when the British system of money was changed. Cash registers all over the

United Kingdom made for the old system had to be changed. All catalogs and advertisements had to be changed. The price of all goods in all stores had to be changed to the new system. There were probably a lot of people who thought the old system was good enough for them!

We are used to a decimal system of money. Many of us, though, are not used to a decimal system for other measures—the **metric system**. Some people think we should not change to the metric system. All the rest of the world, however, is now using the metric system. We must begin to think with metric measures, too. And some day we might have a special day to change weights, lengths, temperatures, and many other measures from our present system to the metric system.

REMEMBER, THINK, DO

1. What advantages did Great Britain have that helped it become a leading steel producer?
2. Assign the old British values to our coins and attempt to make change. Use the values as follows:

hapenny (½ pence)	use a penny
1 pence	use a nickel
5 pence	use a dime
1 shilling (12 pence)	use a quarter
1 pound (20 shillings)	use a dollar

3. Use a globe or the map of the world on pages 542-543 to answer this question and the next one (item 4). Which of the following is farther from the equator:
    a. London, United Kingdom, or Montreal, Canada?
    b. Glasgow, United Kingdom, or Juneau, Alaska?
    c. Southampton, United Kingdom, or Hobart, Australia?
    d. Belfast, United Kingdom, or Calgary, Alberta, Canada?
    e. the northern tip of Scotland or the southern tip of South America?
4. Suppose that you flew straight west from Edinburgh, Scotland, around the world. You would fly over the oceans for about half the journey. You would fly over two continents. What oceans would you fly over? What continents would you fly over? What countries on these continents would you fly over?
5. How many cities in the United Kingdom have more than 500,000 people? Name the cities that have more than two million people.

The United Kingdom

WHAT HAVE YOU LEARNED?

1. Where is Northwestern Europe located compared to the 48-touching states of the United States of America?

2. Why do the western parts of the United Kingdom have more rainfall than the eastern?

3. Why are the mountains in the United Kingdom not as high nor as rugged as many other mountains in Europe?

4. What are the four main parts of the United Kingdom? Do you think these parts should become independent countries?

5. What does a British monarch do? Does the monarch rule? Should the United States of America have a monarch? Why or why not?

BE A GEOGRAPHER

1. Make a bar graph showing the population of the ten largest cities on the Earth. Use a recently published atlas or almanac to find your information. How many of these cities are in Europe?

2. Use the map on page 60, or a globe, to answer the following questions. What direction is it:
   a. from Birmingham to London?
   b. from Hull to London?
   c. from Dover to London?
   d. from Cardiff to London?
   e. from Cardiff to Edinburgh?

3. Give the name and location in degrees and minutes of each of the following cities. Use atlases to get the data you need.
   a. Great Britain's largest city
   b. a main cotton manufacturing center in the United Kingdom
   c. a main port on the English Channel
   d. a linen-producing city
   e. a city producing fine knives and scissors

4. Find the following on the map on page 60 and be prepared to point to each on a large wall map:
   a. Fenland
   b. Isle of Wight
   c. Orkney Islands
   d. Strait of Dover

QUESTIONS TO THINK ABOUT

1. Some parts of the world have many people. Other parts have only a few. What reasons can you give that helps to explain why the United Kingdom has a large population for its size?

2. The United Kingdom is smaller in land area than the state of Oregon. Yet, the United Kingdom is a leading nation of the world. How is that possible? Why did it happen?

3. Some day, a long time in the future, most of the better deposits of iron ore will be used up. Somewhat later most of the low-grade deposits will probably also be used up. What will people do then? Suppose that all the iron ore would be used up in ten years. What would we need to be doing in the next ten years? What changes would be brought about?

4. Why is it possible for members of the Commonwealth of Nations to work together better than the members of the United Nations? Or do they?

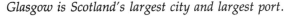
Glasgow is Scotland's largest city and largest port.

Chapter Review

Chapter

FRANCE

Southeast of Great Britain across the English Channel is the country of France. Find it on the map on page 39. France has more land than any other country in Europe, not counting the U.S.S.R. It has more than twice as much land as the United Kingdom. As you can see, France looks quite large on the map. However, it has less land than the state of Texas.

France has a long coastline on the Atlantic Ocean and on the English Channel. It also has a southern coast on the Mediterranean Sea. France's location has helped it to become a nation with much influence. Its location has also made France a battleground for wars between European nations. France has borders with eight other countries. See if you can find all of them on the map. Six of them are easy to find. They are Spain, Italy, Switzerland, West Germany, Luxembourg, and Belgium. The other two, Andorra and Monaco, are so small that they look almost like cities on the map.

Why France Is So Important

France is one of the leading nations of the world. It is of special importance to people who live in the United States and Canada. Here are some of the reasons why:

● The history of France is closely related to the history of the United States and Canada. The French were among the first explorers and settlers in the South Central States and along the Great Lakes. Until 1803 France claimed most of the land west of the Mississippi River. For many years they controlled much of eastern Canada. The province of Quebec is still mainly a French-speaking province. The French helped the American colonies during the War for Independence. This war against Great Britain might not have been won without their help. In this century, both the United States and Canada helped France during World Wars I and II.

● France carries on a large trade with other countries. Exports include wines, perfumes, clothing, automobiles, and weapons of war. France imports mainly raw materials.

● France still has a number of overseas possessions. These are mainly islands scattered over the world. Two of them are only a few miles south of the island of Newfoundland in Canada. These are St. Pierre (san *pyehr*) and Miquelon (*mihk* uh lahn). You can probably find them on a large map of North America or Canada. Some countries which once were French colonies still do most of their trading with France. Many raw materials needed in France are purchased from these countries. France has rich natural resources of its own, too. Together, these resources have made it possible for France to become an important industrial nation.

● French leaders take an active part in international affairs. They often meet with officials from other countries to discuss world problems.

● Paris, the capital of France, is one of the world's most important cities. For many years, Paris has been a great center for art, music, fashions, and learning.

● French is one of the best known languages in the world. For many years, most educated persons in Europe have learned to speak French. Many Canadians learn to speak French as their first language. Many Americans and Canadians also learn to speak French as their second language. The language is often used by representatives from different countries when they meet for conferences. It is one of the official languages of the United Nations.

Paris is considered to be one of the most beautiful cities of Europe.

History of France

France is one of the oldest nations in Western Europe. No other nation has played a more important part in the history of the region. Unlike Great Britain, which has not been invaded since 1066, many battles have been fought on French soil. Some of the fiercest battles of both the First World War and the Second World War were fought in northern France.

France now has a parliamentary type of government. They have a president who is elected by the people to a seven-year term. The

France

president has much power. The president appoints the prime minister and the cabinet. The French have had this kind of government since 1958.

The French Parliament has two houses, as in the United Kingdom. The **National Assembly** is the more powerful house in the Parliament. Its members are elected to five-year terms by the people. The members of the **Senate** serve nine-year terms. They are elected, also, but by regional and city groups.

France was ruled by a monarch for centuries. Then, in 1789, soon after helping in the American Revolution, the people of France had their own revolution. The success of the new American nation probably encouraged the French people to change their government. Two years after the revolution in France started, the king and queen tried to escape from the country. They were captured, put in prison, and killed in 1793. Then a representative form of government was started under a constitution.

The present government of France is known as the "Fifth Republic." The present constitution is the fifth under which France has been governed as a *republic*. A **republic** is a form of government in which the people elect their lawmakers and the head of state. Usually the head of state in a republic is called president.

Climate

Much of France has a mid-latitude marine climate like that of Great Britain. As a result, winters are warmer and summers cooler than one might expect. Enough rain falls along the coast in northwestern France to make the area a fine one for dairying. Pastures grow well in the moist, cool climate of the Brittany Peninsula. (See map, page 87.) Farther inland in France, the climate is slightly drier than near the coast. Differences between summer and winter temperatures there are somewhat greater.

Southern France has a Mediterranean climate. Summers are hot, long, and dry. Winters are mild, with some rain. The rainfall map on page 166 shows that enough rain falls during the year for crops. Much of it, however, falls during the winter months. The rain usually comes in the form of heavy showers. Much of the farmland in southern France, therefore, has to be irrigated. Because winters are mild, vegetables are ready for market in early spring. They are shipped to northern France where they bring higher prices.

Highland areas in France generally have cooler temperatures than the lowlands. In the high Alps, much snow falls during the winter. Glaciers cover the valleys between the highest peaks. Skiing and other winter sports are popular. Many tourists visit the cool Alps during summer months, also. In the lower Jura (*joor uh*) and Vosges (*vohzh*) mountains winters are also long and cold. (See map, page 87.) Much snow falls. Summers are warm, however, and there are even some hot days in the valleys.

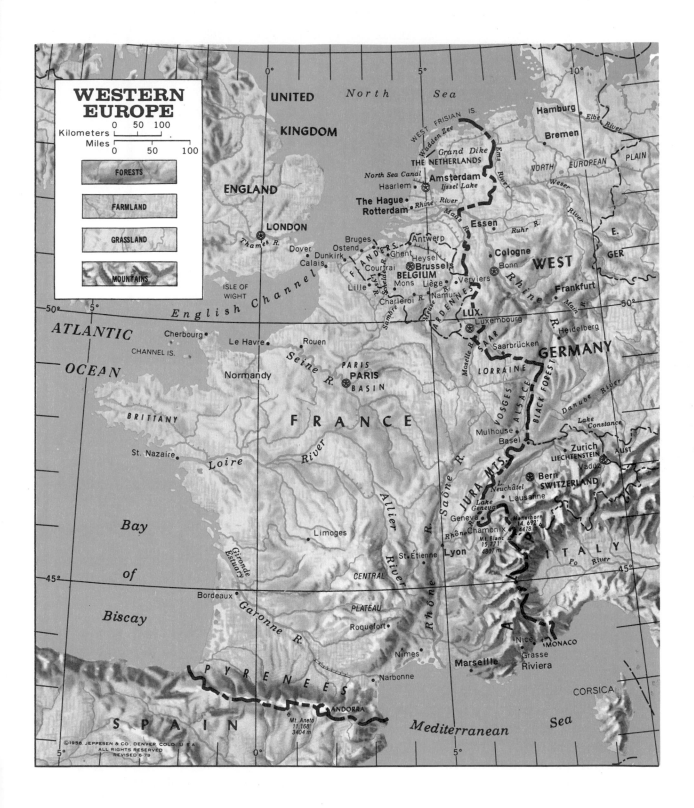

France

1. Look at the land use map on page 202. If you wanted to be a farmer who raised crops, would you prefer to have land in France or in the United Kingdom? Why? If you wanted to be a dairy farmer where would you prefer to have land? Why?
2. What part of France has land in west longitudes?
3. What differences do you think it makes in school to have the prime meridian drawn through your country?
4. Use the map on page 87 to answer the following questions:
   a. What direction is it—
      (1) from Lyon to Paris?
      (2) from Narbonne to Lyon?
      (3) from Paris to Bordeaux (bor *doh*)?
      (4) from Paris to Lille?
      (5) from Paris to London?
   b. How far is it—
      (1) from Paris to London?
      (2) from Paris to Marseille?
      (3) from Paris to Brussels, Belgium?
      (4) from Paris to Bonn, West Germany?

The Land

Since France is a fairly large country, it has several natural regions. These regions can usually be classed either as lowlands or highlands. The lowlands of France usually are *river basins*. A **river basin** is a low area drained by a river and its tributaries. As the map on page 45 shows, there is a lowland in southwestern France and in the Rhône-Saône (rohn-sohn) valley. Most of the lowland, though, is in northern France. The *Paris Basin* is located in this region. It is the best farming area in all of France. All the lowland areas, as the map shows, are joined by canals.

Highlands are generally found along the southern and eastern borders. As the map shows, however, there is an area of highland in southern France west of the Rhône-Saône valley. This area is called the *Central Plateau*.

High mountains form some borders of France. They include the Pyrenees on the border with Spain and the Alps on the Italian border. Lower mountains include the Jura range along the Swiss border and the Vosges near Germany. The two large peninsulas in northwestern France are upland areas. A more detailed description of both areas follows.

Lowland France

The Paris Basin

As you have learned, the plain that covers much of northern France is called the Paris Basin. The land slopes mostly toward Paris on the Seine (sayne) River. That is why it is called the **Paris Basin**. Numerous hills rise above the lower levels of the plain.

The Paris Basin has the richest and most productive farmland in France. It is also the greatest manufacturing region in France. As a result, most people live in this part of the country.

The land in the Paris Basin is carefully farmed, and many kinds of crops are grown. Wheat is the main crop of the region. Much of the wheat is used to make the famous French bread. Another large grain crop is oats. Barley is grown mainly to use as feed for farm animals.

People in the cities need a great deal of milk and vegetables. These are provided by dairy and *truck farms*. A **truck farm** is a small farm or garden where vegetables are grown for sale. These farms are usually located close to main cities. Other important crops of the Paris Basin include potatoes and sugar beets. Most of the sugar used in France comes from sugar beets.

Grapes are the most important fruit raised in France. Grapes grow in many areas including the southeastern part of the Paris Basin. A choice wine, called **champagne** (sham *payne*), is made from grapes raised there. The special flavor of

Although France is an industrialized nation, many of the traditional methods of farming are still practiced. The growing, harvesting, and processing of grapes is done mostly by hand.

Many oil tankers, such as the one pictured above, deliver their cargos to the Port of Fos. Much of the oil that is used in France comes through this port.

champagne is caused by limestone in the ground. It changes the flavor of both the grapes and the water used to make the wine. The wine is bottled and stored for years in caves before it is marketed.

Southwestern France

Southwestern France is a plain drained mostly by the Garonne (gah *rawn*) River. Find it on the map on page 87. Note that a canal links this river with the Mediterranean Sea.

The westerly winds bring the Garonne Basin enough rain for agriculture. Farming is the leading occupation of people in this region. The crops raised include grains, hay, vegetables, and fruits. This area is the only part of France where corn can be grown easily. Many grapes are raised and made into wine. Much of it is exported through Bordeaux. Bordeaux is the leading sea-port of the Gironde (jih *rawnd*) *estuary*. An **estuary** is the wide mouth of a river where fresh water and salt water mix. An estuary has tides, as do all shores along oceans.

The Rhône Valley and Mediterranean France

The Rhône River, with its northern *tributary*, the Saône, flows through a long lowland in southern France. A **tributary** is a stream that flows into a river. Usually, but not always, a tributary is smaller than the river it flows into. The Rhône flows southward to the Mediterranean Sea. Find it on the map on page 87. You will see that the river flows between the Central Plateau and the Alps. This lowland forms an important north-south transportation route. The main roads and railroads between Paris and the southern coast use this valley. Note that farther north along the Saône River, canals join it with the Seine River. A new canal will soon join the Saône and the Rhine rivers. The canal will end at Mulhouse (see map, page 87).

The northern part of the Saône Valley is very wide. It has good farmland. At Lyon (lee *awn*) the Saône and the Rhône Rivers join. South of Lyon the Rhône Valley narrows. There the Alps and the Central Plateau are closer together. In southern France, near the Mediterranean, the valley widens again.

Lyon. The largest city in the Rhône Valley is Lyon. It is the second largest French city. Lyon has more than 1 million people, but Paris is about nine times larger. Lyon is famous for its silk industry that started there more than 400 years ago. Factories in Lyon also manufacture cloth made from synthetic yarns and many other products. The city also serves as a market and trade center for the region around it.

Marseille and Mediterranean France. Marseille (mahr *say*), the third largest city in France, is located on the Mediterranean coast near the Rhône *delta*. A **delta** is land formed at or near the mouth of a river. The delta is made as the river drops the soil it has been carrying from higher ground. Marseille is France's main seaport on the Mediterranean. The port handles large quantities of products brought from countries in Africa and Asia. Especially important is the petroleum brought from Algeria, a former French colony, in Africa. Algeria supplies most of France's petroleum needs.

Many manufacturing industries have grown up in and near Marseille. A new industrial port has been built at Fos (faws) about 48 kilometers (30 miles) west of Marseille. It has been built so that the huge supertankers can unload their oil there. Pipelines from Fos deliver oil to many refineries in France, Switzerland, and West Germany. New factories, including oil refineries, chemical plants, and a steel plant have been built at Fos.

Along the Mediterranean citrus fruits and olives grow. There are also many vineyards and peach orchards. Some of the Rhône Valley is planted in rice. There are many truck gardens, too. West of the Rhône is a rather dry region. It is located between Nimes (neem) and Narbonne (nahr *bawn*). A huge irrigation project has been built to bring water to this area.

East of Marseille and south of the Alps is the French *Riviera* (rihv ee *ehr* uh). The **Riviera** is a narrow strip of land along the Mediterranean coast. It has very mild winters because it is sheltered by the Alps. Many famous winter resorts have been located there.

The warm, sunny climate in southern France is a good one for raising flowers. Cut flowers are flown daily to many large cities in Europe. Some flowers are used to make perfume. The cities of Grasse (grahs) and Nice (nees) serve as centers for this industry. French perfumes are sold in many countries, including the United States. They usually are quite expensive.

France

Highland France

The Central Plateau

The Central Plateau covers much of the southern half of France. It is highest in the southeast. The land slopes steeply downward from the Central Plateau to the Rhône Valley. Swift-flowing streams have cut deep valleys into this side of the plateau.

Many streams flow westward from the Central Plateau. Find some on the map on page 87. These streams show that the Central Plateau is a region that gets much rain. See the rainfall map on page 167. The valleys make it fairly easy to go into and out of the highland region. Some of the rivers are important waterways. The broad valleys of the Loire (lawhr) and Allier (ah *lyay*) rivers have very fertile soil. They are centers for truck gardening.

Many uses are made of the natural resources of the Central Plateau. Stone is cut for homes and other buildings. Lumber is obtained from trees growing in the forests. Coal is mined and steel manufactured at St. Etienne (san ay *tyehn*). (See map,

page 87.) The city of Limoges, (lee *mohzh*) near the western edge of the plateau, is famous for its excellent china. It is made from **kaolin** (*kay* uh luhn), a fine clay found there.

Along the southern border of the plateau is a region with deep, narrow canyons and caves. It is a poor region for farming because the soil is dry. The dryness is not due to lack of rain. The rain sinks rapidly into the ground because of the large amounts of limestone there, which makes the soil **porous** or full of holes. The few people who live in this region raise sheep and goats which they milk. Have you ever heard of *Roquefort* (*rohk* fert) cheese? **Roquefort** cheese is made from sheep's milk in this part of France. The cheese is aged in limestone caves.

The Pyrenees

The Pyrenees Mountains rise steeply from the southern plains of France. They are a real barrier to travel between France and Spain. There are no good, low passes through the range. Streams flow swiftly down the steep, narrow valleys.

Moist winds from the Atlantic Ocean bring rain to the western Pyrenees. Forests cover the slopes. The eastern Pyrenees region has little rain, however. There are few trees and shrubs. (See the rainfall map on page 167.

Many years ago railroads were built around the ends of the Pyrenees near the coasts. Now, two railroads also cross the Pyrenees farther inland. They run through tunnels dug through the mountains. The Pyrenees themselves furnished the power that was used to make them less of a barrier. Electricity made from falling water—**hydroelectric power**—drove the machines used to dig the tunnels. Now this electricity is used to run the trains.

The cheesemakers of Roquefort, France, have been aging wheels of cheese in these caves for more than 2,000 years.

The Mountains of Eastern France

Three mountain groups are found in eastern France. The Alps, in the southeast, are part of the high, rugged range extending through several countries. The Alps are noted for their beautiful valleys and high rocky peaks. The Alps also have huge snowfields, glaciers, and waterfalls. The highest Alpine peak, Mont Blanc (maw *blahn*), is in France. It is 4746 meters (15,771 feet) high.

North of the Alps are the lower Juras which extend into Switzerland. The Juras are much like the Appalachian Highlands in eastern North America. Still farther north are the older Vosges Mountains. Most of them have rounded tops. The Vosges Mountains slope steeply down to the Rhine River on the east.

Forests grow on the slopes of many of the eastern mountains. Lumbering and papermaking are two of the more important industries. Summer pastures, often above the forests, are used for grazing animals.

Swift streams come tumbling down the steep slopes. There are many hydroelectric plants which use the swift-moving water to make electricity. Factories in the valleys use this hydroelectric power.

France

Living in a Mountain Village in France

Life in many mountain villages in the Alps and Jura mountains in France is interesting. People live almost the same today as they did hundreds of years ago. Some differences exist, of course. Most homes now have electricity, for instance. Yet daily living goes on much as it has for centuries.

Little land is available for crops. Much of the land is too steep for anything except grass, trees, and wild flowers. Villages are located in valleys where water and more level land is found. There, during the summer months, women and girls take care of the vegetable gardens. These are located on the outskirts of the village. They let grass grow in some of their small fields. When the grass is quite tall, they cut it by hand. To do so they use long knives with handles, called **scythes** (syths). They then place the grass on carts and take it to the village. There they store it in sheds and in the attics of their houses. The grass is used for insulation and for cattle feed during the winter months.

During this same time, men and boys live in the nearby mountains. Good summer pastures for the cattle are found there. The boys watch the cattle during the long summer days. The men milk the cattle and make cheese. When autumn comes, the cattle are driven back to the village. They are allowed to graze in the small fields where the grass had been cut. The cattle can graze there until the winter snows come. The men and boys then go back to their mountain hut and carry down the cheese they have made. Some of it is sold. Most of it, though, is kept to provide food for the family during the winter.

During winter months, life is very different. The family is together again. Both girls and boys go to school. When snow is on the ground, they often ski to school. Would you like to be able to do that? They can do it for several months! In most mountain villages, families have workshops. There, during the long winter months, the women and men make articles for sale. Some of them make watches. Some carve figures out of wood. Some weave cloth or make articles of clothing to sell. From the money they get from these sales, they buy what they need. Most of their food, though, comes from their gardens and their cattle.

The cattle are also used as "furnaces" during the winter! They are not really furnaces of course. But most villagers in the Alps keep the cattle in one side of the same building as the house. The cattle and the extra wall help to keep the house

Much of the work on mountain farms must be done by hand.

warmer than it would otherwise be. And, during very heavy snows, the cattle are nearby to feed and milk. Of course, there is a good door between the "cattle home" and the "people home"! The grass that was cut during the summer, and is now hay, is fed a little at a time to the cattle. And the cattle provide rich milk all winter long.

Of course, some mountain villages are very different from the one just described. Some mountain villages are tourist resorts. There almost everyone works to take care of visitors. During winter months people come to ski. During summer months they come to enjoy the cool mountain air. They hike to high places—sometimes climbing even the highest peaks. Villagers in such places are usually too busy taking care of people to have workshops. They probably make more money, though.

Would you like to live in a mountain village in France? Which kind of village would you prefer? Why?

Corsica

The island of Corsica (*kawr* sih kuh) is in the Mediterranean Sea off the southern coast of France. This island is one of the **departments** (or states) of the French Republic. About 250,000 people live on the island. The land of Corsica is rocky and mountainous. The main crops are olives, grapes, grains, tobacco, and vegetables. Many sheep are raised. Many of the people make their living by fishing in the Mediterranean.

95

The Industries of France

Agriculture

Farmers in France usually are able to raise enough food for the French people and have some left over. France, therefore, exports food to other countries. About two-fifths (⅖) of the land is cultivated, and one-fifth (⅕) of the people are farmers. There are many more farmers in France than in the United Kingdom. Farmers in France were slow to adopt modern methods of agriculture. Most of them now have tractors, however. For many years French farmers have cultivated their land with great care. As a result, crop yields per hectare (or acre) are higher than in the United States.

Most French farmers own the land they cultivate. Many of them live in villages. The farmland is located outside the villages. They go to work on their land almost every day during the summer months. Most of the farms are small.

More land is planted in wheat than in any other crop. Most of it is raised in the Paris Basin. More wheat is grown in France than in any other country in Europe, not counting the Soviet Union. France also raises more grapes and produces more wine than any other country!

Many French farmers now have joined *cooperatives* to help them make more money. A **cooperative** is much like a corporation. Many people own a corporation, and many people also own a cooperative. The difference is that persons who own part of a corporation usually have bought shares of stock in the corporation. By contrast, persons, such as farmers, *join* a cooperative. They buy much of what they need (such as fertilizer, machinery, seed, tools) from the cooperative. They sell most of what they produce through the cooperative. The cooperatives process the farm produce and store it until it is sold. Cooperatives usually get higher prices than the farmers could get if they sold the products directly. Also, the farmers can buy supplies that they need at a cheaper price. Thus, the cooperatives usually can save the farmers money that way, too.

The perfumes manufactured in France are sold all over the world. The many small manufacturers blend their special (and secret) ingredients to get just the right fragrance.

Manufacturing and Mining

Through the years France has had many small factories making special kinds of products. Most of these factories are family owned. The industries are kept small on purpose. Some of the products made in these small factories are wines, perfumes, and jewelry. Much of the fine cloth and fashionable clothing is also made in small factories.

Compared with Great Britain, France was slow in developing *heavy industries*. Factories which make iron and steel, locomotives, automobiles, ships, and other large goods are examples of **heavy industries**. Less effort was made to develop large industries here than in most western European countries. A possible reason is that France did not have to rely on trade in order to feed its people. Industries were developed to meet French needs and to make special products for export.

Today, however, France is an important industrial nation. France has large deposits of iron ore. This ore is used to make steel. There is some coal in France, but not enough. Moreover, much of the French coal is not good for making *coke*. **Coke** is coal that has been processed to give off more heat than ordinary coal, and is used in making iron and steel. Coking coal has to be imported from West Germany.

Iron ore mining and steel producing are big industries in Lorraine (loh *rayn*), a region near West Germany. The largest deposits of iron ore in France are found in Lorraine.

France

Find Lorraine on the map on page 87. Chemical industries are also located in Lorraine and Alsace (*al* sas). Alsace is an area along the Rhine River. Both Alsace and Lorraine have large deposits of salt. Alsace also has potash which is used in making fertilizers.

Tractors, farm machines, locomotives, airplanes, automobiles, and trucks are some of the larger manufactured products. Many French automobiles are exported. The French also make many war materials. They sell planes, tanks, guns, and many other materials to many countries. Paris is the main center for such industries.

The *Concorde*

Many of you probably have seen pictures of the *Concorde* on television news broadcasts. There is a picture of one below. You probably know that the *Concorde* is a jet airplane called a supersonic transport (SST). You may know that it flies the fastest scheduled passenger flights in the world. It flies much higher than other passenger airplanes. It also flies faster than the speed of sound. You can leave Paris or London and be in New York or Washington, D.C., before you left—according to clock time. How wonderful that seems. But is it?

The *Concorde* is a wonderful achievement. The British and French engineers who designed and built it are proud of it. They should be. They were the first to design a comfortable plane that would fly so far and so fast, safely. At one time, the Boeing company in Seattle, Washington, was also working on a supersonic airplane for the airlines. Congress decided not to spend the billions of dollars that Boeing needed to finish the plane. Work on the airliner stopped. The Soviet Union also built a supersonic airliner. They finished theirs before the *Concorde* and flew it to Paris for an air show. It crashed shortly after takeoff. The British and French governments kept on building the *Concorde*. And it began passenger service in 1976.

The problem is that the *Concorde* is noisy. In fact, it is quite loud! For a long while the New York City Port Authority would not permit the *Concorde* to land there. Yet that is just the place the British and French airlines using the *Concorde* wanted it to land. The greatest amount of business an airline can get is on the route between New York City and London or Paris.

Many people who live near the airports in New York did not want the plane to land there. They staged protests against its landing. Among the things that they did was to create a traffic jam on the highways leading to the airport. The people got in their cars and drove slowly toward the airport. They filled all the lanes of the highway. People who needed to get to the airport could not do so in time to catch their flights.

Not permitting the *Concorde* to land in New York City caused some bad feelings for awhile. The British and French did not understand why their plane should not be permitted to land in New York City. People in the United Kingdom and France seemed to think that the United States' government was just trying to protect American airlines.

The British and French also think the *Concorde* should be bought and used by American airlines. But our airlines seem to think that the *Concorde* will not be a success in making money. The plane costs too much for the number of people that it can seat. It can carry only about 108 passengers. Large planes, such as the Boeing 747, cost much less and carry more than four times as many passengers as the *Concorde*.

Finally our government gave the French and British permission to land the *Concorde* in New York City. As so often happens, the people found out that the *Concorde* was not much worse than other noisy planes.

How would you feel about a situation like this if you lived near an airport? Is your school near an airport? Some schools are, and almost everything stops whenever a plane flies over. Should new airports be built farther from cities?

Would you like to fly in the *Concorde*, the fastest commercial passenger plane in the world? Why or why not?

Tourists

Millions of people visit France each year. Most of them visit Paris. There they can eat in world-famous restaurants or climb the Eiffel (*eye f'l*) Tower. Some people visit rural France. During winter months many visitors go to the Riviera. All during the year people go to the French Alps. Caring for visitors from European countries and from other continents is big business. Hotels, cafés, and travel agencies employ many people just to take care of the many visitors.

Paris

France's capital began as a small settlement on an island in the Seine River about 2,000 years ago. It was easy to cross the river there. They built two short bridges from the river's banks to the island. About 1,000 years ago, Paris was chosen as the capital of France. Since that time it has almost never stopped growing. Today *metropolitan* Paris is among the world's ten largest cities. **Metropolitan** is a name given to an area that includes a main city and the smaller cities around its edges. Sometimes the word "Greater" is used to mean the same thing.

Besides being the center of government, Paris is the leading city of France in many other ways. It is a great center for international travel. It is also the main manufacturing and transportation center in France.

Paris is perhaps best known, though, as a cultural center. Visitors from all over the world come to see famous cathedrals, museums, and universities. Many persons move to Paris to study art, music, or literature. For many years it has been the home of great artists and writers.

EUROPE

ARCTIC CIRCLE

Stockholm
Copenhagen
Birmingham
Hamburg
Berlin
Warsaw
London
Cologne
Brussels
Prague
Paris
Munich
Vienna
Budapest
Turin
Milan
Bucharest
Madrid
Rome
Barcelona
Naples
Athens

Scale

Kilometers	0	300	600
Miles	0	300	600

Population

People Per Square

kilometer	mile
uninhabited	uninhabited
under 1	under 3
1 to 10	3 to 25
10 to 51	25 to 130
51 to 102	130 to 260
over 102	over 260

• Cities over 1,000,000 population

Paris began as a village on an island in the Seine River.

REMEMBER, THINK, DO

1. Try to explain reasons for the following related facts:
   a. France has less land than the state of Texas.
   b. In France 20 out of 100 people are farmers.
   c. In the United States only 4 out of 100 people are farmers.
   d. French farmers raise more crops per hectare (or acre) than farmers in the United States.
   e. Yet farmers of the United States raise much more food.
2. What is a cooperative? What does a farm cooperative do?
3. Plan a trip from your home to France. Plan to spend about 21 days in France. Show on an outline map where you plan to go. Be prepared to explain how you will travel, what you will visit, and where you will stay.
4. Find out what you can about the Eiffel Tower. A recent newspaper story said that the elevators in the tower no longer worked well. It will cost millions of dollars to put in new elevators. Do you think that this should be done? Why or why not?

France

WHAT HAVE YOU LEARNED?

1. Why is France, a nation with less land than the state of Texas, so important in world affairs? (Give several reasons.)

2. What are winters like in areas that have a Mediterranean climate? What are summers like?

3. What is a river basin? Do we have river basins in our country? Do you live in one?

4. What is a truck farm? Are there such farms near where you live?

5. What region of France has the most productive farmland? What region of France has the greatest amount of manufacturing?

6. What are the main mountain ranges in France?

BE A GEOGRAPHER

1. List five major rivers which flow across or along the borders of France. How are these rivers connected?

2. If you flew around the world, flying directly south from Le Havre, without turning to the right or to the left:
   a. What country in Europe would you fly over early on your journey?
   b. What country in Europe would you fly over shortly before returning to Le Havre?
   c. In what direction would you be flying when you returned to Europe?
   d. On your flight from Europe, what countries in Africa would you fly over?
   e. Near what meridian would you be flying on your southward journey?
   f. In what direction would you be flying after passing the South Pole?

3. Look at the map on page 87. In what general direction does each of the following rivers flow:
   a. Rhône/Saône
   b. Garonne
   c. Loire
   d. Seine
   e. Rhine
   f. Allier

4. What does the information that you discovered in question three tell you about the land of France?

QUESTIONS TO THINK ABOUT

1. Paris is one of the most expensive cities in the world. It costs more to eat in a restaurant or stay in a hotel in Paris than almost any other city. Why are some cities more expensive than others? Why are some restaurants and some hotels more expensive than others? Why should things cost more in France than they do in the United Kingdom?

2. If the French rather than the British had won the early wars in our country, what differences might this have made? How might your life be different? What differences might there be in France today?

3. If France had not sold the United States the vast area west of the Mississippi River, what might have happened? Write a story about what might have happened if President Jefferson had decided not to buy, or if France had decided not to sell, this territory.

4. France's government is known as the Fifth Republic because France has had five constitutions. The United States has had only one constitution since 1789. Why do you suppose the United States has not had to start over several times as France has?

5. The island of Corsica is more than 27 times larger than Malta, which is also in the Mediterranean Sea. (Find both on the map on page 87.) Malta is an independent nation and a member of the United Nations. Corsica is one of the departments, or states, of France. Why do you suppose the people of Corsica have not formed their own nation?

6

Chapter

SWEDEN

Find Sweden on the map of Eurasia on pages 8 and 9. It is located on one of Europe's many peninsulas. Sweden and Norway are on the Scandinavian Peninsula. When the Romans conquered parts of Northwestern and Central Europe, they named this land *Scandia* (*skan* dih ah). The Romans thought it was a large island north of the Baltic Sea. The Scandinavian Peninsula is not an island, however. It is a peninsula that separates the Norwegian (nawr *wee* jan) Sea and the Baltic Sea.

Sweden, an Old Country

Sweden is one of the oldest of the present European countries. Ancestors of the present Swedes lived on the land at least 5,000 years ago. At first, Sweden was made up of a number of small states under different rulers. About A.D. 700 all the states were brought under one ruler. The kingdom of Sweden was established.

In the early history of Sweden there were many wars. Partly because of its location, however, the country has not been invaded often.

During the world wars of this century, Sweden did not fight on either side. The Germans did not try to invade Sweden because Sweden sold needed iron ore to Germany.

Today the Swedish people have a very high standard of living. The average income in Sweden is higher than in the United States.

Climate of Sweden

Westerly winds from the Atlantic Ocean have less effect on Sweden's climate than on Great Britain's. As the map on page 45 shows, Sweden is east of the Scandinavian Highlands. Its summers are slightly warmer and its winters colder than those in Norway to the west. Less rain and snow falls in Sweden. Stockholm has about one-fourth (¼) as much rainfall each year as Bergen, located on Norway's west coast.

Sweden is a long country. From north to south it extends for 1600 kilometers (1,000 miles). The climate is, therefore, much milder in the southern part of the country. Winters in the northern section are long and very cold.

Burrowing Underground

Have you ever dug a deep hole in a vacant lot, beach, or field? Where were you planning to get? (When the authors were young, they were always digging to China.) Look at a globe and see where you'd come out if you dug a hole through the Earth.

Do you know of any animals that **burrow** (dig) homes underground? Name some of them. Digging is fairly easy if the soil is moist or loose. But digging through hard rock is something else. The Swedes are probably the best hard-rock diggers in the world. They have done a great deal of it. The question is why?

Well, at first the Swedes started digging underground in hard rock for protection. Some of the fighting during the Second World War took place in Finland to the east and Norway to the west. Germany, a short distance to the south, was one of the main countries in that war. The Germans controlled Denmark and Norway, both close to Sweden. They also had troops in

This subway station beneath Stockholm, Sweden, has been carved out of solid rock.

northern Finland. Sweden remained **neutral**, which means "not taking sides." But the Swedes also wanted to defend their land.

To do this, the Swedes started to build some important projects in the solid rock underground. They started with hydroelectric power plants. They learned how to use drills and dynamite to make large rooms underground.

The Swedes continue to burrow in the hard rock of Sweden. Today they store oil and gas in tanks underground. (You don't see "tank farms" in Sweden as you do in our country.) They have built many tunnels under the cities. Some of them are used by subways to provide transportation. Others are used for water and sewage pipes. Some sewage treatment plants are completely underground. In Stockholm garbage trucks are stored in an underground garage. And many commuters and shoppers park their cars in underground garages, too. Instead of building a needed bridge in Stockholm, the Swedes built a tunnel. It was cheaper and it looked better!

Other interesting ways of using underground storage have been developed. Ice cream is stored there. The temperature is always the same—so it is easier to control. Wine is also stored in large tanks underground. Trucks drive in and out of the underground storage areas just like the cars do in underground garages. And now the Swedes are talking about storing air underground. They will use the air to improve the output of generators that are above ground.

One of the best ideas of using underground tunnels has been adopted in the town of Västerås (vehs tuh *rohs*). This city is about 113 kilometers (70 miles) west of Stockholm. Almost every house and building in Västerås is connected to a central heating system. The heating system works with the electrical system. The town burns oil in one large power plant to heat water. The resulting steam is used to produce electricity. After the steam has cooled a bit and has become water again, it is very hot. Pipes from the power plant carry this hot water to almost every home and building where it is used to supply heat. The hot water pipes are, of course, underground. The water, as it is cooling off, is returned to the plant to be reheated. On the way back to the plant the water travels through pipes that are located just below the surface of the streets. Thus, all winter long, Västerås has streets that are free from ice and snow! And, it costs less to heat the city this way than if all homeowners had their own system.

The Aurora Borealis, as it is seen in Northern Sweden.

The Midnight Sun and the Northern Lights

Northern Sweden is so far north that during the summer months the sun shines at midnight! Which lines on globes and maps show where the sun shines at midnight at least once each year? How is that line drawn on the map on page 109?

If you lived in Kiruna (*kee* ruh *nah*), Sweden, you would be able to see the sun at midnight from early June to the middle of July. For an equal time during the winter, however, you would not see the sun at all. During the long winter nights you might often see the *northern lights*. The **northern lights** are really named the *aurora borealis* (uh *rohr* uh bohr ee *al* uhs). They are often seen on dark nights. These rays of light from the north flicker and usually look a little green. The northern lights are caused by rays from the sun. These rays are drawn toward the *magnetic poles* of the Earth. The **magnetic poles** are the places toward which a compass needle points. The north magnetic pole is at about 76° N. and 101° W. The magnetic poles seem to shift position a little bit from time to time.

Vasalöppet

If you were Swedish and also knew English, you would know that *Vasalöppet* (*vah* suh *low* peht) means the "Vasa Race." What do you suppose the Vasa Race is? A dog race? A horse race? A race between runners? A turtle racing a hare? No. The Vasa Race is a cross-country ski race. Except for two years, the race has been held each year since 1922. (In 1932 and 1934 there wasn't enough snow!) The race is always held on March 1. It is, believe it or not, 86 kilometers (53 miles) long!

How many people do you think might enter a cross-country ski race 86 kilometers (53 miles) long? Would you? If you grew up in Sweden you might like to try it. You would have grown up skiing. You would have started to ski almost as soon as you could walk. You would use skis after school and on weekends almost all winter long. You might even ski to school. In a recent year almost 10,000 persons entered the race! Some of them came from other countries. Of course, not everyone finished, but thousands of the skiers did. The fastest time ever made in the race was 4 hours, 35 minutes, 3 seconds!

Why would so many people enter Vasalöppet? Because the race is run to celebrate an important event in Swedish history. In 1520 Gustavus Vasa was leading a revolution against the Danes. The Danes controlled Sweden at that time. Vasa lost a battle and decided that the cause was lost. He fled on skis toward the Norwegian border. Some friends from the town of Mora raced after him. They caught up with him at Sälen—86 kilometers (53 miles) away. The friends talked Vasa into continuing the fight. He did, and the Danes were finally defeated in 1523. Vasa became King Gustavus I of the independent country of Sweden. It has been independent ever since.

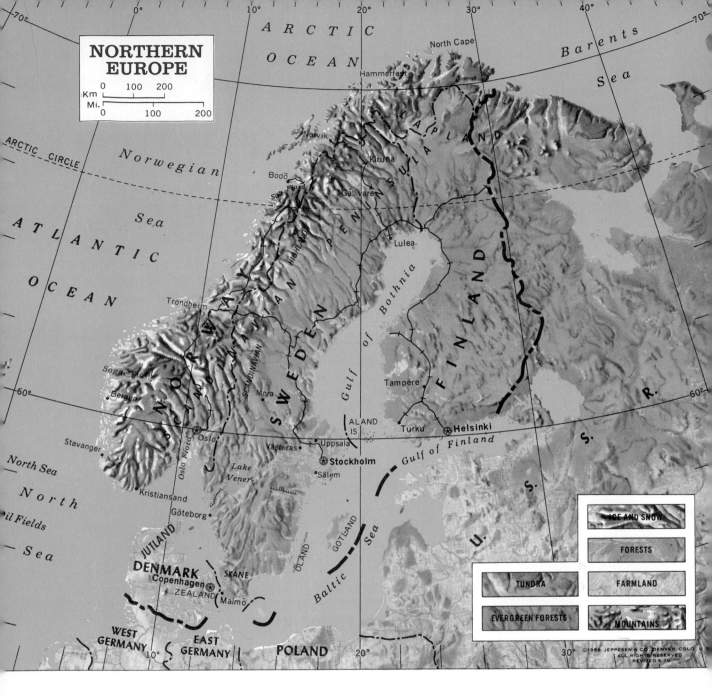

The Land of Sweden

The western part of the Scandinavian Peninsula is an old mountainous region. It was once much higher above sea level than it is now. The glaciers of the Ice Ages scraped soil from the high land. Much of the highland is, therefore, bare and rocky. The soil was carried to lowlands to the south and east. As a result of the glaciers' work, there are few sharp mountain peaks in Sweden. The sides of the higher moun-

Sweden

tains, especially in the north, are tundra. Low places among the highest peaks have snow and ice.

The eastern slope of the Scandinavian Highlands is long and gentle. The western slope, which is in Norway, is much steeper. As the map on page 109 shows, the eastern slope is drained by a number of rivers. These rivers flow into the Gulf of Bothnia (*bawth* nih uh). In the upper part of the river valleys there are many lakes. In fact, there is a lake in every one of the larger river valleys. Near the coast the river valleys are broad. The *divides* between the rivers are low. **Divides** are ridges of land between two river basins.

Near the center of the peninsula there is some higher land, but much of southern Sweden is lowland. This broad lowland extends across southern Sweden. Several large lakes cover much of the land in this area. (See map, page 109.) These lakes are connected by a series of rivers and canals. Small ships can, therefore, cross the country by water from Göteborg (*yuhrt* uh *bor* ee) to Stockholm.

The most southern part of Sweden is part of the North European Plain. This part of Sweden is known as *Skåne* (*skah* neh). It has fertile soil and is Sweden's best farming area.

Sweden is separated from Denmark by a narrow strait. Find it on the map on page 109. Two islands in the Baltic Sea near the southeastern coast are also part of Sweden. These islands are Gotland (*gawt* lund) and Öland (*uhl* ahnd).

Most of the people of Sweden live on the plains near the coast. Very few of them live in the western and northern upland areas. (See population map, page 100.) About nine out of ten Swedes live south of 61° N. Stockholm is the capital and largest city in Sweden. (It is about the size of Seattle or Washington, D.C.) Göteborg is Sweden's busiest port city. The harbor there is ice-free most winters.

REMEMBER, THINK, DO

1. Why are winters in Sweden colder than they are in Great Britain?
2. What is a divide? Is there a divide near where you live?
3. What can be found in the upper part of the river valleys in Sweden? What reasons can you think of for this fact?
4. Where is Sweden's best farming area? What is it called?
5. What two large cities in or near the east and west coasts of Sweden are connected by lakes, rivers, and canals?
6. Find out how locks work in a canal. Use library resources. If there is a canal with locks near where you live, see if you can visit it.

For convenience, Sweden can be divided into two parts. The larger part is northern Sweden which makes up about two-thirds (⅔) of the country. Southern Sweden, the smaller part, has the most people. The dividing line between these two areas is about 61° N. Only about one-tenth (1/10) of the people live north of this line. This vast area, however, is very important to the country's economy. On most of the streams that can be seen on the map on page 109, the Swedes have built dams and hydroelectric plants. About half of the electricity used in the country is produced in this area.

Iron Ore. Northern Sweden is also the source of much raw material that is exported. A major source of income for Sweden is high-grade iron ore. The best ore is found at Kiruna and Gällivare (*yehl* lih vah reh). Find these towns on the map on page 109. The Swedish govern-ment has built railroads to and from Kiruna. One is built to Narvik (*nahr* vihk) in Norway. The other goes through Gällivare to Luleå (*loo* leh oh) on the Gulf of Bothnia. During the summer months the port at Luleå can be used. For several months each winter, however, the Gulf of Bothnia is blocked by ice. Then all the iron ore must be shipped through Narvik.

The iron ore found at Kiruna and Gällivare is of a very high grade. And there is lots of it. The Swedes believe that the supply of ore at Kiruna will last for at least another 100 years.

For many years the ore was taken from the mountain at Kiruna using **strip mining** methods. Huge power shovels were used above ground to dig out the ore. Now much of it is mined from shafts below ground. Some of the shafts are more than 900 meters (3,000 feet) deep. During the long winter that's a good place to work.

Modern equipment has increased the amount of ore that can be taken from the mines in Kiruna and has also made mining safer.

Life for the people of Lapland has changed greatly in the past few years. Modern products are now common in most homes. What modern items do you think these Lapps would find most useful?

Kiruna has a population of only about 30,000 people. Nevertheless, the people there boast that it is the world's largest city. They are talking, of course, about *area*. We are not sure if Kiruna is the world's largest city. There is little doubt, however, that Kiruna's city limits provide it with room to grow. One of the largest cities in area in the United States is Los Angeles, California. It has about seven million people in an area about 1166 km² (450 sq. mi.). Kiruna, by contrast, has about 12 950 km² (5,000 sq. mi.) within its city limits! Most of the ''city'' is, of course, without people.

Home of the Lapps. *Lapland* is the name given to the extreme northern part of Europe north of the Arctic Circle. The map on page 109 shows that Lapland stretches across all of northern Norway, Sweden, Finland, and parts of the Soviet Union. These areas are the homeland of interesting people called *Lapps*. The Lapps are small people, usually not over one and one-half meters (five feet) tall. They moved from Asia to this part of Europe thousands of years ago.

Until recently, the Lapps were *nomadic*. People who move from place to place, rather than having a permanent home, are **nomadic**. For centuries, the Lapps depended mainly upon their herds of reindeer for their living. They also did some hunting and fishing. But reindeer were the most important property of the Lapp families. The reindeer were the source of meat, milk, and hides. The hides were used to make tents, shoes, and clothing. The reindeer also provided a means of transportation. They were used during winter months to pull sleds. During summer months the Lapps used the reindeer to carry tents and supplies.

Reindeer feed largely on a small plant called reindeer moss. This moss grows on tundra lands in the far north. When the moss is eaten from an area, it takes a few years for it to grow again. So, in summer months the reindeer move from one place to another in search of food. They also move to get away from mosquitoes or bad weather.

The Lapps, in turn, followed the reindeer. The homes of the Lapps during these summer months were tents. They made the tents from reindeer skins or heavy cloth. The skins or cloth were stretched over a framework of poles much like a tepee. An opening at the top of the tent let out smoke from a fire.

Today there are about 10,000 Lapps in Sweden. Most of them have permanent homes of wood. Some of them have become fishers or farmers. In the far north some of them still hunt animals for fur. Some Lapps still raise and care for reindeer. But most of the Lapps in Sweden work in the mines or on the railroads.

For many Lapps who still care for reindeer, ways of doing so have changed greatly. These Lapps may have regular jobs in Kiruna or elsewhere. They use motorcycles, cars, boats, snowmobiles, and even airplanes to locate and follow the reindeer. They herd the reindeer into corrals to sort them. About one-fifth

(⅕) of the herd is killed each year for meat and hides. The reindeer are killed in well-equipped slaughterhouses. The Lapps sell the meat to special restaurants. The money they make is used to buy products that are needed.

The Lapps still have the right to use the northern third of Sweden as grazing lands for their reindeer. And, only Lapps may breed reindeer in Sweden. Nevertheless, many Lapps, like many Indians in North America, believe that much of their land has been taken from them unfairly. They frequently complain when reindeer are struck and killed by automobiles and trains.

Most Lapps in Sweden are able to read and write. Some speak two or more languages. Lapp children, like other children who live in Sweden, go to school. Some of them have to live near the school during the week. These children sometimes ski home for the weekends.

In the past, the nomadic Lapps created problems for the countries where they lived. They did not understand why they could not pass freely from one country to another following their reindeer. This did not matter much when traveling through Sweden, Norway, and Finland. People move easily across these borders anyway. Crossing the border into or out of the Soviet Union was a different matter, however. To do that, even the Lapps had to have **passports** and **visas**—documents permitting them to enter and leave the Soviet Union—approved in advance.

Sweden

REMEMBER, THINK, DO

Work with a friend to do the following:

1. Estimate quickly, using the map on page 109, about how far it is by air from Stockholm to Kiruna. Write your answer on a sheet of paper. How did you do it? Now measure and see how close your estimate was.

2. Using the map of Eurasia on pages 8 and 9 or a globe, compare the location of London, United Kingdom, and Stockholm, Sweden. Then answer the following questions:
   a. Which city has more daylight on December 22? Why?
   b. Which city has more daylight on June 21? Why?
   c. Which city sees the sun rise first on April 21? Why?
   d. Which city probably has a warmer average climate? Why?
   e. In which city would the sun set the earliest on March 21? Why?

3. Figure how far it is by air from:
   a. Stockholm to London, United Kingdom
   b. Stockholm to Paris, France
   c. Stockholm to Bordeaux, France
   d. Stockholm to Rome, Italy

Forests and Forestry. One of Sweden's greatest natural resources is its timberlands. About half of the country is forested. Much of northern Sweden is covered with needle-leaf evergreen trees. In the far north and on the upland slopes, however, many of the trees are birch. The trees become smaller and more scattered north of the Arctic Circle. The forests of Sweden are well cared for. Young trees are planted as older ones are cut.

In northern Sweden the logs are cut during the winter months. Most of the logs are brought to mills by trucks or trains.

The logs are trimmed and cut in the sawmills. Some of them are made into timber for shipbuilding or railroad ties. Some are cut into lumber for making houses and furniture. Many mills, instead of cutting logs into lumber, shred the wood into small pieces and make wood pulp. Wood pulp is needed to make paper and many other products. Sweden produces a great deal of high-grade wood pulp. Some of it is exported. Much of it is made into paper that is exported. About one-fourth (¼) of Sweden's income from exports comes from forest products.

Much work has been done in Sweden to use the trees more completely. The bark is burned with oil in many mills to provide power. Sawdust and scrap pieces from the sawmill are used to make hardboard. Also, much has been done to

The Swedish lumber industry is highly mechanized. Machines do most of the work that people used to do.

lower pollution. The Swedes have learned that, as they control wastes, they improve profits. What had been thrown away or allowed to escape into the air or water is now used. New products are made from the waste or it is used for fuel.

Southern Sweden

Southern Sweden is the real heart of the nation. This part of Sweden has long been a center for agriculture and for making steel products.

Steel and Steel Products. Most of the steel made in Sweden is produced northwest of Stockholm and south of 61° N. Small towns and cities in this area are located near another important source of iron ore. The steel and steel products made in this region are known throughout the world for their high quality. Because Sweden does not have good supplies of coking coal, most of the steel is made by using electric furnaces. This way of making steel costs more than the method using coke. Therefore, Swedish steel makers produce small amounts of high-quality steel rather than a big amount of low-quality steel.

Many things made of steel are produced in the lowland area between Stockholm and Göteborg. Perhaps the best known products are knives, ball bearings, surgical instruments, automobiles, trucks, buses, and electrical appliances. Telephones, sewing machines, milking machines, needles, tractors, motors, and tools for woodworking are also made. Most of these products are exported from Göteborg's harbor.

Use a globe or a world map to answer the following:

1. What city in the United States is at about the same latitude as Stockholm, Sweden?
2. If you flew straight west from Stockholm, would you fly over any part of the world's largest island?
3. If you flew straight north from Stockholm around the world:

   a. What country would you fly over before reaching the Norwegian Sea?

   b. What island would you fly over before reaching the North Pole?

   c. What part of North America would you fly over?

   d. What direction would you be flying when you crossed part of North America?

   e. What continent would you fly over before reaching the South Pole?

   f. What direction would you be traveling when you reached the shores of Africa?

   g. What city in Africa would you first see?

   h. Would you fly over the Mediterranean Sea after leaving Africa?

   i. What sea would you fly over just before reaching Stockholm again?

Glass. One of Sweden's best-known products is beautiful glass. Not many people work in the glass industry. And they do not make a lot of glass. But what they do make is among the world's finest. The products of the Swedish glass industry are sold in fine stores almost everywhere. They are expensive because they are handmade with great care. The glass products are also famous because of their design. Swedish glass blowers are very skilled and spend many years learning their craft.

Do you know how glass is made? You may want to read some more about it in encyclopedias. The basic ingredient of all glass is sand. It is mixed with soda ash and lime and heated in a very hot furnace. The materials melt together and form a syrupy liquid. When still hot, the glass can be shaped in different ways. One way is by blowing. A person, known as a glass blower, blows into a long tube containing the hot liquid. The artist then hand-shapes the glass as desired using simple tools. When it has cooled, the glass can be cut, polished, and engraved. The final product, if made by skilled workers, is a true work of art.

Agriculture. Farming and dairying have long been important occupations in Sweden. About 5 out of every 100 Swedish workers are farmers. These people are able to grow most of the food needed by the people except for tropical products like coffee and oranges. Swedish farmers use only about one-tenth (1/10) of the land for crops. They are able to raise so much food because they use scientific methods. Until recently most Swedish farms were small in size. Much of the work was done by hand or by using horses. That is changing now. Most farmers in Sweden who have small farms are growing old. As they retire, the small farms are joined with farms nearby. In this way the younger farmer has a larger area to care for.

As a result it is possible for the farmer to care for the larger area with machinery that is too expensive for the small farmer.

Dairying and truck farming are carried on throughout southern Sweden. Potatoes, sugar beets, and hay are grown widely. Skåne, the southernmost province, is known as the granary of Sweden. The soil there is fertile, and the growing season is a bit longer than it is farther north. The leading grain crops are oats, wheat, rye, and barley.

Stockholm, the Capital

Stockholm has a beautiful and unusual location. It is built on about a dozen islands and along the shores of a beautiful lake. Stockholm has

Glassblowers spend hours making a single item. They must be both artistic and skillful in their craft to create useful and beautiful objects.

Sledding and skiing are not only for enjoyment in Sweden but are also ways to get from place to place.

been the capital of Sweden for several hundred years. Modern apartment and office buildings stand tall against the sky. As in most other large cities, traffic jams are frequent.

Two subway systems help people move from the suburbs to the central city and back home. The newer of the two systems has its stations decorated by modern art. Constructing subways in Stockholm was not easy because of its location on so many islands. The subway trains in Stockholm now are equipped so that they are controlled by **automatic pilots**. The "pilots" increase or decrease the speed of a train to keep it a safe distance from the train in front. The automatic pilots stop the train at a particular place in each station. They slow down the train for sharp curves. Only one employee is needed on each train. This person closes the doors and gives an "all-clear" signal to the automatic pilot.

Government and Education

The government of Sweden, like that of Great Britain, is a constitutional monarchy. In Sweden the legislature, or law-making group, is called the *Riksdag* (*rihks* dahg). The **Riksdag** is much like the House of Commons in Great Britain. Members of the Riksdag are elected by the people. The king, prime minister, and members of the cabinet are all responsible to the Riksdag.

Sweden has a good educational system. Every boy and girl in Sweden must attend school for nine years. Nearly every adult can read and write. There are four state universities. The one at Upsala (*up* suh lah) was established in 1477, 15 years before Columbus sailed for America.

A Young King and a Royal Wedding

For many years the Swedes were used to having old kings. Gustaf V, for instance, became king in 1907 at the age of 49. He was king until 1950 when he died at the age of 92. His son, Gustaf VI, then became king. He was 68 years old when he became king! He lived until 1973—or until he was almost 91 years old.

Then things changed. Because Gustaf VI's oldest son had been killed in a plane crash in 1947, his grandson became king. King Carl XVI Gustaf was only 27 years old. What a change! He was also unmarried.

Some people in Sweden had thought that there might not be another king. Many politicians wanted to cut the cost of having one. So a new constitution was adopted that took all power away from the king. King Carl has a lot of work to do anyway. He receives newly appointed ambassadors. He opens new factories. He talks with political leaders. And he visits and talks with school children. King Carl also tries to answer the many letters he gets from people.

In June, 1976, three years after he became king, Stockholm was full of flags and bands. It was having its first royal wedding since 1797. The king had met his future wife, Silvia Sommerlath, at the 1972 Olympics. That year the summer Olympics were held in Munich, West Germany. Ms. Sommerlath was working there as a guide.

King Carl XVI Gustaf and Ms. Sommerlath came to the United States shortly before they were married. The couple spent 26 days in the United States. They visited New York City and Boston. They stopped for three days in Washington, D.C., and were guests of President and Mrs. Gerald Ford. They visited Chicago and Detroit and other cities in the Midwest. They went to Houston to visit the Lyndon B. Johnson Space Center. And they visited Hollywood. Like good Swedes, they also spent two days skiing at Vail, Colorado.

Stockholm had a big celebration after their marriage. The king and queen drove part of the way from the cathedral to the palace in an open carriage. Then they traveled the last part of the short trip by boat, passing by ships of the Swedish navy. Swedish army units paraded beside the palace. Fiddlers and dancers in folk costumes entertained. The evening before the wedding there had been a special performance at the Stockholm Opera House. The Royal Mint issued a Jubilee Coin. And

After their marriage ceremony, Queen Silvia and King Carl greeted some of the Swedish people.

the Swedish post office printed two new stamps showing the royal couple.

What a thrill it must have been for Silvia Sommerlath of West Germany. Her father is a retired business executive. She did not grow up in palaces or as part of a royal family. She was a commoner—as we are. The new queen was able to speak six languages when she met the king. She is now able to speak seven. Guess what the new language is. If you said Swedish—you are right!

REMEMBER, THINK, DO

1. What are Sweden's leading exports? Where are they obtained?
2. How is living changing for the Lapps?
3. What kind of steel is made in Sweden?
4. What are some of the products made in southern Sweden?
5. What is unusual about the location of Stockholm?
6. Would you like to live in Sweden? Why or why not?

LEARN MORE ABOUT NORTHWESTERN EUROPE

There are a number of other interesting countries in Northwestern Europe. In many ways, Norway, Finland, and Denmark are much like Sweden geographically. In other ways, however, they are quite different. Iceland and Ireland are two small island countries. Near France are three smaller countries: Belgium, Luxembourg, and The Netherlands. The Study Guide on page 31 may be of help to you if you plan to learn more about one or more of these other countries.

Belgium

Even though it is small, the people of Belgium speak two quite different languages. In the northern part of Belgium, the people speak **Flemish**. This language is much like Dutch, the language spoken in The Netherlands. In southern Belgium the people speak a language much like French. It is known as **Walloon**. Belgium has been famous for textiles since the 12th century. One of their most famous exports is lace. Antwerp is one of Europe's busiest port cities. Brussels is the capital.

Denmark

South of Sweden is an interesting small country named Denmark. As you can see by looking at the map on page 109, Denmark is on a peninsula that extends into the North Sea.

Denmark also has many islands—more than 500 of them. Most are too small to be seen on the map. Denmark is mainly low country with some rolling hills. The Danes have worked for many years to make their sandy land better for farming. With only a few natural resources they have developed a high standard of living. Denmark is noted for the quality of its farm products. Copenhagen (*koh* pen *hay* guhn), the capital, is one of the largest and best seaports in Northwestern Europe.

Life for the people of Denmark is in some ways similar to life in the United States. The style of clothing is only one example.

Sweden

On January 23, 1973, a volcano erupted on the island of Heimaey in Iceland. When the eruption stopped in June, 1973, one-third (⅓) of the town of Vestmannaeyjar had been buried under ash and lava.

Finland

Finland also is mainly lowland. Like the eastern part of Canada, Finland has thousands of lakes. And they were caused in the same way as Canada's—by glaciers that scraped the land thousands of years ago. Finland has a long border with the U.S.S.R. Although a little country, the Finns are very independent. They cooperate in many ways with the Soviet Union. But they refuse to be controlled by their large neighbor. Forests cover more than two-thirds (⅔) of the land of Finland. Helsinki (hell *sing* kee), the capital, is a beautiful, clean city.

Iceland

The state of Virginia has a little more land than the island of Iceland. Iceland was built by volcanoes. Some of them are still active. Reykjavik (*rayk* yuh veek), the capital, is located near hot springs. Hot water is piped from these hot springs to the city. It is used to heat homes and other buildings. As the map on page 39 shows, Iceland is located just south of the Arctic Circle. Much of the land is covered by glaciers. Warm ocean currents provide a climate that is mild for a land so far north. Only very hardy crops, such as hay, potatoes, cabbage, and turnips can be grown outdoors. Other vegetables are grown in greenhouses heated by the hot water. One of the main occupations of the people is fishing. Iceland is a world leader in exports of fish each year.

Ireland

Ireland did not become an independent republic until 1949. It had been a self-governing part of the British Empire since 1921. The mild, moist climate in Ireland is perfect for the growth of grass. Perhaps that is why Ireland is often called "The Emerald Isle." Emeralds are precious stones that are green in color.

Most of the Irish people make their living from farming, from manufacturing, or from providing services. The government has been trying to increase the number of manufacturing plants to provide more jobs for the people. Unfortunately, except for fertile soil and abundant rainfall, Ireland has few natural resources. Manufacturing, therefore, is related mostly to agriculture. Exports include clothing made from woolens and linen, livestock, meat, dairy products, and alcoholic beverages.

In the 1840s, the potato crop—which is the basic food crop in Ireland—failed because of a plant disease. Thousands of people starved to death. Many people left Ireland and moved elsewhere, including to the United States. That is one of the main reasons why there are so many people of Irish descent in the United States today.

Luxembourg

The little country of Luxembourg is even smaller than Rhode Island! It is only about 60 kilometers (37 miles) across the country from east to west. And it is only about 100 kilometers

Many of the farmhouses of Ireland still have roofs made of thatch—tightly packed hay.

(62 miles) from north to south. French and German are the main languages spoken. About half the land is used for farming. The mining of iron ore and the manufacture of steel are the largest industries.

The Netherlands

One of the most interesting countries in Northwestern Europe is The Netherlands. You may think of it as Holland—but its real name is The Netherlands. It is one of the most crowded countries in the world. More than two-fifths (⅖) of the country is below sea level. The Dutch have worked for many years to build dikes to keep out the North Sea. Then they have reclaimed the land behind the dikes as farmland.

Windmills have been used for centuries in Europe to perform tasks such as pumping water, grinding grain, and sawing lumber.

They are still at work taking land away from the sea! The most famous crop of The Netherlands is tulip bulbs. Most of the Dutch live in cities, however. Rotterdam, near the mouth of the Rhine River, is the busiest port in Europe. Amsterdam is the capital, but the government buildings are at The Hague (haig).

Norway

Norway is often called the "Land of the Midnight Sun." Only in northern Norway does the sun shine at midnight, however. Norway has many fiords. Do you remember the definition of a fiord given earlier? Some of them reach inland almost across the country. For a land so far north, Norway has a mild climate due to the North Atlantic Current. Most ports are open all year long. The land is mountainous, and little of it can be farmed. For that reason, Norwegians have turned to the sea for much of their living. Many of them fish for a living. Many more operate ocean ships that take people and goods all over the world. Discovery of oil under the North Sea off Norway's shores has recently provided a new power source for Norway. New jobs may, therefore, be created. Oslo (*aws* loh), the capital, is the largest city. Norway is a favorite country for tourists. The mountain scenery with fiords, glaciers, and waterfalls is among the best in the world.

A SUMMARY OF NORTHWESTERN EUROPE

Northwestern Europe is one of the most productive areas on Earth. Many people live there, yet it still has areas with few people. People from this region were largely responsible for the early exploration and settlement of North America. Many people now living in the United States and Canada have ancestors from Northwestern Europe. At one time the countries of Northwestern Europe controlled much of the land areas of North America, Africa, and southern Asia.

The land area of most countries in Northwestern Europe is quite small when compared to the United States and Canada. France, the largest country, is only about the size of two Colorados. The Netherlands is only about the size of Massachusetts, Connecticut, and Rhode Island combined. Belgium and Luxembourg are smaller yet. Luxembourg is even smaller than Rhode Island.

Mineral wealth, manufacturing, and trading have had much to do with making Northwestern Europe influential. Coal, iron ore, and limestone made this area an early leader in steel production. The coal has been used to produce electricity. The power was used to drive machines in factories. Ships took the manufactured products all over the world. Recently good supplies of oil have been discovered in the North Sea. This discovery is giving Northwestern Europe a new power source.

Although small in land area, much food is raised by the farmers of Northwestern Europe. Farms are much smaller than in the United States. Crop yields are generally higher, though, because of rainfall and careful farming methods. Nevertheless, food has to be imported by most countries.

The people of Northwestern Europe suffered greatly during the last two world wars. Now the peoples of the area seem to want to work together to solve their problems. You should continue to be interested in Northwestern Europe. It is an area with much influence and many interesting people.

WHAT HAVE YOU LEARNED?

1. How are the United Kingdom, France, and Sweden alike? How are they different?

2. What countries in Northwestern Europe have at least one day when the sun is not seen above the horizon?

3. What two countries in Northwestern Europe are likely to be exporters of oil in the future? Where was the oil found? What effect may the development of the oil have on these two countries?

4. Which countries in Northwestern Europe have a royal family? What advantages and disadvantages are there to having a monarch?

5. Which one of the three countries in Northwestern Europe just studied—
   a. mines the most coal?
   b. raises the most farm crops?
   c. has the largest city?
   d. helped the United States during the War for Independence?
   e. did not take part in the Second World War?

BE A GEOGRAPHER

1. List the countries of Northwestern Europe from largest to smallest according to land area.

2. List the countries of Northwestern Europe from largest to smallest according to population. What differences are there in rankings?

3. List the ten largest cities in Northwestern Europe from largest to smallest according to population.

4. Using the maps on pages 60, 87, and 109 locate all of the following. Be prepared to locate them on a wall map:
   a. Gulf of Bothnia
   b. Shetland Islands
   c. English Channel
   d. Bay of Biscay
   e. Baltic Sea
   f. North Sea
   g. Jura Mountains
   h. Rhône River

5. Without looking at a map, guess whether each of the following cities is located in east longitude or west longitude:
   a. London, United Kingdom
   b. Paris, France
   c. Oslo, Norway
   d. Dublin, Ireland
   e. Reykjavik, Iceland
   f. Helsinki, Finland
   g. Antwerp, Belgium
   h. Edinburgh, United Kingdom
   i. Bordeaux, France

6. Without looking at a map, guess which of the following cities is located north of 60° N.:
   a. Paris, France
   b. Marseille, France
   c. Reykjavik, Iceland
   d. Stockholm, Sweden
   e. Kiruna, Sweden
   f. Bergen, Norway
   g. Aberdeen, United Kingdom
   h. Copenhagen, Denmark
   i. Helsinki, Finland
   j. London, United Kingdom

7. Guess which of the following cities in North America are located north of 60° N.:
   a. Montreal, Quebec
   b. Augusta, Maine
   c. Anchorage, Alaska
   d. Quebec, Quebec
   e. Whitehorse, Yukon
   f. Fairbanks, Alaska
   g. Edmonton, Alberta
   h. Goose Bay, Labrador
   Use the maps in this unit to check your guesses for items 5, 6, and 7.

————————— QUESTIONS TO THINK ABOUT —————————

1. Wars in this century have been very destructive and very costly. Have any benefits resulted from wars and preparation for wars? Do you think there will ever be a time when wars between nations will not occur? Why or why not?

2. Mont Blanc, the highest mountain in the Alps, has a number of large glaciers on its slopes. In the Colorado Rockies, no peak is quite as high as Mont Blanc. Many peaks are almost as high, though. Nevertheless, large glaciers are not found on the Colorado peaks. Why do you think this is so?

3. When the people in Northwestern Europe use the term "Scandinavian," they are usually thinking about the Danes, the Swedes, and the Norwegians. When they use the term "Nordic," they also include the Finns and the Icelanders. Why do you think they do that?

4. Why do few trees grow in the far north? Are there any areas in the United States where trees do not grow? Where?

5. If you wanted to plan an overland-oversea journey to the North Pole, where would you start? How many times has someone reached the North Pole *over* the ice? When was the last time? Has anyone ever reached the North Pole *under* the ice? Who and when? Has anyone ever reached the North Pole *through* the ice? Who and when?

Unit Review

Industrial development, the exploitation of natural resources, and traditional economic activity often exist side by side in Central Europe.

Unit

CENTRAL EUROPE

Unit Preview

- How many countries can you see in Central Europe?

- On a sheet of paper write the names of the countries. Start with the country farthest to the northwest and work toward the southeast.

- Parts of three seas are shown. What are their names?

- Two main rivers are shown. What are their names?

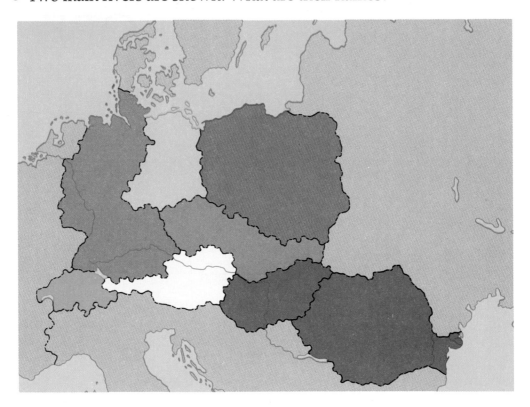

PREDICT

- Where are winters the coldest in Central Europe?

- Where are summers the hottest?

- Where are mountains the highest?

- Which country produces the most manufactured products?

- Which countries have governments controlled by a Communist party?

- Where do you think you would prefer to live in Central Europe? Why? (You may change your mind of course!)

DO YOU KNOW?

- What major differences there are between the governments of the nations of West Central Europe and East Central Europe?

- Why the city of Berlin is divided between East Germany and West Germany?

- Why tourism has been—and is—such an important business in Switzerland?

- Why the boundaries of Poland have changed so often?

- How COMECON aids the nations of East Central Europe?

Unit Introduction

Central Europe

The region called Central Europe in this book is made up of two smaller regions. These are called *West Central Europe* and *East Central Europe*. West Central Europe contains three interesting countries and a *principality* (*prihn* suh *pal* uh tee). A **principality** is a country that has a prince as its head of state. The countries are West Germany, Switzerland, and Austria. The principality is Liechtenstein (*lick* tuhn stine). Two of these countries, West Germany and Switzerland, are presented in this book in detail. Short coverage is given to Austria and Liechtenstein.

East Central Europe contains the countries of East Germany, Poland, Czechoslovakia (*check* uh sloh *vahk* ee uh), Hungary, and Romania. The countries of Poland and Romania are presented in detail in this book. Brief information is given about the other nations. Find all of these countries on the map on page 138.

In the countries of West Central Europe, the people are allowed to criticize their government. They can openly support differing political views. At election time several political parties put up candidates. The people may leave their own countries to travel or to live in other lands.

By contrast, the people living in the countries of East Central Europe are not as free. These countries all have political systems that are controlled by a Communist party. People living in East Central Europe may not leave their countries without special permission. This permission is sometimes hard to get. All borders are carefully patrolled by police and/or by army troops. Many borders are "protected" by high barbed-wire fences or walls. In general, people in these countries vote. But usually only one candidate for each office is on the ballot. The only candidates on the ballot are those approved by that country's Communist party.

In some countries of East Central Europe, some criticism of the government is permitted. That condition is not widespread, however. Czechoslovakia, which had permitted some criticism, was invaded in 1968 by troops from the Soviet Union. All open criticism of the government was stopped. Any person criticizing the government was jailed. Twelve years before Czecho-slovakia's invasion, Hungary had also been invaded by Soviet troops. Control by the Soviet Union seems to be lessening now, but the threat of invasion by Soviet troops is still there.

Pressure for changes in the Communist-controlled governments of East Central Europe will probably continue. These nations have made progress in developing their industries. Educational opportunities for their people have been greatly improved. Educated people, however, think and act on their own. Desire for more freedom cannot be kept down forever. The governments will have to change—and it appears that they are gradually doing so.

Things You Might Like To Do As You Study About Central Europe

1. Prepare a large outline map of Central Europe. On it locate the main rivers and mountains of this region. Draw in the boundaries of all countries. Then, locate and label their capital cities.

2. On your large outline map of Central Europe, shade the areas where coal is produced. Then locate the cities that have more than 500,000 people. Finally, think about where cities tend to grow. Why, for instance, do you suppose the largest city in West Germany grew where it did? The largest city in Poland?

3. Make a series of maps, each drawn to the same scale, showing the size of Poland in 1770, 1772, 1793, 1795, 1918, 1939, and 1945. Display these maps in the room and prepare a report explaining why Polish borders have changed over the years.

4. Learn all you can about the European Coal and Steel Community, the European Free Trade Association, Euratom, and COMECON. What differences have these organizations made in Europe's economy? How are they likely to affect European nations in the future?

5. Make a scrapbook or prepare a display in the classroom of short biographical sketches of some famous men and women of Central Europe, such as:
  a. Albert Schweitzer
  b. Annette Kolb
  c. Gregor Mendel
  d. Rosa Luxemburg

e. Johann Gutenberg
f. Wanda Landowska
g. Thaddeus Kosciusko
h. Wolfgang Mozart
i. Marja Sklodowska (Marie Curie)
j. William Tell
k. Karol Wojtyla

Be sure to name the country of Central Europe from which each person came.

6. Prepare and give a report on one of the following topics:
a. The Ruhr Valley
b. The Rhine River in Story, Song, and Fact
c. The Brenner Pass
d. Adolf Hitler and the Second World War
e. Watchmaking in Switzerland
f. The Black Forest
g. The North European Plain
h. The Danube River
i. Manufacturing in Poland
j. Winter and Summer Sports in Switzerland
k. The Black Sea Coast

This machine is used to extract coal in an open pit mine in the Ruhr Valley in West Germany. It is the world's largest digger.

Unit Introduction

Chapter

———— WEST GERMANY ————

West Germany is the only country of West Central Europe with a seacoast. Much of the land of Switzerland and Austria is in the Alps (see map, page 138). A little of the southern part of West Germany also extends into the Alps.

Before World War II Germany was one country. Now it is divided into two parts. These parts operate as separate countries and are called *West Germany* and *East Germany*. The official name for West Germany is the Federal Republic of Germany. (The official name for East Germany is the German Democratic Republic.) West Germany is about the size of New York and Pennsylvania combined and has about 64 million people. Its capital city is Bonn (bahn). Berlin, its largest city, is not located in West Germany. Rather, it is surrounded by another country—East Germany!

Probably no other country has had as much influence on the history of Europe since 1900 as Germany. In 1914, and again in 1939, world wars began when armies from Germany attacked other countries. In order to understand how West Germany came to be, we need to learn some history.

The Growth of Germany

Until 1871 the area that became Germany was divided into many independent cities and kingdoms. At the close of a war with France in 1871, these kingdoms and independent cities were united to form the German Empire. Germany soon became a strong and productive nation. In less than 40 years, it became a leader in manufacturing. Its foreign trade was second only to Great Britain's. Germany's leadership in music, education, and science was recognized throughout the world.

Coal was one of the main reasons for Germany's rapid rise to importance as a nation. Large deposits of high-quality coal gave the Germans, like the British, a source of power for manufacturing. As a result, Germany became one of the world's great industrial nations.

When World War I started in Europe in 1914, Germany had a large, well-equipped army and navy. Opposed by France, Great

Many kinds of vessels can be seen on the Rhine River. Large ocean-going ships, barges, ferries, and private pleasure craft make the Rhine River Europe's greatest artery for waterborne traffic.

Britain, and the United States, Germany was finally defeated in 1918.

The treaty ending the war gave about one-seventh (⅐) of the land of Germany to nearby countries. All of Germany's overseas colonies were taken away. The areas of Alsace and Lorraine, rich coal regions, were returned to France.

After Germany was defeated in World War I, a form of government new to Germany was started—that of a democratic republic. But 15 years later, in 1933, a group of people under the leadership of Adolf Hitler gained control of the government. These people and their followers were called *Nazis* (*naht* sees). They began to build a strong army, navy, and air force. After a few years of preparation, the Nazis led Germany into war.

Austria was taken without any fighting. Next, Germany's troops took Czechoslovakia, also without bloodshed. In September of 1939, German troops invaded Poland and the Second World War began. France and Great Britain joined Poland in the war against the Nazis. In time the war spread to include countries in all parts of the world. Germany had as its major **allies** (helpers or partners), the nations of Italy and Japan. Together, these nations were known as the Axis powers. France and Great Britain had many allies, including the United States, Canada, and the U.S.S.R. Together they were known as the Allied powers.

West Germany

Adolf Hitler—a Lesson from History

Between 1933 and 1945 Adolf Hitler was *dictator* of Germany. As you may know, a **dictator** is one who rules with absolute power. Under his leadership Germany developed a huge army of trained and well-equipped soldiers. Fighter and bomber planes were built, and pilots and mechanics were trained. Navy ships, including many submarines, were built. Having built a huge military machine, Germany set out to conquer the world. At the height of the German conquest during World War II, Hitler's armies controlled most of Europe. The armies advanced far into the Soviet Union. They controlled much of the land along the Mediterranean Sea in North Africa.

Millions of people were killed as Hitler's armies conquered Europe. Many of them were killed in the war. But, in addition, it is estimated that about twelve million people were starved, beaten, and murdered in death camps. About six million of these people were Jews. Hitler hated all Jews. He blamed them for most of what was wrong with the world. Both in Germany and in the countries his armies conquered, he had his troops imprison, torture, and kill Jews. People who tried to help Jews escape were also imprisoned or killed.

Why would any leader act this way? Why would any person want to kill and destroy? Why would any person want to conquer the world?

Ever since World War II, historians have been trying to understand why it all happened. No one knows for sure why the German people followed Hitler. No one knows for sure why they permitted him to be so cruel, especially to Jews. No one knows for sure why destroying cities and killing millions of people in war was seen by many as glorious. Yet it did happen!

Many reasons have been given to explain how Hitler got his power. The people of Germany, after being defeated in World War I, were poor. Some of their land had been taken by other countries. There were few jobs in Germany, and there was little food. The new government that had been forced upon Germany seemed unable to do much to improve conditions. The German people were defeated in spirit and had little hope for the future. The people seemed to want a strong leader.

Hitler moved in and took over. Under his leadership, the Nazi Party gained control of the government. Hitler was a powerful speaker. He told the German people that they were the best people on Earth. He told them that they were members of the

It made no difference if they were young or old, male or female, sick or healthy. Millions of people, all over Europe, were sent to concentration camps by the Nazis. Only a few survived.

master race and that Germany should rule the world. He killed or put in prison anyone who disagreed with him. Many German people thought that Hitler could solve their problems. They overlooked the evil things he did.

For a time it looked as if Hitler's dream of conquest might succeed. But he made some very costly mistakes. Most people think that he became increasingly mad, crazy, or insane. Gradually the rest of the world saw what life would be like if Germany was not defeated. Beginning in 1942 the war began to turn around. By 1945 the "glorious dream" of conquest was over. Germany's armed forces surrendered. Hitler took his own life shortly before he would have been captured.

Could something like Hitler's madness happen again? One hopes not. One hopes that people everywhere have learned a lesson from history—especially the "lesson" taught by Hitler.

Germany Divided

After six long years of war, Germany surrendered to the Allied powers in 1945. The United States, France, the United Kingdom, and the Soviet Union divided the land of Germany into four parts. These four areas—each to be occupied by one of the Allied powers—were termed

West Germany

CENTRAL EUROPE

© 1980 JEPPESEN & CO. DENVER COLO. U.S.A.
ALL RIGHTS RESERVED
REVISED 6-79
PUBLISHED BY THE H.M. GOUSHA CO.

Kilometers
Miles

FORESTS
FARMLAND

EVERGREEN FORESTS
GRASSLAND

MOUNTAINS

U. S. S. R.

SWEDEN

Baltic Sea

DENMARK

Copenhagen

North Sea

JUTLAND PENINSULA

Kiel
Lübeck
Hamburg
Bremen
Emden

THE NETHERLANDS

Amsterdam
Rotterdam

BELGIUM
Brussels

EUROPEAN PLAIN

Gdynia
Gdansk

Warsaw
Vistula River
Lubin

POLAND

Lodz
Wroclaw

Krakow

NORTH
EAST GERMANY

BERLIN
Oder River
Neisse R.
Elbe
Leipzig
Dresden
Chemnitz
Jena
Elbe River
Weser River

HARZ MTS.
THURINGIAN FOREST
ORE MTS.
SUDETEN MTS.

Saxon Gate

Prague

CZECHOSLOVAKIA

BOHEMIAN FOREST

Pilzen

CARPATHIAN MTS.

Ostrava

TATRA MTS.

WEST GERMANY

Dortmund
Essen
Düsseldorf
Cologne
Bonn
Koblenz
Rhine R.
Frankfurt
Mainz
Heidelberg
Mannheim
Ruhr R.
Ems River
Luxembourg
LUX.
Saarbrücken
Saar
Moselle R.

FRANCE

ALSACE LORRAINE

BLACK FOREST

Main R.
Nürnberg
Regensburg
Danube

BAVARIA
BAVARIAN ALPS

Augsburg
Ulm
Munich
Inn River
Linz

AUSTRIA

Vienna
Bratislava
Morava R.
Danube

BOHEMIAN PLATEAU

Graz
Klagenfurt

HUNGARY

BUDAPEST
R.
Dunaújváros
Lake Balaton
Tisza R.

HUNGARIAN PLAIN

Miskolc
Debreczen
Szeged
Timisoara

ROMANIA

BIHOR MTS.
TRANSYLVANIAN PLATEAU

Cluj
Oradea

MOLDAVIA
Bacau
Galati
Braila
Buzau
Ploesti

Transylvanian Alps
Stalin
Craiova

WALLACHIA
Bucharest
Giurgiu
Constanta
Danube
Iron Gate

BULGARIA

YUGOSLAVIA

Belgrade

Black Sea

SWITZERLAND

Basel
Zürich
Luzern
L. Lucerne
L. Zürich
Vaduz
LIECHTENSTEIN
Bern
Neuchâtel
L. Neuchâtel
Lausanne
L. Geneva
Geneva
Rhône R.

JURA MTS.

ALPS

Monte Rosa
4638 m
15,203
Mt. Blanc
4807 m
15,781

SIMPLON PASS
Brig

ITALY

Trieste
Venice
Padua
Bologna
Florence
Genoa
Po River

SAN MARINO

Adriatic Sea

Marseille

ISTRIA

For almost a year the people of West Berlin watched and waited for the airplanes that brought the food, fuel, medicine, and other supplies that they needed to survive during the blockade of West Berlin in 1948-1949.

occupation zones. Each zone was placed under the control of a military officer from one of the four countries. The officer was given the responsibility for governing the zone. The city of Berlin, which was the capital of Germany, was also divided into four zones.

In 1948 France, Great Britain, and the United States agreed to form a new nation out of their three zones. The nation was named the Federal Republic of Germany. It is usually called West Germany. The zone governed by the U.S.S.R. then became the German Democratic Republic. It is usually called East Germany.

Three parts of the city of Berlin—the parts occupied by the Allied powers—wanted to be a part of West Germany. Even though West Berlin

is about 175 kilometers (110 miles) inside East Germany, it is part of West Germany.

In the agreement reached among the Allied powers, it was promised that transportation of people and goods from West Germany to West Berlin would be permitted. But when the Federal Republic of Germany was started in 1948, the U.S.S.R. stopped all ground traffic. They wanted all of Berlin to be under Soviet control and part of East Germany. They would not let trucks carrying food or fuel cross from West Germany into West Berlin.

The other Allied powers, of course, protested this blockade. On some days, a few trucks would be permitted to travel to West Berlin. But not enough food and fuel could

West Germany

be provided that way for the two million people in West Berlin. The Soviets hoped that the people in West Berlin, as they began to starve, would ask to join East Germany. They thought that the other Allied powers would not start another war over Berlin. The blockade almost worked.

The United States, France, and the United Kingdom, however, decided not to start another war. They decided instead to supply West Berlin by air. From June, 1948, until May, 1949, hundreds of cargo planes flew each day from West German airports to West Berlin. The people of West Berlin had to learn to live with less food, clothing, coal, and petroleum than usual. After almost a year of the airlift, the U.S.S.R. decided to re-open the borders.

Times were tense during the months of the airlift and blockade. For many years after those tense times, people who did not wish to live under communism left East Germany. They went to West Berlin and then flew to West Germany. Finally, in 1961, the Communist-led government of East Germany built a wall through Berlin. They put guards along the wall to stop people from leaving East Germany. They also built high fences along the borders with West Germany.

Living conditions have improved a great deal in East Germany since those days. The two nations now carry on quite a bit of trade with each

In many sections of Germany the population density is high. However, the people have been able to adapt. They have developed comfortable housing that uses the available space efficiently.

other. At times, during holidays, people in West Berlin are permitted to visit relatives in East Berlin. The wall and the fences are still there, however.

Progress in West Germany

At the end of World War II, most cities in West Germany had been badly damaged. British and American planes had dropped thousands of bombs on the cities, bridges, and docks. Most factories had been destroyed. Most roads and railroads had been made unusable.

Now, that damage has been repaired. The Allied powers, except for the U.S.S.R., helped speed West Germany's recovery from the war. They provided money and needed materials for rebuilding. Cities, industries, and roads were rebuilt. Production of manufactured goods is now greater than before the war. West Germany is again one of the world's leading industrial nations.

REMEMBER, THINK, DO

1. What are the two main parts of Central Europe? What is the main difference, politically, between these parts?
2. What mineral resources helped Germany achieve a position of power before World War I?
3. What were the followers of Hitler called? What did they do?
4. What nations in World War II were known as the Axis powers? What nations made up the Allied powers?
5. Why was a wall built in the city of Berlin and high fences put up along the borders of East Germany? What do you think about fencing people in? Would you like to be forced to stay in your town, city, or country?
6. Locate Berlin on the map on page 138. What problems do you suppose exist when a large city that is part of one country is located inside another country?

The Land of West Germany

The northern part of West Germany is part of the North European Plain. This vast plain extends from France eastward into the Soviet Union. In West Germany this area is gently rolling land. Thousands of years ago melting glaciers left deposits of earth and stone here. Much of this area was once wooded marshland. It has since been cleared and drained. Along the North Sea coast, dikes have been built to hold back the waters. The land behind the dikes can now be farmed.

South of the plains in West Germany is an uplands region. The Rhine River divides the uplands into an eastern and a western section. On

West Germany

Although large cities and industrial centers are located on the banks of the Rhine River, areas of great beauty can be seen along much of the river.

the higher parts of the uplands, the soil is generally poor. The hills, moreover, are swept with winds and rain or snow during the winter. A majority of the people live, therefore, in sheltered valleys. There it is drier and the soils are better.

The Ruhr (roor) River flows from the eastern part of these uplands westward to the Rhine (see map, page 138). Europe's best coal fields are in the Ruhr Valley.

The southern part of West Germany is highland. Most of the land is from 300-600 meters (1,000-2,000 feet) above sea level. In the extreme south, the Alps form the border with Austria. The Alps in West Germany are called the Bavarian Alps. This part of the country is known as Bavaria.

Two of the world's most important rivers flow through West Germany. Both get much of their water from the Alps. (See map, page 138.) The rivers are the Rhine and the Danube. Although both are important to Europe, the Rhine is of greater importance to West Germany.

The Rhine River

Two main branches of the Rhine flow from high in the Alps near the border between Switzerland and Italy. The two branches join in Switzerland to form the Rhine. The river then flows northward along the border of Liechtenstein to Lake Constance (*kahn* stans). Then the Rhine flows westward to Basel (*bah* zehl),

Switzerland. Near Basel the borders of France, Switzerland, and West Germany meet. The Rhine then turns northward and flows along the border with France. Then it flows through West Germany, The Netherlands, and into the North Sea.

The Rhine is about 1100 kilometers (700 miles) long. It is navigable as far south as Basel. Thus, it provides a river port for Switzerland. It is even more important, however, for West Germany and France. Barges move coal, ore, and other products up and down the Rhine. Railroads have also been built along its banks. The Rhine is one of the world's busiest rivers.

Many other smaller rivers flow into the Rhine on its journey to the North Sea. On a number of these rivers canals have been built. These canals connect the Rhine with the Rhône and Seine rivers in France. Thus, bulk cargoes can be moved cheaply between West Germany and France.

The Danube River

The Danube River flows eastward from southern West Germany for almost 2900 kilometers (1,800 miles). As the map on page 138 shows, the Danube empties into the Black Sea. The Danube is the longest river in Europe outside the U.S.S.R. It is navigable throughout most of its length. On the map on page 138 trace with your finger the course of the Danube River from its source to its mouth. How many countries does it flow through? Note how a canal at about 49° N. 11° E. joins a tributary of the Danube with the Main River. The Main flows into the Rhine west of Frankfurt.

Climate of West Germany

As in countries to the north and west, the part of West Germany near the North Sea has a mid-latitude marine climate. Summers are cooler and winters warmer than would be expected this far north. Look at a globe and see again what part of North America is at about the same latitude. Inland, farther from the North Sea, the climate is more like the humid continental type. There, winters are a bit cooler, and summers are a bit warmer.

Throughout West Germany enough rain falls for crops. The rainfall is usually distributed throughout the year. As the map on page 167 shows, the heaviest rainfall occurs in the Bavarian Alps.

Industries of West Germany

West Germany has coastlines on both the North Sea and the Baltic Sea. Trade with other countries is made easier, therefore. Before World War II, Hamburg, on the estuary of the Elbe (ehl buh) River, was the greatest commercial port on the mainland of Europe. Since the division of Germany, not as many ships use it. Nevertheless, Hamburg is West Germany's main port city. It is still one of Europe's busiest port cities.

Hamburg is also West Germany's largest industrial city as measured

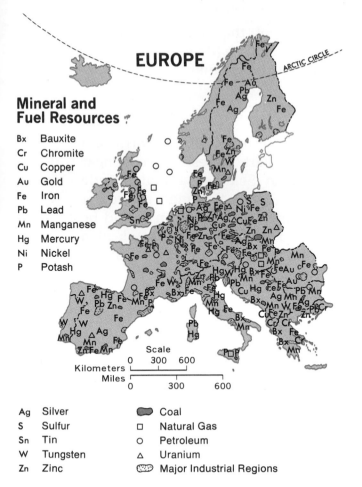

EUROPE

ARCTIC CIRCLE

Mineral and Fuel Resources

Bx	Bauxite
Cr	Chromite
Cu	Copper
Au	Gold
Fe	Iron
Pb	Lead
Mn	Manganese
Hg	Mercury
Ni	Nickel
P	Potash

Scale
0   300   600
Kilometers
Miles
0   300   600

Ag	Silver		Coal
S	Sulfur	□	Natural Gas
Sn	Tin	○	Petroleum
W	Tungsten	△	Uranium
Zn	Zinc		Major Industrial Regions

Factories in Hamburg make iron and steel products, chemicals, and wood products. Other factories process food, coffee, tobacco, fish, and other imported goods. Hamburg is a beautiful city, too. It has lakes, parks, golf courses, and lots of interesting places to visit. Shops and stores have goods from all over the world.

The Kiel (keel) Canal has been built across the Jutland Peninsula north of Hamburg. (See map, page 138.) This canal joins the North Sea and the Baltic. Much bulk cargo is transported through the Kiel Canal to the Elbe River and Hamburg.

Mineral Resources and Manufacturing

West Germany has a number of important mineral resources. Some of them are coal, potassium salts, salt, iron ore, copper, lead, zinc, and petroleum. Some ores and petroleum have to be imported because not enough of these resources are available in West Germany. Much of the iron ore used, for instance, is imported from Sweden and France.

Some of West Germany's leading manufacturing centers are located near the Ruhr Valley coal mines. A number of large cities are located in this region. During World War II, many of the buildings and industries in the Ruhr Valley were destroyed by bombs. Before they were destroyed, the iron and steel mills of the Ruhr turned out vast amounts of steel. The steel was used to make tanks, trucks, and guns. When the

by products. Find Hamburg on the map on page 138. It is located about 100 kilometers (about 65 miles) up the Elbe River from the North Sea. Hamburg has about 1.75 million people and a land area almost as large as New York City. It is treated as a state in West Germany. Hamburg has many factories, but it is the port that is of most importance to the city. Hundreds of trainloads of goods reach Hamburg every day to be placed on ships. Thousands of people work on the docks and in the warehouses, running the machinery to load and unload ships.

mills were damaged or destroyed, Germany's ability to continue the war was weakened. Factories in the Ruhr Valley have been rebuilt. Today, they make tools, automobiles, and structural steel used in buildings. Ships, bridges, and many other products are also made.

The Ruhr Valley also has many chemical and textile plants. Large salt mines made the development of the chemical industry possible. Great beds of rock salt are found at several locations. The salt often is found mixed with potassium salts. Potash, that is made from these salts, is used in making glass and fertilizers.

West Germany now produces much more than the whole of Germany did before World War II. Only the United States, the U.S.S.R., and Japan now produce more goods. For its area and population, it is surely one of the most productive nations in the world. Almost completely destroyed at the end of World War II, West Germany's recovery is a true success story.

More than 1,100 American firms have established branches or agencies in the Federal Republic of Germany. Many firms have huge plants turning out industrial goods worth millions of dollars each year. The German people themselves are the main reason why so much progress has been made.

REMEMBER, THINK, DO

1. The northern part of West Germany is part of what vast plain?
2. Where are Europe's best coal fields located? What difference have these fields made in the lives of the people of West Germany?
3. Why is the Rhine of more importance to West Germany than the Danube?
4. What advantages does Hamburg have that helped to make it West Germany's most productive industrial city?
5. What mineral resource, other than coal, has helped make the Ruhr Valley West Germany's main industrial region?
6. Do a ministudy to discover what products can be made from coal and salt. Use encyclopedias to locate the information you need.
7. What factors have made it possible for some nations to become world leaders in the production of goods? Why have other nations with more natural resources not become world leaders?

Agriculture in West Germany

The best way to describe the soils in West Germany is "productive" rather than "fertile." In most areas, and especially on the sandy plain in the north, the soils have been made productive through use of fertilizers. Even so, not enough land can be farmed to supply the people's needs for food. Almost half of the food supply has to be imported.

Potatoes and sugar beets are two of the most important crops in West Germany. Potatoes are one of the most common foods of the people. They are also fed to animals and are a source of flour, starch, and alcohol.

The northern part of West Germany, with its sandy soil and cool, moist climate, is better suited to raising rye than wheat. Considerable wheat is raised in the central and southern parts of West Germany. There, both the soils and the climate are more suitable for wheat. Even so, some wheat must be imported.

Oats, barley, and hay are other large crops; these are used mainly for feeding livestock. In the southern part of West Germany, grapes and hops are also leading crops. They are

Almost 90 percent of the potatoes harvested in the world are grown in Europe. Potatoes are an important crop in Germany.

used in making wine and beer. Most grapes are grown along the Rhine River.

Many cattle are raised in West Germany. Most of them are dairy cattle. The farmers take good care of their stock. Many pigs and poultry and a number of horses, sheep, and goats are also raised.

More than one-fourth (¼) of West Germany is forested. Trees are planted on the poorer soils of the plains in the north. The rugged hills in the central and southern parts of West Germany are also covered with forests. Forested areas in West Germany are treated almost as if they were cropland. The forests are carefully tended to increase growth and prevent fires. For each tree that is cut, another is planted.

Berlin—a Divided City

For about 30 years the largest city of West Germany has been about 175 kilometers (110 miles) inside East Germany. Berlin is a divided city. As you already know, it has a wall through it. The wall keeps people in East Berlin from moving to West Berlin.

West Berlin

West Berlin is a busy city of more than 2 million people. It has an area of almost 500 km² (190 sq. mi.). West Berlin is treated as a state by the country of West Germany. (Our smallest state, Rhode Island, has more than six and one-half times as much land as West Berlin.)

During World War II, most of the larger buildings and factories of Berlin were damaged or destroyed by bombs. A famous church was almost totally destroyed. The people of West Berlin decided to rebuild the church to help people remember what happens in war. The old bombed-out tower was kept, and the new church was built around it. The Kaiser Wilhelm Memorial Church is now a famous landmark in downtown West Berlin.

Much of the land of West Berlin is still wooded and beautiful. There are a number of small lakes and a world-famous zoo. Within the city itself the main streets are wide, and most buildings are quite modern. Most of them have been built since World War II.

As in most large cities, many people in West Berlin work in stores, offices, and factories. The raw materials needed in the factories of West Berlin are brought, for the most part, from West Germany. Most of these raw materials come by canals, trains, or trucks. Many factories in West Berlin make electrical products.

As you know, people who live in West Berlin may now visit relatives and friends in East Berlin several times each year. The opposite is not possible, however. Usually, only people who are members of special groups, such as athletes, musicians, or artists, may cross from East Berlin into West Berlin. Sometimes, by special permission, relatives may spend a day in West Berlin. The East German government sets these regulations.

The Kaiser Wilhelm Memorial Church stands amid the new buildings that have been put up since the end of World War II.

East Berlin

East Berlin is the capital of East Germany. It has about half as many people, but almost as much area as West Berlin. East Berlin was much slower to recover from the damage done in World War II. It now is probably the most modern city in all of Europe. East Berlin, of course, does not have the same problem that West Berlin faces. It is surrounded by East Germany. Goods flow easily and without delay to and from East Berlin.

East Berlin is an important industrial city. Factories there make many electrical and mechanical products. It is also a center for textiles and food processing. As contrasted with most cities in Europe and the United States, there are few traffic jams. Not nearly as many people own automobiles in East Berlin as in West Berlin. A huge television tower with a restaurant in it is a main landmark in downtown East Berlin. The tower is 360 meters (about 1,180 feet) high.

East Berlin has been largely rebuilt according to a master plan. About half of the city, for instance, is heated by one of the three heating centers. One of these heating centers serves all the buildings in the city center. As the city was rebuilt, many sports centers and theaters were built. Many areas were left as open spaces. As a result, East Berlin now is an attractive place to live. But the people, although proud of their accomplishments, do not have the freedoms that West Berliners enjoy.

Other Cities of West Germany

Bonn, the capital of West Germany, is located where the Rhine Valley begins to widen. It is the birthplace of the famous musician, Ludwig Van Beethoven (*bay* toh vuhn). Several large industrial cities are north of Bonn in the Rhine and Ruhr valleys. Find Cologne (koh *lohn*), Dusseldorf (*doos* uhl dorf), Essen (*ehs'*n), and Dortmund (*dawrt* muhnd) on the map. Many old and interesting cities are found along or near the Rhine River. The largest city in southern Germany is Munich (*myoo* nihk). As you know, however, the largest city of West Germany isn't in West Germany.

Education

All boys and girls between 6 and 14 years of age must attend school in West Germany. After finishing elementary school, they may attend a secondary school. In West Germany, schools are much like those in the United States, but there are differences. Children in Germany go to school six days a week. In most schools, children study two foreign languages. West Germany has a number of world-known universities. Training in science has been especially good in these universities for many years.

REMEMBER, THINK, DO

1. How are soils in West Germany best described?
2. What are some of the important crops grown in West Germany?
3. Where in West Germany are some of the main manufacturing cities located? Why?
4. If you have not done so, learn to say "Good Morning," "Excuse me," "Hello," and "Goodbye," in German. Practice using these words (and others) in your classroom.
5. What difference would it make in the United States if most people in one state used a language different from that used in neighboring states?

CHAPTER 7 REVIEW

---------------- WHAT HAVE YOU LEARNED? ----------------

1. Which has more people, West Germany or the two states of New York and Pennsylvania combined? Which is more crowded?

2. What four countries divided Germany into zones after the Second World War?

3. What kind of climate does West Germany have?

4. On what two seas does West Germany have seaports? How have these two seas been joined?

5. What mineral resources must West Germany import? Why?

6. How is Berlin different from all other cities?

---------------- BE A GEOGRAPHER ----------------

1. Make a bar graph showing some comparisons of the United Kingdom, France, and West Germany. You might show population of largest cities, length of major rivers, or the production level of a product. You may want to use encyclopedias to get the information you need.

2. Develop several maps that show the boundaries of Germany at different times in its history. Be sure to include one map that shows both West and East Germany as they are today.

3. Look at a globe and figure out answers to the following:
   a. If you flew straight south from Berlin, what sea would you cross before reaching Africa?
   b. If you kept flying straight south, would you be west or east of the tip of Africa?
   c. If you kept flying south what continent would you fly over next? What direction would you be flying after reaching the South Pole?
   d. Would you fly over the Ross Ice Shelf before or after reaching the South Pole?

e. If you kept flying straight north, would you fly over any part of North America?

f. If you kept flying straight north, what direction would you be going when you reached Europe?

g. What countries in Northwestern Europe would you fly over before reaching Berlin again?

4. If it is 11:00 P.M. Tuesday in Berlin, what time and day is it in:
   a. London?
   b. New York City?
   c. Honolulu?

—————— QUESTIONS TO THINK ABOUT ——————

1. Germany had much to do with starting both World War I and World War II. Why do some countries seem to become more involved in wars than others?

2. How do you suppose the people in Alsace and Lorraine feel about wars? What do you suppose it is like to be part of one country for awhile and then to be a part of another country?

3. Some people say that history repeats itself. Do you think that is true? If you think it is true, does that mean that France and West Germany will be on opposite sides in World War III?

4. What are the advantages and disadvantages of having a royal family, as the British have? Would the West Germans be better off if they had a royal family? Would the United States?

5. If the Rhine River, for some reason, stopped flowing what differences would that make to West Germany?

6. Should East Germany and West Germany reunite under one government? Why or why not?

Chapter

——— SWITZERLAND ———

Switzerland is an inland country surrounded by France, Germany, Italy, and Austria. Find it on the map on page 138. Switzerland is about the size of the states of Vermont, Connecticut, and Rhode Island combined. Nearly seven million people live there.

The Land

Switzerland is the most mountainous of the European countries. The central part of the great mountain system called the Alps is in the southern part of this small country. The Alps cover more than one-half (½) of the land of Switzerland. The Jura Mountains, along the border with France, are in the northwestern part of the country. Between these mountain ranges is a rolling, hilly plateau. The plateau area lies between Lake Constance on the German border and Lake Geneva (juh

nee vuh) on the French border. Most of Switzerland's farming land and its cities are located on this plateau.

The Climate

Although small in area, Switzerland has a climate that is different from place to place. The reason for this is Switzerland's mountains. The best way to describe the climate in most mountainous places is to call it a *highland climate*. In a **highland climate**, the climate depends largely upon *altitude*. (**Altitude** is the height of the land above sea level.) Much of Switzerland has such a climate. High in the Swiss Alps, much of the land is covered with glaciers. Heavy snows fall during the winter. In the summer these highlands also have a great deal of rainfall. The temperature in the mountains is cool.

At lower altitudes in Switzerland, the climate is warmer and drier. In the plateau area summer days are

Water melting from the Stein Glacier has formed a small pond. It is from glaciers such as this one that many rivers in Europe have their source.

quite warm. Nights are always cool, however. Winters are long and cold. Less rain and snow fall on the plateau than in the mountains.

In southern Switzerland at low altitudes, as in the Rhône River Valley, the climate is like that of southern France. Little rain falls there during the hot summer months. Winters are generally mild with some rain but little snow. Do

you remember what such a climate is called?

Europe's Watershed

The Alps are Europe's *watershed*. A **watershed** is a large area of land which supplies water for a river or lake. Because so much rain and snow falls in the Alps, they are the source of several of the most famous rivers of Europe. The Rhine, the

Rhône, the Po (poh), and the Danube's main upper tributaries begin in the snowfields of the Alps. Find all these rivers on the map on page 138.

Switzerland has many lakes. Most of them are long and narrow because they are located in valleys between mountains.

Hay is stacked for drying before it is stored for the coming winter. Most of the farmers who work in these fields live in the village in the valley.

Agriculture

Nearly one-fourth (¼) of the land in Switzerland is unproductive. Much of this land is either steep and rocky or covered with ice. Another one-fourth (¼) of the land is forested. The forests grow on the ridges in the plateau region and on the slopes of the Juras and the Alps.

Most of the rest of the land is pasture or used for growing hay. As we have learned, the climate is cool and moist, and most of the usable soil has an uneven surface. Therefore, the land is better suited for grazing animals and growing grass for hay than it is for producing food crops. About one-tenth (¹⁄₁₀) of the land is used for growing fruits and vegetables.

The crops which are grown by the Swiss are used mainly in their own country. About half of the food eaten by the people is imported. Potatoes, sugar beets, grapes, apples, pears, and cherries are some of the crops grown. The grains raised are wheat, rye, barley, and oats.

Dairy farming is an important industry in Switzerland. Enough meat and milk are produced to supply the people's needs. Most of the milk is taken to factories near the cities and towns to be made into butter and cheese. Some of the butter and cheese is made high in the mountains during the summer months. Some of it is sold in the markets of the nearest town. Swiss cheese and Swiss milk chocolate, the most famous of all the dairy products of Switzerland, are exported.

Using Mountain Pastures

Pastures high in the mountains can be used during summer months to graze cattle. These mountain pastures are called **alps** and the mountains were named for them. The way the alps are used is interesting. In the spring the men and older boys drive young cattle to these high pastures. They remain there during the summer, milking the cattle and making butter and cheese. The women and younger children remain at home in the lower valleys. They plant and care for crops, cut hay, and care for the older cattle which are left behind. In the fall the men and cattle return to the valley. All the cattle are sheltered in barns during the winter. They are fed hay and other crops that have been stored for the winter.

A High Standard of Living

The Swiss people live in a very rugged land. The country has few mineral resources to use for manufacturing. Yet the Swiss have one of the highest standards of living in Europe. How can this be?

Tourists Are a Big Business

One reason is that the Swiss, for many, many years, have been caring for tourists. Visitors come from all over the world to view the high, snow-capped mountain peaks, glaciers, lakes, and waterfalls. The Swiss have worked hard to make sure that tourists enjoy their stay.

Even in the places far from towns or cities, Swiss hotels are good. Hotel managers and workers are carefully trained to give their guests good service. Many of the workers speak several languages. More than two-thirds (⅔) of the people in Switzerland speak German as their main language. Those who live near the border with France speak French. Those who live near the border with Italy speak Italian. Many people who deal with tourists speak English. Many Swiss people speak all these languages.

Railroads and good roads have been built through Switzerland. They make it easy for travelers to view the scenery. They also provide main routes for moving freight across Europe. Because of the mountains, the roads and railroads were difficult and expensive to build. Deep gorges had to be bridged in many places. High mountain ranges, especially in the southern part of Switzerland, had to be crossed. Some of the longest railroad tunnels in the world—such as the one under the Simplon Pass—cut through the highest ranges. Swiss engineers have become noted for their skill in constructing roads, bridges, and tunnels. The Swiss have also developed fine international airports at Zurich (*zoor* ihk) and Geneva.

Switzerland attracts many tourists during summer months. They hike in the mountains, go boating on the lakes, and enjoy the beautiful scenery. Swiss mountain guides are available to help mountain climbers up the famous peaks.

A strong and stable economy has made the Swiss franc one of the world's most valued currencies.

Banking by Number

Do you have your own bank account? If not yet, you probably will have one before too much longer. Banks are safe places to keep money. Banks have two main kinds of accounts. Probably the most used is a *checking account*. People deposit (or put in) money in their checking account on payday. Then they can withdraw (or take out) money when they need it. The other type of account is a *savings account*. People use a savings account when they think they have more money than they need for a while. The bank pays *interest* on money in savings accounts. **Interest** is money paid by the bank to encourage people to save.

To open an account in an American bank, you have to sign your name. Then you are given an account number. Checks and deposit slips all carry that number. Computers use the number to keep track of the amount of money you have in the bank. Most banks today require a person's name and number to be printed on checks. That way there are fewer mistakes or attempts to get money from someone else's account.

In Switzerland, however, a person may have a bank account that is only numbered. Only the banker and the person know the number. The banker may not tell anyone, even the government, the name of the person. Such accounts make it possible for persons to keep money secretly. And so, Swiss banks have become favorite places to keep large sums of money.

Why would anyone want a bank account with a secret number? Perhaps you are worried about whether the government in your country is going to *devalue* your money. **Devalue** means to change the value of money so it is worth less. You might want to send some of your money while it was worth more to a safe place. One of the best places in the world to do that is Switzerland. Banks there are regarded as being very safe. The country does not take part in wars. And Swiss bankers don't ask questions. They accept your money, paying you interest for as long as you keep the money there. Thus, Switzerland has become a center for international banking.

Hydroelectric Power

Another reason for the Swiss high standard of living is the development of hydroelectric power. The large amount of water from rain and snow is used to make electricity. Many dams have been built to store water during the spring months when the snow melts rapidly. The water is used throughout the year to run generators which produce electric power.

Electricity is used in many ways. Almost every home has electricity. All the trains are run by electricity. Most factories use electricity for all of their power. Because they have no coal, the Swiss have learned to use the power of falling water.

Peace

A third reason why the Swiss have been able to develop a high standard of living is that they have lived in peace for many years. While other nations around them have used their resources in costly wars, the Swiss have remained neutral. One of the reasons why Switzerland was not involved in the two world wars is probably the Alps. They are fairly difficult to cross and can be rather easily defended against land armies. Still, airplanes could easily have bombed Swiss cities during World War II. They did not, mostly because both sides needed and wanted Switzerland to stay at peace. During wartime and during peacetime, Switzerland has served as an international banking center. The railroads of the country were needed, moreover, to move arms from one country to another.

Also, Switzerland is too small a country to be dangerous to any of its more powerful neighbors. But the Swiss do have a good, well-trained army for the defense of their own country. During the world wars in Europe, they were ready to fight any invading forces. The Swiss were ready to destroy railroad tunnels and bridges to keep armies out of the country. In their mountain strongholds they could have made it costly for any country trying to conquer their land.

Switzerland has been a neutral nation for many years. It has, therefore, become known throughout the world as a good meeting place. Representatives of countries that are having problems often meet in Geneva to try to solve them. The good care given visitors, the beautiful scenery nearby, and the central location of Switzerland have helped attract such meetings. The international headquarters of the Red Cross is located in Geneva. Many other international organizations or agencies are also located in Geneva.

REMEMBER, THINK, DO

1. About how much of Switzerland is covered by the Alps?
2. What kinds of climate does Switzerland have?
3. Where are most of Switzerland's cities located?
4. Why was Switzerland able to remain at peace during the two world wars?
5. Using the map on page 138, figure the distance between the following places: (a) Zurich and Geneva; (b) Basel and Frankfurt; (c) Bern and Bucharest; (d) Geneva and Warsaw; (e) Hamburg and Krakow.

What can you see in this picture of an outdoor cafe that shows Switzerland is the host to visitors from all over the world?

Manufacturing

A fourth reason for the Swiss high standard of living is manufacturing. Although Switzerland has few mineral resources, it is mainly an industrial nation. It manufactures many goods to sell to other countries. In this way the Swiss obtain money to buy what they need and want. Most of the raw materials needed in the industries have to be imported.

The Swiss have long been known as skilled workers. Years ago they became noted for certain articles that they made in their homes. Many of these articles are now made in factories. One product that requires great skill to make is watches. Years ago each watch was made entirely by one person. Now watchmaking is a factory industry with different people making different parts of the watches. Geneva and a number of cities in the Jura Mountain region are watchmaking centers. Large numbers of watches and clocks are exported.

Switzerland's main industrial development began with the manufacture of cotton and silk cloth. Woolens, rayons, linens, and other fibers were later included. As the textile industry grew, the Swiss made machines and tools that were needed in manufacturing textiles. In this way, another important industry grew— the **machine-tool** industry. Water wheels, turbines, diesel engines, agricultural machines, steam engines, locomotives, looms, and various aluminum articles are now made. Coal tar dyes, drugs, paper, and chemicals are also manufactured. Coal, **bauxite** (from which aluminum is made), and other metals used in making these products have to be imported. Many of the manufactured articles and textiles are exported.

Another industry for which the Swiss are famous is woodcarving. Beautiful carved clocks and other wooden articles are exported. Small articles, such as jewelry, toys, music boxes, and metal tools and instruments are also made in Switzerland.

Government and Education

The Swiss systems of government and education have also made possible a high standard of living in Switzerland. Switzerland is the oldest republic in Europe. In 1291, men living in three valleys on the borders of Lake Lucerne (loo *surn*) met and formed a union. They helped defend each other against the rulers of surrounding kingdoms. In this way they kept their independence. The land controlled by each valley group was called a **canton** (*kan* tuhn). As time passed, other cantons in surrounding valleys joined the union. Now there are 22 cantons and 3 half-cantons which function as states. Each canton has its own local government and elects persons to represent it in a National Assembly. The president is also elected. In 1971 women were permitted to vote in national elections in Switzerland for the first time.

The Swiss believe in giving everyone a good education. Almost every person in the country can read and write. As in the United States, education is controlled by the states, not by the national government. Thus, there are differences in the schools from place to place. All children have to attend school, though, until they are at least 14 years of age. In some cantons everyone must attend school until the age of 16. Boys and girls study two or more languages, arithmetic, geography, history, and other subjects.

An Inland Nation with Ships

As you can see by looking again at the map, Switzerland has no sea-coast. Nevertheless, the Swiss own a number of ships that carry goods from one country to another. The Rhine River is the only waterway the Swiss have to the ocean. Barges are used to move goods up and down the Rhine. Swiss manufacturing depends almost entirely upon imported raw materials. Thus, keeping goods flowing on the Rhine is especially important to Switzerland.

Switzerland also helps its neighbors in the transportation of their goods. The shortest route from northern Italy to Germany is through Switzerland. One of the

Trains are an important means of transportation in Switzerland. Because they all run on electricity produced by falling water, they cause little pollution.

best routes from Paris, Brussels, or Rotterdam to Italy is through Switzerland. The country also serves as a route from France to Austria.

Cities

Most of the cities of Switzerland are located on the shores of lakes. Some are located along the larger rivers in the plateau region. Zurich, the largest city, is a center for textiles and for machinery of many kinds. Basel is Switzerland's river port on the Rhine River. Textiles, dyes, and chemical products are manufactured in Basel. Bern, the capital, is also a manufacturing center. Textiles, machinery, scientific instruments, and chocolate are made there. Geneva, at the western end of Lake Geneva, is best known as a place for international meetings and conferences. It is also a center for watchmaking.

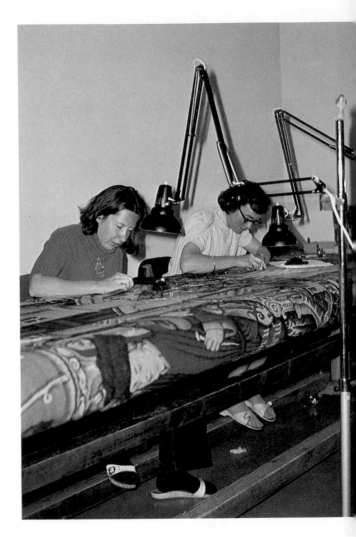

Repairing old tapestries and cloth garments is a difficult skill that some Swiss have learned.

REMEMBER, THINK, DO

1. What are several of the main products of Swiss industry?
2. How does Switzerland get the raw materials which are needed for industry?
3. What is the name of Switzerland's port city? What kind of port is it?
4. Compare the height of the mountains in Switzerland with the height of the mountains in Colorado. Which are higher? Which have more ice and snow in summer months? Why?
5. Do you think a country should permit bank accounts with secret numbers? Why or why not?

WHAT HAVE YOU LEARNED?

1. What is a watershed? What four large European rivers begin in the little country of Switzerland?

2. Why is it important for the people of Switzerland to attract many tourists? Why do tourists want to visit Switzerland?

3. What are the three main languages used in Switzerland? Which of them is spoken by the largest number of people?

4. Why is electricity of special importance to the Swiss?

BE A GEOGRAPHER

1. Compare Switzerland with the state in which you live. Which is larger? Which has more people? Which has higher mountains?

2. Five states in the United States have a northern border that is at about the same latitude as Switzerland. What are the names of those five states? Is any of the five states as small as Switzerland? Does any of the five states have as many people as Switzerland?

3. Look at the Landforms Map of Europe on page 45. Where else in Europe besides Switzerland would you expect there to be ski resorts? Why? Be prepared to defend your answer.

4. Look at the Population Map of Europe on page 100. How many cities are there in Europe with more than 500,000 people? (Be sure to include in your count all cities that have more than 1 million, of course!)

5. Look at the Annual Rainfall Map of Europe on page 167. Based on what you already know, how much annual rainfall do you think is needed for crops to grow well?

6. Look at the Land Use Map of Europe on page 202. Does the map agree with what you have learned about Switzerland?

QUESTIONS TO THINK ABOUT

1. What differences might it make in your life if people in the northeastern part of the United States spoke English, those in the southeast French, those in the southwest German, and those in the northwest Swedish? What differences would there be in radio and television programs? In school books? In road signs? Do you know of any other country that uses more than one language?

2. Why do you suppose women were not permitted to vote in Switzerland until 1971? Are there countries even today where women may not vote?

3. Would you prefer to live in a mountain village, in a city on a river or lake, or on a farm located on a rolling, hilly plateau? Why?

4. Why is much milk made into cheese by farmers in the Alps?

5. Are you able to speak more than one language? What advantages does that give you? What difficulties has it caused for you?

6. What do you think of "not taking sides" as Switzerland did in the world wars? Is there ever a time when a person, or a nation, should fight? Do you think the Swiss would fight if their country were invaded? Would you?

Chapter

POLAND

The word *Poland* means "country of the plain." Almost all of the land of Poland is part of the North European Plain. Poland is one of the larger countries of Europe. It has almost as much land as the state of New Mexico. About 35 million people live in Poland. Only about one million people live in New Mexico. The southern border of Poland is at about the same latitude as the southern border of western Canada.

A Brief History

During the 1500s Poland was a country with great power and influence. But it had few natural borders on the east and west. As a result, Poland was involved again and again in wars. Poland lost portions of its territory at different times. In 1795, after a war, the whole country was divided among the Germans, Austrians, and Russians.

For more than 100 years the Russians, Germans, and Austrians tried to keep the land they had taken in 1795. But the Poles dreamed and planned for the day when they would again be an independent country. The Poles who moved to other countries helped to keep the dream alive. In 1918 the Polish dream came true. At the close of World War I, land was taken from Austria, Germany, and Russia. It was returned to the Poles. Poland again became an independent nation.

The boundaries of Poland set up in 1918 were farther east than the present boundaries. In 1939 the armies of Germany and the Soviet Union invaded Poland. These two countries, again, divided Poland between themselves. When the Germans attacked the Soviet Union in 1941, they seized the whole country of Poland. Poland was held by the Germans until armies from the Soviet Union drove the Germans out in 1945.

After World War II the Soviet Union and Poland signed a treaty of peace. Poland gave up a large section of its eastern territory to the U.S.S.R. In the west, however, Poland was allowed to shift its boundaries to the Oder (*oh* der) and Neisse (*ny* suh) rivers. The Germans living east of these rivers had to make a choice. If they wished to live in Germany, they had to move westward.

Government, Education, and Religion

Today the government of Poland is controlled by the Polish United Workers [Communist] Party. Two other political parties also operate in Poland. They support communistic policies and probably would not be allowed to exist if they did not. Only about 7 out of every 100 adult Poles belong to the Communist party. Yet they control the government. If the Poles decided to get rid of the Communist-controlled government, armies from the Soviet Union would probably move in to prevent it.

Education is controlled by the government. All boys and girls between 7 and 15 years of age must attend school. They are taught government (their own), the Polish and Russian languages, and science.

Over 90 out of every 100 Poles are members of the Roman Catholic church. They have stubbornly remained so even though the government has tried for many years to discourage religious worship.

As archbishop of Krakow, Karol Cardinal Wojtyla was an important leader of the Catholic Church in Poland. In October 1978, he was elected Pope John Paul II, the first non-Italian pope in more than 450 years.

Poland

The Climate

Poland has a humid continental climate with cold winters. Close to the Baltic Sea the climate is milder and more even than farther inland. Away from the coast, the weather often changes quite rapidly. The people of Poland say they have six seasons: early spring, spring, summer, autumn, early winter, and winter.

Snowfall is not heavy on the plains during the winter, but it lies on the ground from two to three months. Streams are frozen for about the same length of time. Generally, winters are colder in the southern and eastern parts of Poland. The mountains in the south have more snowfall in winter months and more rainfall in the summer than the plains. As the rainfall map on page 167 shows, most of the country has enough rainfall for crops. Most of this rain falls during the summer months.

The Land

The land of Poland slopes gently from the Carpathian (kahr *pay* thee uhn) Mountains in the south toward the Baltic Sea. The northern part of Poland at one time was probably quite flat. However, the glaciers that once covered the land left deposits when they melted. These are now low hills. Some of these hills are about 250 meters (820 feet) above sea level. Many lakes are scattered throughout northern Poland. See the map on page 138. The lakes are located both east and west of the Vistula (*vihs* tyoo luh) River.

The river valleys are broad and shallow in northern Poland. The rivers used to overflow their banks and flood the lower land. Now low banks of earth, called **levees**, have been built along the many rivers.

The Polish people have made the rolling hills of northern Poland into productive farmland.

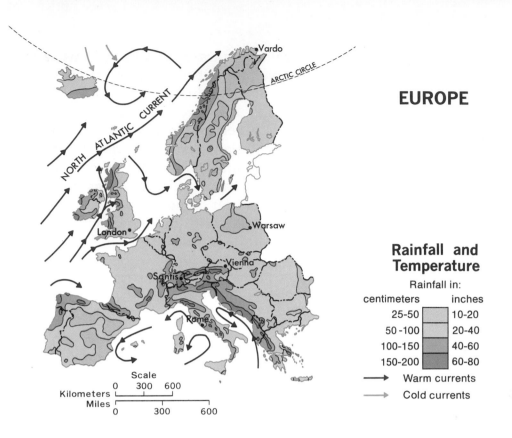

EUROPE

Rainfall and Temperature

Rainfall in:

centimeters		inches
25-50		10-20
50-100		20-40
100-150		40-60
150-200		60-80

→ Warm currents

→ Cold currents

Scale

Kilometers 0 300 600

Miles 0 300 600

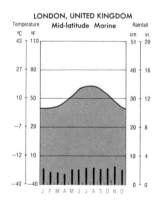

LONDON, UNITED KINGDOM
Mid-latitude Marine

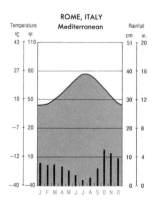

ROME, ITALY
Mediterranean

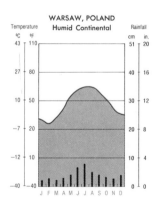

WARSAW, POLAND
Humid Continental

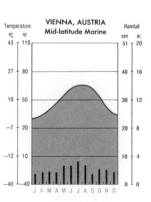

VIENNA, AUSTRIA
Mid-latitude Marine

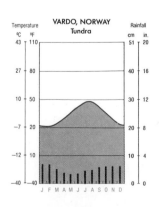

VARDO, NORWAY
Tundra

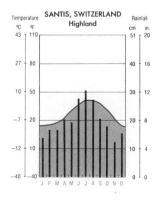

SANTIS, SWITZERLAND
Highland

Poland

The levees usually keep the rivers from flooding the land.

At one time many of the rivers in this section were bordered with marshes. Most of these have now been drained.

In the southwestern part of Poland are the Sudeten (soo *day* t'n) Mountains. These form the boundary between Poland and Czechoslovakia. As the map shows, the Oder River and its tributaries drain the Sudetan highland. The Oder River Valley also provides an easy route into Czechoslovakia. In southeastern Poland are the High Tatra Mountains (*tah* truh), a beautiful mountain range. Poland has no natural border along most of its eastern boundary with the Soviet Union.

REMEMBER, THINK, DO

1. What landform region extends through most of Poland?
2. Poland has few natural borders to its east and west. Why has this been a problem for this nation?
3. What seasons do the Polish people say they have?
4. What church do most people in Poland belong to?
5. What have the Poles built to prevent floods in northern Poland?

During the summers many vendors sell cold drinks and ice cream on the streets of Warsaw.

Jan Paderewski in concert. Inserts: Frederic Chopin; Artur Rubinstein.

The Piano—and Poland

Three of the greatest pianists that ever lived were born in Poland. All three are known and their work admired throughout the world.

The first of the three—and regarded by many as the best composer—was Frederic Chopin (shoh *pan*). Most people think of Chopin as French, because for most of his adult life he lived in Paris. Chopin was born, however, near Warsaw, Poland. Chopin studied music at the Warsaw Conservatory. A **conservatory** is a school that teaches young, talented people in the fine arts. Chopin began to play the piano in public when he was only eight years of age. Most of his adult life was spent writing music for the piano. You probably have heard pianists play some of his music. Many people think that, in writing for the solo piano, he was the most creative musician who ever lived. Chopin died in 1849 of tuberculosis. He was only 39 years of age when he died.

By contrast, Ignace (ehn *yahs*) Jan (yahn) Paderewski (pah deh *rehf* skee) lived to be 81 years of age. He was born in 1860 in a part of Poland that is now in the Soviet Union. He began taking piano lessons when he was six years old. When he was 12, he entered the Warsaw Conservatory. And when he was only 18, he was made a professor at the Conservatory! Paderewski was probably the world's most famous pianist in the early years of this century. He was also one of the people who kept alive the dream of an independent Poland. After the First World War he entered politics and served for a short time as premier of Poland. But his government did not remain in power very long. He returned to the concert stage and played for many more years. His last concert tour in the United States was made when he was 78 years of age. Paderewski died in New York City during World War II. At the time of his death, German troops were occupying his beloved homeland.

As this book is written, the third famous pianist from Poland is still alive, but he is very old. You may have seen him on television. Arthur Rubinstein (*roo* bihn stine) was 91 years of age in 1978. Like one of his teachers, Paderewski, he played the piano on concert stages through most of his adult life. Rubinstein was born in Lodz (looj), a city southwest of Warsaw, in 1886. He was a soloist with the Berlin Symphony Orchestra when he was only 12 years of age. Rubinstein made his first American tour when he was 20 years old. He became an American citizen when he was 60. And he was still giving concerts at 90!

Isn't it interesting that three of the world's greatest pianists were born in Poland! And, all of them left Poland because of wars and for opportunities elsewhere. Chopin, Paderewski, Rubinstein—three pianists—all with Poland in their hearts and in their fingers.

Industries in Poland

In giving up the eastern territory to the Soviet Union in 1945, Poland lost some petroleum and potash deposits. However, the southern part of the region—which was taken from Germany—is rich in coal. It also contains copper, sulphur, and iron ore.

Much effort has been made to increase industrial production in recent years. And the effort has been successful. In East Central Europe, only East Germany produces more goods.

Coal is Poland's most valuable mineral fuel. It is also an important export. Poland ranks fourth as a coal-producing nation. Poland produces slightly more coal than Great Britain. It exports more coal than any nation except the United States.

Until recently the main coal fields were west of Krakow (*krah* cow). A new field has been found, however, near Lublin (*loo* blihn) in eastern Poland. This field is nearer to Warsaw and other industrial centers. It can also be mined more easily by using machines. Since the world faces a shortage of energy, Poland's exports of coal will probably increase.

Coal is used in the chemical industry. It is also used in the smelting of ores and in the manufacture of goods made of steel. Poland has a large supply of low-quality iron ore. About two-thirds (⅔) of the iron ore used is imported, however. The ore comes mainly from Sweden and the Soviet Union.

Poland also has an important supply of uranium ore. The uranium mines are controlled by the Soviet Union, and the ore is shipped to that country.

Manufacturing

Many products, including tools needed in industries, are made in Polish factories. Trains, ships, motors, cranes, tractors, bicycles, and mining machinery are all made in Poland. Wroclaw (*vrot* slahf), in southwestern Poland, is an important center for such manufacturing. Lodz, between Wroclaw and War-

Coal mining is an important industry in Poland. These students are preparing for careers in mining by studying science. Why do you think there are no girls in this classroom?

saw, is a textile center with linen, cotton, and woolen mills.

The government controls industry as in other countries that are governed by Communists. The Poles are probably better off today, however, than they were between the two world wars. Then more than half the people were farmers and almost all were poor.

Agriculture

Poland's best farmland is found in the central and southern parts of the country. In the northern sections the soils are generally poor. Much of the land in the north is poorly drained, too. In central and southern Poland, the land is better drained and the soils are richer. Even so, large amounts of fertilizer must be used to obtain good crops.

After World War II the govern-

Warsaw was destroyed by the German army in World War II. The Old Town section of the city has been rebuilt.

ment in Poland tried to start a great many large *state farms*. **State farms** are farms owned and managed by the government. But most of the people in Poland refused to give up their land. They wanted to farm their own small farms. They did most of the work by hand or with the help of horses. The state farms that were formed did not succeed. Today about four-fifths (⅘) of the farmland is still farmed by owners. And they still do most of the work by hand.

The main crops raised by Polish farmers are rye, potatoes, and oats. These crops grow well on the poor soils. In the southern part of the country, wheat, sugar beets, barley, and some corn are raised. Some hemp and flax are grown, too. Until recently enough food was raised that some could be exported.

Almost all Polish farmers raise livestock as well as crops. Poland is a leading nation in milk and butter production. Butter, canned hams, and bacon are exported most years to other countries.

Warsaw and Krakow

Warsaw is Poland's capital and largest city. It was nearly destroyed by the Germans during World War II. A large section of the city known as the Jewish quarter was completely destroyed. Every building was leveled. So much rubble was left lying on the ground that the new city was built right on top of it. Most buildings and industries destroyed during the war have been rebuilt.

Students arrive by bus for school. What differences are there between your school building and the one you see in the picture?

Krakow, in southern Poland, was the capital of Poland until about 1600. It is an old and favorite tourist city. Its university was founded more than 100 years before Columbus sailed from Spain. Krakow is near the mountains in southern Poland and is the doorway to this favorite vacation area for the Poles. Old parts of the city are being restored. And attempts are being made to stop air pollution. In the future, only factories that do not pollute will be permitted in the area. Automobiles will not be permitted in old parts of the city. A new and modern city is being built near the old part of Krakow. Thus, the Poles will have both the old and the new side by side.

REMEMBER, THINK, DO

1. Where is the best agricultural land in Poland?
2. What is a state farm?
3. Who operates and controls industry in Poland, private owners or the government?
4. Why may Warsaw be properly named a "new city"?
5. Is it higher above sea level at Krakow or at Warsaw? How can you tell?

WHAT HAVE YOU LEARNED?

1. What does the name "Poland" mean? Why is that a good name for the country?

2. What kind of climate does Poland have? Where in the United States is that kind of climate found?

3. What is Poland's most important mineral fuel? What advantages does having a good supply of that fuel give the Poles?

4. What are the main crops grown by Polish farmers?

5. What political party controls the government of Poland? What might happen if that party lost control?

BE A GEOGRAPHER

1. Is Warsaw farther north or farther south than:
   a. London, United Kingdom?
   b. Montreal, Quebec?
   c. Seattle, Washington?
   d. Anchorage, Alaska?
   e. Berlin, East Germany?
   f. Paris, France?

2. How far is it by airplane from Warsaw to:
   a. Geneva, Switzerland?
   b. Paris, France?
   c. Moscow, U.S.S.R.?
   d. Washington, D.C.?

3. Make a chart on which you list the capitals of the European countries we have studied. Put the largest city first on your chart, and the smallest city last. Then choose a scale to show population, and draw lines by each capital city to show how large each is. You should have six capitals on your chart when you have finished.

4. If you wanted to ship a bulky product like coal from Krakow, Poland to Stockholm, Sweden, what would you do? Suggest several possibilities.

5. If it is 6:00 P.M. in Warsaw, Poland, what time is it in:
   a. Stockholm, Sweden?
   b. New York, New York?
   c. Denver, Colorado?
   d. Honolulu, Hawaii?
   e. Reykjavik, Iceland?
   f. Moscow, U.S.S.R.?

6. How far is it from the border east of Warsaw to the western border where the Oder and Neisse rivers join? How far is it from the northern coast to the southern border?

7. How far is it from the northern to the southern borders of Poland along 20° E.?

8. Are Berlin and Warsaw in the same time zone? London and Warsaw? If it is 12:00 noon in London, what time is it in Warsaw?

QUESTIONS TO THINK ABOUT

1. How do you think you would feel if you lived in a country where armies from nearby countries had fought across your land several times? Why?

2. How can a political party that has so few members (only 7 out of 100 adults) control the government of Poland?

3. Why should one small country in central Europe produce three of the world's greatest pianists? Is there any reason why something like that happens, or is it just chance?

4. The Communists have tried to stifle religion in Poland, yet more than 90 of every 100 Poles still belong to the Catholic church. Is that a bit hard to understand? Why do you suppose it is true?

5. Which of the six countries we have studied about would you choose to visit if you could? Which would you choose to live in for five or more years? Why?

10
Chapter

─── ROMANIA ───

Romania is the farthest south and the farthest east of the countries of East Central Europe. It has about as much land as the states of Indiana and Illinois combined. Romania has about 22 million people. On many maps and in some books, the name of the country is spelled "Rumania." Romania is spelled with an "o" in this book. Both spellings are correct. Spelling the name with an "o" helps explain how important the Roman Empire was in the early history of this land.

A Brief History

Romania was conquered by armies from Rome during the second century A.D.. Since then the land has had many wars. The people, at different times in history, have been under the control of a number of great powers. A glance at the map of Eurasia on pages 8 and 9 will help you understand why.

Much of Central Europe is mountainous. Until this century, armies had to cross the land mainly on foot or on horseback. Usually they followed low passes through the mountains or traveled through the broad river valleys. The plains of southern Romania, near the Danube River, are a natural pathway for armies. These pathways are easy to locate on the map, page 138. Thus, Romania has been invaded many times and has had many changes in its borders.

Romania was a battleground in World War II. It became one for a special reason. Romania has large petroleum fields. Modern armies must have gasoline and oil to run their planes, tanks, and trucks. Germany did not have enough petroleum for a long war. Therefore, German armies struck through Hungary and the Danube Valley to take the Romanian oil fields. Germany also wanted the food crops that grew well on the plains near the Danube.

A German-controlled government ruled Romania for several years during World War II. Then, near the end of the war, armies from the Soviet Union moved westward and helped defeat Germany. As in other countries in East Central Europe, the Soviet Union set up a communistic government. In recent years the Romanians have been making changes in the government that the U.S.S.R. started. They have been doing some things different from what the Soviet Union would like. For instance, in 1967 they started trading with West Germany. Romania also refused to place all the blame on Israel for the war that broke out in the Middle East in 1967. In 1968 Czechoslovakia was invaded by her "friendly neighbors in Communism." Some people thought then that Romania would also be invaded. However, Romania was not invaded. It has remained more independent of the Soviet Union than most nations in eastern Europe.

The Land of Romania

Four natural features are very important to Romania. These are: *the mountains, the plains, the Danube River,* and *the coastline on the Black Sea.*

The Mountains. Romania has two major mountain ranges which join near the central part of the country. The Carpathian Mountains extend southeastward into Romania. (See map, page 138.) These mountains extend southward to near the center of the country. The Transyl-

Farmers gather freshly-cut hay in the Transylvanian region of Romania. Compare these hay fields with the ones pictured on page 154. Which area do you think has better farmland? Why?

Sailing a ship through the Iron Gate on the Danube River has been made safer and easier by the construction of a dam and locks.

vanian (*trans* uhl *vay* nee uhn) Alps extend from the Danube Valley in western Romania eastward to the Carpathians. These two ranges formed the eastern rim of the huge lake that once covered the Great Hungarian Plain. (See map, page 138.) The lower Bihor (bee *hawr*) Mountains extend northward from the Transylvanian Alps. They almost surround a fertile plateau near the center of Romania.

Romania's mountains are quite rugged but not very high. The highest peaks are in the Transylvanian Alps and they are about 2500 meters (8,000 feet) high. Many of the slopes are covered with forests, which provide an important natural resource. Since the countries farther south have few forests, they are good customers for Romania's timber. The mountains also furnish small supplies of many minerals and a good supply of bauxite.

The Plains. As the map on page 138 shows, there are plains west of the Bihor Mountains. Romania also has broad plains south and east of the other ranges. These plains are drained by streams that flow into the Danube River. The plains have fertile soils and usually get enough rainfall for good crops.

Between the plains and the mountains there are foothills. In the foothills of south-central Romania, near Ploesti (plaw *yesht*), the oil and natural gas fields are located.

The Danube River. The Danube River flows almost completely across southern Romania. Part of the country's southwestern border with Yugoslavia follows the river. The river also forms the southern border with Bulgaria until it turns northward not far from the Black Sea. Almost all of Romania is drained by the Danube and its tributaries. The river is very important to Romania because it is navigable throughout the country. Two of Romania's larger cities are important river ports: Galati (gah *lahts*) and Braila (bruh *ee* luh). Ocean vessels sail up the Danube River from the Black Sea to both cities.

The Danube River has cut a deep gorge through the mountains on the border between Yugoslavia and Romania. This gorge is called the *Iron Gate*. At one time it was a dangerous channel for shipping. Then rocks were blasted from the channel. A canal near the Yugoslavian side of the river was built. Vessels moving upstream then were pulled through the swift water by locomotives on the river bank. Romania and Yugoslavia now have built a dam at the Iron Gate. There are **locks**—gates in the canals—which make it possible to raise or lower the water level. These locks on both shores make it easy for tugs and barges to move through the Iron Gate today.

The dam was built in an interesting way. After agreeing to build it, engineers from Romania and Yugoslavia planned how to construct it. Then workers from both countries started to build from each shore. They met in the middle of the river! The dam is about 425 meters (1400 feet) long. A lake 80 kilometers (50 miles) long has been formed behind the dam.

The dam not only makes river traffic easier, it is also used to produce hydroelectric power. Each country has a power plant on its side of the river. Together they produce about one-third (⅓) as much electricity as was produced in the two countries before the dam was built.

Like all big rivers, the Danube River is forming a delta at its mouth on the Black Sea. The Danube delta is being extended into the sea about 5 meters (16 feet) each year. As the map on page 138 shows, the Danube has several mouths. The delta area and much of the land along the lower part of the river is marshland.

Every few years floods occur along the Danube in Romania. In 1975 about 100,000 homes were damaged by high water. Thousands of cattle, sheep, pigs, and chickens were drowned. And many crops growing in fields along the river were destroyed.

For many years Romanians have wanted a canal connecting the Danube and the Black Sea. The canal was to be built from where the Danube turns north to near Constanta (kawn *stahn* tuh). A canal there would shorten shipping time on the Danube. Many years ago a canal was started. The workers were political prisoners. Their tools were shovels and wheelbarrows. Not much was done, because most of the prisoners became sick. The canal was never completed.

Now, Romania is again planning to build a canal through the swampy area. But it will be done very differently this time. Workers will use

As in most major cities of Europe, streets in Bucharest are often crowded.

large earth-moving machines. The workers will be paid. And the Romanians have planned the project well. On the Danube River a dam will be built to control the flow of water. The dam will also produce electricity. A new seaport will be built on the Black Sea coast 16 kilometers (ten miles) south of Constanta. Someday tugs will be pushing barges through the canal. Fewer floods will probably occur along the lower Danube. The planners are quite sure that some good farmland will be obtained. The swamps near the canal will dry out or will be drained.

The Coastline on the Black Sea. The Black Sea provides Romania with an ice-free port. Constanta is the only major Romanian city on the Black Sea coast. Ships from Constanta, Braila, and Galati sail through the narrow Bosporus (*bahs* puh ruhs) at the southwestern end of the sea. (See map, pages 200-201.) Then the ships sail through the narrow Dardanelles (dahr d'n *ells*) into the Aegean (ih *jee* an) Sea and then into the Mediterranean. (Both the Bosporus and the Dardanelles are controlled by Turkey.) The Black Sea coast is a favorite vacation spot for tourists who want to swim and sunbathe. Many hotels have been built near the shore.

Bucharest, the Capital

Bucharest, the capital of Romania, is the nation's largest city. It has about 1.5 million people. The city has a number of manufacturing industries and is the leading business and trade center.

The climate of Romania is the humid continental type. It is much like that of the north-central United States. Winters are colder and summers hotter than in Germany and Poland. On the plains near the Danube, the temperature often falls to −30° C (−22° F) during the winter. In the summer temperatures frequently rise above 38° C (100° F). High in the mountains, of course, temperatures are cooler both in winter and in summer.

Most of the country has a little more than 50 centimeters (about 20 inches) of rainfall annually. In eastern Romania near the coast of the Black Sea, even less rain usually falls. Most farmers in that area have to irrigate their crops. A very little difference in the rainfall on the plain makes a large difference in crops. If the rain does not come at the right time, the crops can be ruined.

COMECON

COMECON—the Council for Mutual Economic Assistance—is an organization made up of countries working with the Soviet Union. Eight nations of eastern and central Europe are members of COMECON (*kahm* ee *kahn*). Bulgaria, Czechoslovakia, East Germany, Hungary, Poland, Romania, and the Soviet Union are full members; Yugoslavia is an associate member. In addition, Cuba and Mongolia are members. The Soviet Union, of course, is the most powerful member. Vietnam and North Korea are not members, but they take part in some COMECON activities.

What does COMECON do? COMECON tries to get the member countries to cooperate with each other—especially in the area of economic planning. The agency was first organized by the Soviet Union in 1949. At that time the U.S.S.R. wanted to recover quickly from World War II. It needed resources, tools, and equipment that could be made in the eastern and central European countries. The Council, while founded for economic cooperation, really worked to benefit the strongest nation—the U.S.S.R. For a long time COMECON was not very effective.

In recent years COMECON has become much better at economic planning. Resources now flow both ways—to the Soviet Union and from it. For instance, petroleum and natural gas pipelines have been built to eastern European countries from the U.S.S.R. Iron ore and other minerals dug in the Soviet

Ore from the Soviet Union is used to make steel in COMECON countries. Here molten steel is being processed in Romania.

Union are shipped to the other nations. Products grown or made in one country are shipped to where they are needed in another. COMECON works for the countries with governments controlled by Communist parties. It works in much the same way as does The European Economic Community in western Europe.

Most countries controlled by Communist parties have developed five-year plans. These plans provide goals that the countries try to reach. COMECON countries are trying to coordinate or combine their five-year plans. That way they believe that more progress can be made. Special factories, for instance, are built only in one country. Products from that one factory are shipped to all COMECON nations.

Romania is a bit more independent than the other COME-CON nations. Its leaders often argue that planning should be done by each country. Then each country should seek what it needs wherever it can be found. Romania was the first COMECON nation, for instance, to start trading on a large scale with the United States. Other nations in eastern Europe then began to trade with non-COMECON countries. A great deal of trade now takes place between COMECON and EEC (European Economic Community) nations. More trade, however, still takes place among COMECON nations.

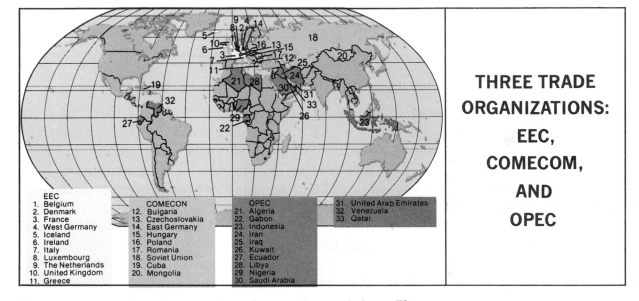

THREE TRADE ORGANIZATIONS: EEC, COMECOM, AND OPEC

EEC
1. Belgium
2. Denmark
3. France
4. West Germany
5. Iceland
6. Ireland
7. Italy
8. Luxembourg
9. The Netherlands
10. United Kingdom
11. Greece

COMECON
12. Bulgaria
13. Czechoslovakia
14. East Germany
15. Hungary
16. Poland
17. Romania
18. Soviet Union
19. Cuba
20. Mongolia

OPEC
21. Algeria
22. Gabon
23. Indonesia
24. Iran
25. Iraq
26. Kuwait
27. Ecuador
28. Libya
29. Nigeria
30. Saudi Arabia
31. United Arab Emirates
32. Venezuela
33. Qatar

There are many international trade and economic associations. The members of the three that are discussed in this book are shown on the map above.

Agriculture in Romania

Many of the people in Romania make their living by farming. Corn grows well in the Romanian climate and is one of the main foods of the people. It also is fed to farm animals, especially cattle and hogs. Wheat is exported to other COMECON nations.

Many other crops are grown. Among the important ones are sugar beets, potatoes, other garden vegetables, tobacco, grapes, and orchard fruits. The orchards and vineyards are usually planted on sloping land along river valleys. Apples, peaches, pears, plums, grapes, and cherries are the main fruits grown.

In hilly and mountainous areas, people farm the fertile river valleys. In addition, they graze many cattle

Romania

and sheep on land that is not good for growing crops.

As in the other countries of East Central Europe, the government has tried to develop state farms. According to the government, state farms use more modern methods of farming, including tractors and fertilizers. As in other nearby countries, however, these farms have not been very successful. Crop yields have not increased greatly.

Manufacturing in Romania

Until recently, manufacturing in Romania was not very important. Most of it had to do with the processing of food. Much of the rest was related to the production and refining of oil and other petroleum products. Pipelines carry the oil eastward to Galati and Constanta for shipment by tankers. Other lines carry oil southward to Bucharest (*boo* kuh rehst) and on to the Danube. From there barges take the oil upriver.

In recent years, however, manufacturing in Romania has grown very rapidly. The manufactured goods have improved in quality, also. As this growth has gone on, the Romanian leaders have often found themselves not agreeing with their COMECON neighbors. Czechoslovakia and Hungary, for instance, wanted Romania to concentrate on agricultural production. The Romanian leaders did not agree. Romania wanted to concentrate on iron, steel, and machinery. They built a huge iron and steel complex.

They also started a chemical industry based upon petroleum. And they used iron and steel to make the tools needed in oil fields. These tools are now sold all over the world. Factories have been built to make automobile tires, tractors, synthetic fibers, and machinery. Romania bought whole factories from countries in Western Europe. It could not find equipment of as good quality in COMECON countries. Romania is working hard to build up trade with nations outside COMECON.

Religion and Education

As you have learned, Romania has been invaded many times by different peoples. These invaders brought to Romania many religious faiths. As a result, the people of Romania worship in many different ways. The majority of the people (about eight out of ten), however, are members of the Romanian Orthodox church. Today all religions and religious teaching are supervised by the government. Also, the government pays the salaries of the ministers and priests.

Before the Second World War, education in Romania was limited to a small number of people. Today education is free to all boys and girls. All children are required by law to attend school. As a result of this policy, almost all of the people in Romania can read and write today. Also, many students are taught a trade so that they will be ready to work in a factory.

The Best in the World at 14 Years of Age

Nadia Comaneci (koh mah *neech*) became the first person to receive a perfect score from Olympic judges. You may have seen her perform at the 1976 Olympics if you watched the "games" on television. Nadia, a gymnast, was only 14 years of age in 1976. She was judged to be the world's best woman gymnast. Seven times during the competition in Montreal, Canada, the judges gave her perfect scores for her performances! People wonder whether that will ever happen again.

Nadia started practicing as a gymnast when she was in kindergarten. For many years she practiced about three or four hours every day, six days a week. In the morning she went to school like most young people in Romania. Afternoons she practiced, practiced, and practiced.

Nadia's mother works at an office, and her father is a mechanic. She has one brother who is four years younger than she is. When she was 14, Nadia liked to ride her bicycle and play with her friends. She also had a hobby—collecting dolls.

Because she is now world famous, Nadia has had many opportunities to travel. She went with the Romanian national gymnastics team all over the world. Nadia Comaneci may be the best-known Romanian citizen. It all started when she was 14 years of age!

REMEMBER, THINK, DO

---

1. Why have the borders of Romania changed many times?
2. Why did German armies invade Romania during World War II?
3. What has been built at the Iron Gate?
4. List all the countries the Danube River flows through or along. Then compare that with the number of states the Mississippi River flows along or through.
5. Plan a trip by ship from Stockholm, Sweden, to Galati, Romania. List all the seas, oceans, straits, channels, and rivers you would sail on.
6. What cities in Romania have more than 500,000 people? What cities in your state have at least 500,000 people?
7. Which has more land, Romania or the state where you live? Which has more people?
8. Imagine that you are the leader of the government in Romania. You face some important decisions. Not many people in your country can afford to own and drive a car. You can decide to raise their wages so that more people can buy cars. Or you can pay them less and spend the money on projects like dams and canals. Which would you choose to do? Why?

LEARN MORE ABOUT CENTRAL EUROPE

There are several other countries in Central Europe. You may wish to learn about them on your own. The Study Guide on page 31 will help you plan how to do so.

In West Central Europe there is one additional country and a principality.

Austria

The land of Austria is much like Switzerland. The main river in Austria is the Danube. Some of eastern Austria is part of a great plain through which the Danube flows. Austria, like Switzerland, is a delightful country to visit. Taking care of tourists is big business there. The main ways of making a living are farming and manufacturing. Steel, aluminum, machinery, and textiles are the main products.

Liechtenstein

The principality of Liechtenstein is not quite as large as the District of Columbia. Only about 20,000 people live in Liechtenstein. This little country is probably best known for its postage stamps. Perhaps you have seen some of them or have some in your stamp collection.

East Central Europe also contains the countries of East Germany, Czechoslovakia, and Hungary. All three of these countries have governments controlled by a Communist party.

East Germany

East Germany is the most prosperous of the countries in East Central Europe. You have learned some facts about it already. Factories in East Germany make many products for export. Among them are cameras, rubber, plastics, fertilizers, and locomotives. East Germany is unusual because the largest city of West Germany, West Berlin, is located within East Germany's borders.

Czechoslovakia

The country of Czechoslovakia is south of East Germany and Poland. It is the homeland of two peoples—the *Czechs* (checks) and the *Slovaks* (*sloh* vaks). Do you see how this country got its name? Czechoslovakia is about the size of the state of New York. It has about 15 million people. Many people work in agriculture, raising sugar beets, wheat, corn, and barley. Manufacturing industries make steel, glass, porcelain, and shoes.

Hungary

Hungary is south of Czechoslovakia and east of Austria. It is about the size of the state of Indiana and has almost 11 million people. Most of Hungary's land is lowland and most of it is farmed. The main farm crops include corn, sugar beets, wheat and other grains, and potatoes. A long, long time ago most of eastern Hungary was probably covered by a large lake. The main manufactured products include processed foods, machinery, and hardware. As in other countries in East Central Europe, the government owns and operates the factories.

Workers unload pipe for the Brotherhood gas pipeline in Czechoslovakia.
This pipeline, 500 kilometers (310 miles) long, links the U.S.S.R. and
Czechoslovakia across the Carpathian Mountains.

A SUMMARY OF CENTRAL EUROPE

Central Europe contains eight countries and a principality. Two of the countries once were one. The boundaries of most of the larger countries have changed many times. Most of the changes have been caused by wars. For a long time Central Europe has often been a battleground. Much of World War II was fought in Central Europe.

Central Europe has many people. Most of them speak more than one language. Central Europe is one of the world's most productive regions. Factories there make almost every product known. Parts of Central Europe are world famous as tourist areas. Maybe you will visit Central Europe some day.

WHAT HAVE YOU LEARNED?

1. Why has Romania been invaded many times?

2. How have Romania and Yugoslavia improved the Danube River for transportation?

3. What are the main parts of Central Europe?

4. How do the governments differ in the two main parts of Central Europe?

5. Which of the countries in West Central Europe produces the most goods? In East Central Europe?

6. Which country in East Central Europe seems to be the most independent of the Soviet Union?

7. What is COMECON and why was it formed?

BE A GEOGRAPHER

1. Using the map on page 138, answer the following questions:
   a. What four large rivers flow either along the borders of West and East Germany or across these countries?
   b. These rivers flow into three different seas. What are they?
   c. How are these rivers joined?

2. What is the highest mountain in the Alps? Is its highest point in Northwestern, Central, or Southern Europe?

3. The capitals of three states in the United States are about the same distance from the equator as Bern, Switzerland, is from the equator.
   a. What are the cities and the states in which they are located?
   b. Which of the three is nearest to Bern's latitude?

4. If it is 1:00 A.M. in Berlin, Germany, on Tuesday, November 1, what time, day, and date is it in the following cities:
   a. Liverpool, United Kingdom
   b. New York, New York

   c. Sydney, Australia
   d. Honolulu, Hawaii

5. What cities in Central Europe have more than 500,000 people? Are any cities in Central Europe as large as London? Paris?

6. Suppose you wanted to ship some goods from London to Budapest, Hungary, by water. How do you think the shipment would be made? Could the goods be shipped by another route? Would the size of the shipment influence the selection of the route?

———————— QUESTIONS TO THINK ABOUT ————————

1. In which country of Central Europe would you prefer to live? Why?

2. Which country of Central Europe would you most like to visit? Why? If your answer to this question is different from your answer to question 1, why?

3. How are the Mississippi River and the Danube River alike? How do they differ?

4. How are Switzerland and Colorado alike? How do they differ?

5. How do you think life would change in Poland or Romania if the Communists lost control of the government? Would it depend on the kind of government that replaced the present one?

6. What do you think life would be like where you live if a Communist party achieved power?

7. When did women get the right to vote in national elections in the United States? In Switzerland? Young people, 18 years of age, were given the right to vote in the United States in 1971. Should persons that age be permitted to vote in national elections? How about persons 16 years of age? Twelve years of age?

Some of the labor on farms in Southern Europe is still performed by hand.

Unit

SOUTHERN EUROPE

Unit Preview

All these photographs were taken in Southern Europe. Based on what you already know about Europe, answer the questions below before you start to read about Southern Europe. Your answers will be *predictions* on your part—what you expect you will learn. Write your answers on a piece of paper.

- The climate in Southern Europe is . . .

- The major languages used in Southern Europe are . . .

- The crops raised in Southern Europe are . . .

- The history of Southern Europe is . . .

DO YOU KNOW?

- What nations are included as part of Southern Europe?

- What is meant by a "Mediterranean climate"?

- What the official state religion of Spain is?

- What nation in Southern Europe has active volcanoes?

- Which is the only Balkan nation *not* controlled by a Communist government?

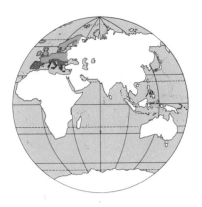

Unit Introduction

Southern Europe

Southern Europe, as the region is called in this book, extends for a great distance from east to west: some 3000 kilometers (1,900 miles). The region has eight nations and three small independent states. All the nations have a seacoast. (See map, pages 200-201.) From west to east, the nations that make up this region are: Portugal, Spain, Italy, Malta, Yugoslavia, Albania, Bulgaria, and Greece. The three small states are Andorra, San Marino, and Vatican City.

The Land

The map of Southern Europe on pages 200-201 shows three large peninsulas in Southern Europe. The one farthest west is called the *Iberian* (eye *beer* ih uhn) *Peninsula*. It extends southwestward from France to the narrow Strait of Gibraltar (jih *brawl*

ter). The strait is at the western end of the Mediterranean Sea. Spain, the largest country in Southern Europe, takes up most of this peninsula. Portugal and Andorra are also on the Iberian Peninsula. In addition to the Pyrenees Mountains, which separate France and Spain, there are several other mountain ranges. Find them on the map on pages 200-201.

The central peninsula of Southern Europe extends like a long boot southeastward from Switzerland. It is called the *Apennine* (*ap* uh nine) *Peninsula*. The mountains with that name extend its full length. This peninsula, and two large islands near it, make up the country of Italy. The tiny republic of San Marino is surrounded by Italy. It lies near the Adriatic (ay dree *at* ihk) seacoast at about 44° N. The Apennine Peninsula and the island of Sicily reach most

193

of the way across the Mediterranean Sea toward Africa. South of Sicily is Malta, a small island-nation in the Mediterranean Sea.

The peninsula to the east is the *Balkan* (*bawl* kuhn) *Peninsula*. It contains the southern European countries of Yugoslavia, Bulgaria, Albania, and Greece. It also contains a small part of Turkey. The Dinaric (duh *nair* ihk) Alps extend northwest to southeast in Yugoslavia. The Balkan Mountains in Bulgaria give the peninsula its name. These mountains form the northern boundary for the peninsula. The Pindus Mountains in Greece and Albania combine with the others to make this peninsula a rugged land.

The Climate

The climate of Southern Europe is much the same on all three peninsulas, except for central Spain. In most places at low altitudes, the region has a Mediterranean climate. You may remember that areas with a Mediterranean climate have cool, moist winters with considerable

The Mediterranean climate of Southern Europe has made it a favorite vacation area for people from Northwestern and Central Europe. Split, Yugoslavia is a popular vacation center.

rain. They have long summers that are quite dry and hot. Throughout Southern Europe more rain falls in mountainous areas than on low-lands. The Mediterranean climate is perfect for olive trees and grapes.

Those crops are grown, therefore, in many places. In river valleys, where irrigation is possible, many other crops are grown. Grains, especially winter wheat, are grown in many places, too.

Things You Might Like To Do As You Study About Southern Europe

As you continue your study of the continent of Europe, you may wish to undertake some of the following activities:

1. Make a survey of the class members to determine where their ancestors came from. If you find that you have parents whose ancestors came from Southern Europe, learn all that you can from them about the "old country." Some parents may have visited relatives in Southern Europe. They may have slides, photographs, clothing, and memories to share.

2. Make a large outline map of each of the three peninsulas in Southern Europe. Place on them the boundaries of countries and the capital cities. Add information to the map as you study Southern Europe.

3. Form a committee to visit the library for information or stories about Southern Europe. If possible, bring the materials to the class—books, filmstrips, or pamphlets. If other classes also need to use the material, ask the librarian to put the materials in a special place within the library.

4. Investigate topics such as the following:

a. The Olympic Games—Past and Present
b. Three small countries in Southern Europe: Andorra, San Marino, and Vatican City
c. Florence and the Italian Renaissance
d. Mythology of Early Greece and Rome
e. The Roman Empire
f. The Great Teachers of Early Greece: Aristotle, Plato, and Socrates
g. The Spanish Armada

5. Make a study of the Greek alphabet to find out how ours was developed from it. Also learn about Roman numerals. Try to do some multiplication and division using Roman numerals!

6. Look around your community for evidence that ancient Greek and Roman architecture still lives.

7. Learn a few words and greetings in Spanish, Italian, Yugoslavian, and Greek.

8. Form committees to do special reports on countries described only briefly in this text. These include Albania, Bulgaria, Malta, and the very small countries of Vatican City, Andorra, and San Marino.

Unit Introduction

11

Chapter

—SPAIN—

Of all the countries in Europe, not including the Soviet Union, only France has more land than Spain. Spain has more than twice as much land as the United Kingdom. However, fewer people live in Spain per square kilometer (or square mile) than in France or the United Kingdom. Only about 73 persons per km² (189 per sq. mi.) live in Spain. Much of central and northeastern Spain, as the population map on page 100 shows, has few people.

Early History

Spain is a very old land with a rich history. It was conquered several times early in its history by invading people. The first important invading armies came from Rome about 200 B.C. For about 600 years, Spain was part of the Roman Empire. The second great invasion was by people from northern Europe. They entered Spain about A.D. 400 and ended the control of Rome. These people ruled Spain until about A.D. 700.

Around A.D. 700 the third, and perhaps most important, group of invaders came to Spain. These people were Arabs of North Africa and were called the **Moors**. They were people who had become **Muslims** (*muhz* luhms)—followers of Islam. The Muslims had conquered all of North Africa. These people brought to Spain a very rich civilization—more advanced than that of Western Europe at the time. Among the contributions brought by the Moors was the figure zero. This made arithmetic much easier than it had been with the Roman numerals, which had no zero. The Moors also taught the Spanish people how to irrigate the land. They knew how to make better steel. They also built fine buildings. For about 500 years Spain was very prosperous under the Moors. Then the Moors, too, were defeated—this time by the Spanish themselves.

For about 100 years, beginning with the defeat of the Moors in 1492, Spain was the world's leading power. (Do you know something else that the Spanish did in 1492?) For a period of time, Spain ruled over Portugal, The Netherlands, and much of Italy. In addition, Spain had many colonies on other continents. Most of the land in the West Indies, Central America, and South America was under the control of Spain. Some of North Africa also was controlled by Spain. Riches from these areas poured into Spain. The country developed the world's greatest navy and powerful armies.

But Spain's glory lasted only about 100 years. In 1588 the Spanish fleet was almost destroyed by the British fleet and a storm at sea. Little by little, revolts in the colonies and problems at home weakened Spain.

Recent History

Spain has never been as rich or as powerful as it was in the 1400s and 1500s. In this century the country has had many problems. The people of Spain fought a long civil war from 1936 to 1939. The war was fought over control of the government. Hundreds of thousands of people were killed. The forces led by Francisco Franco finally won the war in 1939. He ruled Spain as a dictator until his death in 1975.

Spain's recovery from the civil war was a slow and painful process. The country was in ruins. At first, the Franco government placed strict controls on the people. Most personal freedoms were taken away. As conditions improved through the years, some controls were lessened or lifted completely. By the end of the 1960s Spain's economy had recovered greatly. There was new industry and improved agriculture.

The Muslim and Christian heritages of Spain can often be seen in the architecture. Built for use as a mosque, this building in Cordoba is now a Christian church.

The standard of living for the people had risen quite rapidly.

In 1969 General Franco chose as his successor the grandson of the former king of Spain, Alfonso XIII. Alfonso left the Spanish throne in 1939 when the people voted for a republic. Juan Carlos, his grandson, became king of Spain two days after Franco's death in 1975. At once he began to move Spain toward a parliamentary system of government. After 36 years of military dictatorship, however, changing the government was not easy. In 1976 the voters of Spain chose to have a parliament. This parliament, called the **Cortes** (kawr *tehz*), has two houses—an upper house and a lower house. As in the United Kingdom, the lower house is elected directly by the people. The king appoints one-fifth (⅕) of the members of the upper house (the Senate). The rest are either elected or appointed by labor unions, professional organizations, and local governments.

In 1977 national elections were held in Spain for the first time in 41 years. Under the new government more than 6,000 people who had been in jails as political prisoners were freed. Labor unions once again were allowed. Once again the press was allowed to criticize the govern-

In December 1978, the people of Spain voted in favor of a new democratic constitution. The new constitution limits the power of the king. King Juan Carlos and Queen Sophia are shown here casting their ballots.

ment. Political parties began to operate once more.

There were problems in the world economy in the 1970s. Spain's economy suffered, too. Many people who were working in France or Germany returned home to Spain. But jobs were scarce. And, because of inflation, everything cost more. One hopes that the new, more democratic government will be able to solve these and other problems.

The Land of Spain

Three-fourths (¾) of the land of Spain is high tableland called the *Spanish Meseta* (muh *say* tuh). **Meseta** is a Spanish word for "plateau." In most places this highland reaches almost to the shore, leaving a narrow coastal lowland. Several ranges of mountains extend across Spain from west to east. These divide the country into several natural regions. (See map, pages 200-201.) The highest ranges are near the northeastern and southeastern borders. As you have learned, the high Pyrenees in northeastern Spain make travel through them quite difficult. Unlike the Alps, the Pyrenees have no good passes. Through the ages people have gone around them rather than through them. Not until 1928 was a railroad built through the Pyrenees. Fourteen tunnels had to be built to make the railroad possible. The main tunnel on this route is about seven kilometers (four and a half miles) long. It is built at an altitude of more than 900 meters (3,000 feet).

Look again at the map on pages 200-201. Notice that several fertile lowlands extend far into the Meseta in central Spain. In the northeast the Ebro (*ay* broh) River and its tributaries drain the Pyrenees and the eastern Cantabrian (kan *tay* brih uhn) Mountains. In the southwest the Guadalquivir (*gwah* dahl *kee* veer) River drains the large valley north of the Sierra Nevada mountains. Three other main rivers cut the high central plateau. These all flow toward the west through Portugal. Until modern machines were invented, it was difficult to build roads and railroads across Spain. As a result, people in different parts of Spain lived in isolation, with little contact with others. That is why different ways of living developed in different parts of the country.

The Climate of Spain

Most of Spain, like all regions with a Mediterranean climate, has dry summers. Only the northwestern part of the country has plenty of rainfall. See the rainfall map on page 167. Northwestern Spain has a mid-latitude marine climate like that in southwestern France. The hills in this region are covered with oak, beech, chestnut, ash, and birch trees. In the higher mountain areas, coniferous trees grow. The lower slopes and the valleys are used for farming.

In the central plateau region many of the rivers become small streams during the summer. Summer and winter are both resting periods for plants on the central plateau. The

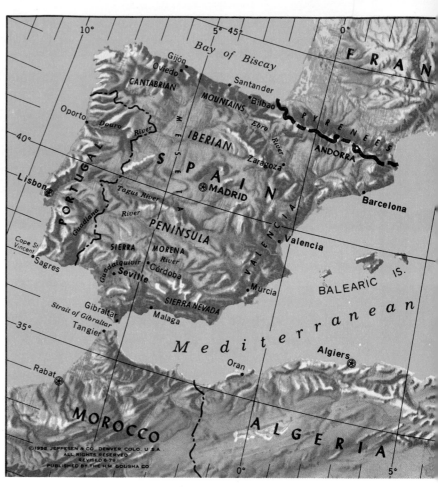

SOUTHERN EUROPE

Km 0 100 200
Mi 0 100 200

EVERGREEN FORESTS

FORESTS

FARMLAND

DRY GRASSLAND

SEMIARID GRASSLAND
AND FOREST

MOUNTAINS

summers are too dry for growing crops, except where irrigation is possible. The winters are too cold for much plant growth. Spring and fall are the growing seasons. At those times enough rain falls, and the weather is warm.

The soil on the Meseta is naturally fertile. Crops that do not need a long growing season do well wherever there is enough water. The rivers on the Meseta have cut very deep valleys through the plateau. Because these valleys are so deep, a good way of using the water for irrigation on the plateau itself has yet to be found. The river valleys are irrigated, and good crops are produced.

On the lowlands of southern and eastern Spain, little or no rain falls during the summer months. Most vegetation dries up during July and August. Only olive and cork trees seem able to survive these hot, dry months. Of course, some of the land is irrigated during the dry summers. Winter temperatures in this area are generally mild enough for agriculture. Enough rain falls for crops during the winter.

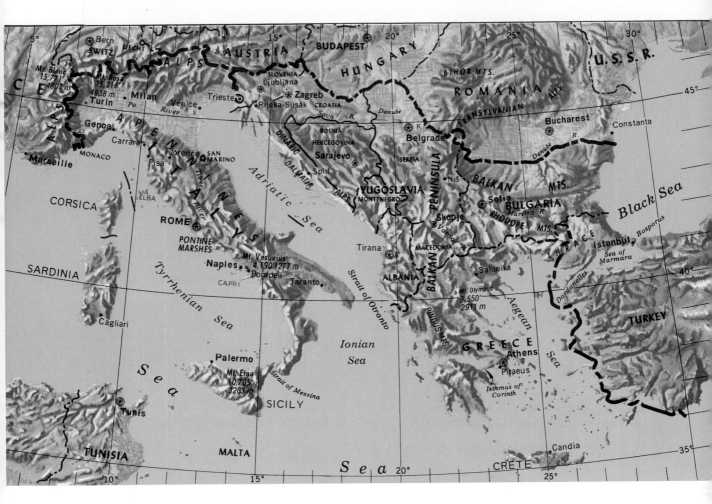

REMEMBER, THINK, DO

1. What three peninsulas extend southward from Europe?
2. What country in Europe has more land than Spain?
3. Describe a Mediterranean climate. Where in the United States would you find that climate?
4. What is a civil war? Why was the Spanish Civil War fought?
5. Why were roads and railroads not built through the Pyrenees for many years?
6. Why is it difficult to use water in the rivers flowing through the Meseta to irrigate land on the plateau?
7. When was Spain the world's richest and most powerful nation?
8. Do you think the new government in Spain will succeed? Would you like to have it succeed?

Spain

Many farmers still use old methods of farming. They have few machines to help with the work. During planting season, much of the grain is sowed by the wasteful method of scattering the seed by hand. Not much fertilizer is used. Sometimes the rain doesn't fall when it is needed. As a result, crop yields are much less than in most other European countries.

Crops in Spain are quite different from place to place. The main crop on the northern half of the Meseta is winter wheat. Because most of the rain falls during winter months, good crops of wheat can be raised. Inland a short distance corn is grown. In addition, cattle are raised in the northwest because of its humid climate.

In the drier parts of the Meseta, barley is often grown. Some rye and oats are also grown in mountainous areas. These crops need less rain to produce grain than wheat does.

Throughout the southern part of the Meseta, olive trees are grown. The land here is too hilly or the soil too poor for other crops. The olive tree grows well in a Mediterranean climate. It can stand long periods of dry weather. Its roots grow deep into the soil. Some trees that are more than a thousand years old still bear olives. The hills along the Guadalquivir River in southern Spain are one of the leading olive-producing regions.

Some olives are eaten in Spain and many are bottled and exported. The most important use made of the olives, though, is in the production of olive oil. The oil that is in the olive is squeezed from ripe olives by presses. Olive oil largely takes the place of butter in the diet of the Spanish people.

In eastern Spain, especially in the province of Valencia (vah *lehn* see uh), citrus fruits are grown. This province grows so many oranges that one type of orange bears the name "Valencia." This type of orange is grown in many other places in the world, too. Many are grown in the United States. So many oranges are grown in Spain that they are a

EUROPE

ARCTIC CIRCLE

Scale

0   300   600

Kilometers

Miles

0   300   600

Landuse

Livestock and food crops
Sheepherding
Mediterranean agriculture
Dairy farming

Non-agricultural, mainly forests
Nomadic herding
Non productive

major export. Other citrus fruits are also exported to northwestern Europe.

Truck gardens in river valleys supply enough vegetables to meet the people's needs. The long, sunny summers make it possible to grow large crops, provided there is water for irrigation.

Another major crop of southern Spain is grapes. Grapevines, like olive trees, grow well where little rain falls during summer months. The grapes are used mainly to make wine. Wine is served with lunch and dinner in most southern European countries. Sherry, a popular wine, was first made in Spain. Some grapes are exported.

Other crops grown in southern Spain include almonds and rice. **Esparto grass**, which is long and coarse, grows in this region. It is used to weave baskets, rugs, and rope. South of the Sierra Nevada range, tropical fruits such as bananas and dates are raised.

On most Spanish farms, work animals are still used to help in plowing and in cultivating the soil. Oxen and donkeys are used more frequently than horses. Horses are more expensive and also need more care. They especially need better pastures. Some farmers now use tractors.

Although there is not much water in this river during the summer, the bridge indicates that there are times when this river on the Meseta is much wider and deeper. What are the reasons for the changes in the amount of water in the river?

Irrigation

A large irrigation project has been built on the Guadiana (gwah *thyah nah*) River in the southwest of Spain. Five huge dams have been built on the river and its tributaries. These dams store water during the spring months, when rain falls. Canals take the water during the dry summer months to valley farms. Electricity is also produced at the dams.

Spain

Every year, during the week of July 9, a festival occurs at Pamplona. The main events are daily bullfights preceded by the "running of the bulls" through the streets of the city.

Bullfighting in Spain

One of Spain's most popular spectator sports is bullfighting. Have you ever seen a bullfight? The sport is not allowed in the United States. But in Mexico and Spain, bullfighting is popular. As with our football, basketball, and baseball games, much of the fun comes from the crowd. Crowds stand and cheer when a **matador** (*mah* tuh dor), or fighter, kills a bull skillfully. They also let a matador know it when he or she is clumsy.

Bullfights are held in more than 400 places in Spain. They are usually held on special days. The bullfights are held in arenas that are built much like football stadiums. The bullfighting arena is somewhat smaller and rounder, however. It is built with strong walls between the arena and the crowd. There are openings in these walls so that the helpers can escape the bull's charges. And there is at least one place where the bulls are let into the arena.

A bullfight starts when the chosen bull is turned loose in the arena. Soon, two or more **picadors** (*pee* kah thors)—persons with long spears and mounted on horses—enter the ring. The bull looks at them and may even charge the horses. The horses are much faster, and can turn more easily than the bull. Seldom is a horse hurt. The picadors stick their spears into the neck of the bull. The spears weaken the neck of the bull so that it will carry its head lower, and they make the bull angry.

After the picadors have done their work, the **banderilleros** (*bahn* thay ree *yay* ros) enter the ring on foot. They also have spears—but theirs are short. These spears are also known as darts. The banderilleros have bright capes to attract the bull. As the bull charges them, they step aside and stick their darts into the neck of the bull. These darts remain in place until after the bull is killed. Usually about four darts are placed in the neck of the bull, two on each side. The banderilleros have to be almost as skillful as the matador. If they are not careful, the bull will injure them. And they only have an instant—when the bull charges—to place the darts properly. The crowd cheers when the darts are placed well and skillfully.

Finally, the matador enters the ring. The banderilleros remain to help attract the bull if the matador is hurt. By the time the matador arrives, the bull is very mad. The matador walks toward the bull and waves a small red cloth at the bull. It is attached to a stick. The bull charges the red cloth, which is called a **muleta** (moo *lay* tuh). The matador skillfully turns so that the muleta is just in front of the bull's horns. The bull turns to follow the muleta. The matador keeps playing the bull until the bull is tired. Finally the bull is so tired he just stands still. The matador then walks toward the bull. If the bull has been judged properly, the bull will stand quietly as the matador approaches. Sometimes, however, matadors are fooled. The bull charges again and may injure or even kill the matador.

At last, if the bull is completely exhausted, the crowd calls for the matador to kill the bull. The best matadors kill the bull with one thrust of a sword. When the killing is done skillfully, after a fine series of passes, the matador is awarded the ears and tail of the bull. If a fair job has been done, the matador may get one ear. If the crowd thinks it has been a clumsy performance, the matador gets paid—but doesn't get the ears or the tail.

In Portugal, Spain's neighbor on the Iberian Peninsula, bullfighting is popular, also. There, however, it is all done from horseback, and the bull is not killed. The bull's horns are padded with leather so that the horses are not often injured.

What do you think of bullfighting as a sport? Which way of doing it would you prefer? Do you think bullfighting should be allowed in the United States? Why or why not?

Spain

Fishers from the Galicia region of northern Spain display their catches. Most of these fish are going to be sold at auction to local fish merchants.

Minerals and Manufacturing

Spain has a good supply of many minerals needed for industry. In spite of this, Spain has been slow in becoming a manufacturing nation. There are several reasons. During the last century, Spain was unable to develop a stable and effective government. Most of Spain's mines, moreover, were controlled by companies from other countries. The mines are located at widely separated places in Spain. Building roads and railroads across the land was difficult until large machines were invented. Finally, because Spain was a poor country, not enough money was spent on education. All of these factors together resulted in very slow industrial growth.

More factories have been built in recent years. Spain is beginning, moreover, to make more effective use of its available resources. The country has good supplies of both iron ore and coal in the Cantabrian Mountains. It has the world's best supplies of **mercury**, the liquid metal. Water from the Pyrenees, the Cantabrian, and Sierra Nevada ranges is available to produce hydroelectric power. Some dams and power plants have been built. Many more good sites are available.

Major manufacturing centers in Spain are located along the northern and eastern coasts. Bilbao (bihl *bah* oh) is located a few kilometers (or miles) up a navigable river from the northern coast. It is a major center for iron and steel production.

Barcelona (bahr seh *loh* nuh), on the northeastern coast of Spain, is the busiest port city. It is also a major manufacturing center. Many textile mills have been built in or near Barcelona. These mills generally use hydroelectric power to run the machines that weave cloth from cotton, wool, and synthetic fibers. Other factories make paper, glass, and leather or metal products. Ships sailing from Barcelona carry exports of olives, wine, grapes, citrus fruits, and cork. They also carry some manufactured products such as shoes. Most manufactured goods are used in Spain, however.

Fishing

Look again at the map on pages 200-201. You will see one reason why people living near the coasts in Spain have for many years harvested food from the sea. The country is almost surrounded by water.

Among the nations of the world, Spain's fishing industry ranks eleventh. That ranking is based on the number of metric tons of fish caught. Some nations, such as Japan and the Soviet Union, send fishing ships thousands of miles away to catch fish. Most of the fish caught by Spanish fishers, however, are taken near their homeland. And most of them are taken by fishers using small boats and nets. Most of the fish catch in Spain is sold fresh in nearby fish markets. Only a little of the catch is frozen or canned.

The Catholic Church

Throughout most of Spain's recent history (since 1492), the Catholic church has been the official state church. The Spanish government guarantees support of the Roman Catholic church. All schools offer religious instruction, and only those marriages that are performed by the Catholic church are considered legal. Other churches may hold services, but they are not permitted to try to get new members. As a result, almost all people in Spain think of themselves as Catholics.

At the time of Spain's greatest glory (1500s), the Catholic church had great influence and power. Every explorer took along priests. These priests not only served the members of the expedition. They also taught their beliefs to the people of the lands explored or conquered. Spain probably had more influence on the spread of Catholicism than any other country. For a time some Catholic leaders misused their power. Nonbelievers were sometimes imprisoned or killed.

Most of the important holidays in Spain today are tied to the Catholic church. Local **fiestas**—religious celebrations to honor saints—are held throughout Spain. These are times of much fun—dancing, eating, athletic contests, and the like—as well as religious celebrations. Much of the finest art, music, and literature of Spain is religious in nature, too.

Madrid

The largest city in Spain is Madrid (muh *drihd*), the capital. It is located in a section of Spain that has limited water and mineral resources. If it were not the capital, Madrid probably would not have become such a large and important city. It is, however, a modern and beautiful city. Railroads and roads go out from

Madrid in all directions. Madrid also has a major international airport. Many flights from both South and North America land there.

Millions of tourists visit Spain's capital city, Madrid, and other parts of the country each year. Tourism is one of Spain's most important industries. About one-tenth ($\frac{1}{10}$) of the people in Spain work in the tourist industry. Many tourists come from northern Europe to get away from the long, cold winters. Many others from Latin America visit Spain because their language and history are related. A popular tourist attraction is Madrid's El Prado, Spain's best-known art museum. Resorts on the Mediterranean coast attract millions of vacationers to Spain each year. The money made caring for tourists is used by Spain to buy needed products from other countries.

Gibraltar

On the southern coast of Spain, near the Strait of Gibraltar, a narrow peninsula extends into the Mediterranean Sea. It is the British *Crown Colony of Gibraltar*. A **crown colony** is an area governed by a person appointed by the British government. Gibraltar includes only 6 km² (2.3 sq. mi.) of land, but it has much more importance than its size would indicate. Years ago the British built a naval base and a fort on the limestone rock of Gibraltar. In case of war, they can control the passage of ships to and from the Mediterranean Sea through the Strait of Gibraltar. At its narrowest, the strait is only 13 kilometers (about 8 miles) wide. At its widest, it is only about 37 kilometers (about 23 miles) across. In underground caves in the rock are huge guns. These guns can fire across the narrow Strait.

The Spanish government claims Gibraltar. For many years the Spanish government has been trying to get the United Kingdom to withdraw from Gibraltar. In 1966, for instance, Spain closed the border to all vehicles. Then, in 1968, Spain stopped permitting tourists to cross from Spain to Gibraltar. And, in 1969, all traffic was stopped with Gibraltar. More than 4,000 Spanish people who worked during the day on Gibraltar lost their jobs.

There seems to be some signs that the new, more democratic government of Spain may lift some of the restrictions. Several years ago a vote was taken by the people who live on Gibraltar to see whether they wished to be under Spanish rule. Almost all of them at that time voted to remain a British colony. It may be that the people on Gibraltar may soon be asked again whether they would prefer to be part of Spain. At present the people there are almost "prisoners" on "the rock." Check to see what has happened on Gibraltar recently.

The Canary Islands

Another possible "hot spot" for Spain is a group of islands located off the northwest coast of Africa. These islands—called the Canary

The rock of Gibraltar guards the entrance to the Mediterranean Sea.

Islands—are about 1300 kilometers (800 miles) from Spain. (Find the Canary Islands on the map on pages 542-543.) They are officially provinces of Spain, but they are much closer to Morocco.

For many years Spain had a small territory on the west coast of Africa called Spanish Sahara. Much trade took place between the Canaries and Spanish Sahara. In 1975 Spain gave part of Spanish Sahara to Morocco.

Part of it was given to Mauritania (mawr uh *tay* nee uh). Some of the people who lived there wanted their independence. Fighting broke out and trade with the Canary Islands almost ended. Some people living in the Canary Islands are not happy with Spain's actions in Africa. These people are asking for independence for the Canary Islands. Check to see what has happened to the Canary Islands since 1980.

REMEMBER, THINK, DO

1. Many more sheep than cattle are raised on the Spanish Meseta. Why would this be so? Where are cattle raised in large numbers in Spain?
2. How is olive oil used by most Spanish families? Why?
3. What agricultural crops or products are exported from Spain?
4. Why was Spain slow in becoming a manufacturing nation?
5. Should the United Kingdom give Gibraltar back to Spain? Why or why not?
6. List reasons for and against independence for the Canary Islands.

Spain

WHAT HAVE YOU LEARNED?

1. After the invasion by peoples of northern Europe, who invaded and ruled Spain for about five hundred years?

2. Who was Francisco Franco? Who was named as his successor?

3. What is Spain's parliament called?

4. What does the word "meseta" mean?

5. What part of Spain has a mid-latitude marine climate like that in southwestern France?

6. In what part of the Meseta is winter wheat grown? Are olive trees grown?

7. What are two of Spain's chief industrial centers?

8. Why do many tourists visit Spain during the winter months?

BE A GEOGRAPHER

1. How does France, Europe's largest country, compare with Spain, Europe's second largest? You might like to make a chart with several items (*e.g.,* population, land area, rainfall, topography, main crops, etc.) listed down the left side of the chart and columns for data about France and Spain to the right.

2. How does Spain compare with Mexico in land area, population, natural resources, and industry? Where, if you had to make a choice, would you build a shoe factory? An automobile factory? Give reasons for your choices.

3. How do the Pyrenees compare with the Alps?

4. Is southern Spain south or north of southern Florida? What main cities in the eastern United States are at about the same latitude as Madrid? How do the climates compare?

5. Is Madrid east or west of:
   a. Paris, France?
   b. London, United Kingdom?
   c. Dublin, Ireland?
   d. Edinburgh, United Kingdom?
   e. Lisbon, Portugal?
   f. Stockholm, Sweden?
   g. Reykjavik, Iceland?

QUESTIONS TO THINK ABOUT

1. Plan and conduct a debate on the following subject: *Resolved*: That the United Kingdom should return Gibraltar to Spain. (If you have never conducted a debate in your class, find out how to do one. A speech teacher from a nearby high school might be a useful person to ask for help.)

2. Look for interesting articles about Spain in *National Geographic*. If your library has one, consult a copy of the *National Geographic Index*.

3. Develop a class notebook of "Famous Spaniards." Have a small committee compile a list of possible people to include. Then have volunteers do short biographies of the persons selected. Be sure to include some present-day persons in your notebook.

4. Compile another class notebook entitled "Places to Visit in Spain." Be sure that you include places within Madrid as well as other cities. Don't forget to check with travel bureaus. They often have good information about places to go and things to see.

5. Invite a student from Spain, or one from the United States who has studied in Spain, to visit the class. Prepare for the visit by developing several questions you are anxious to ask.

6. Using encyclopedias or other sources, learn what you can about Spain's Balearic Islands. You can find them on the map on pages 200-201 in the Mediterranean Sea east of Valencia. Do people live on these islands? If so, what do the people do for a living?

❧ 12 ❧
Chapter

ITALY

The country of Italy covers almost all of the Apennine Peninsula, the central of the three peninsulas in Southern Europe. Two large islands nearby—Sicily (*sihs* uh lee) and Sardinia (sahr *dihn* ee uh)—are also part of this nation. Italy has a little less land than Poland, but many more people. Italy has about 190 people per km² (492 per sq. mi.). Poland only has about 112 people per km² (290 per sq. mi.). As you can tell by looking at the map, much of Italy is mountainous. (See page 45.) Therefore, the lowlands, where most of the people live, are quite crowded.

Two generations come together on the streets of Milan.

A Brief History

The Roman Empire

Rome, the capital city, was started on the banks of the Tiber (*ty* ber) River about 3,000 years ago. Seven low hills in this location furnished good sites for villages. The hills offered good protection against attack. In addition, villages on the seven hills were not flooded when the Tiber overflowed its banks. In time, they were united to form Rome. Rome is still known as the "City of the Seven Hills."

With Rome as its center, this small *city-state* slowly expanded. A **city-state** is an independent state that includes a city and surrounding territory. It became one of the world's

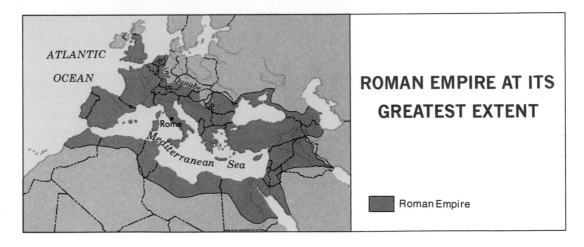

ROMAN EMPIRE AT ITS GREATEST EXTENT

ATLANTIC OCEAN

Rome

Mediterranean Sea

Rhine Danube

▦ Roman Empire

greatest empires. By A.D. 100 the Roman Empire controlled all the land around the Mediterranean Sea. Most of Europe west of the Rhine River and south of the Danube River was also conquered by Rome.

Ancient Rome had a strong central government. Roman citizens took an active part in the government. Problems were debated in public, and there were written laws. Under the Emperor Justinian (juh *stihn* ee uhn), about A.D. 550, these laws were carefully organized. This code of laws—the **Justinian Code**—was, perhaps, the Roman Empire's greatest gift to the world. These laws were more reasonable and just than any laws the world had known up to that time. Many of them still are in use today.

The government of Rome, nevertheless, finally weakened. People from the north invaded the Italian peninsula and ended Roman power.

Recent History

During the centuries that followed the decline of Rome, the peninsula had many small kingdoms. Finally,

in 1870, these kingdoms were united under one king to form the nation of Italy. Once again Rome became the capital of Italy.

In 1922, eleven years before Hitler and the Nazis seized power in Germany, Benito Mussolini (*moo* suh *lee* nee) gained control over Italy. Mussolini was backed by a group of Italians known as *Fascists* (*fash* ihsts). **Fascists** believed in a strong central government headed by a dictator. Mussolini was able to take over the government because many people were unemployed and starving. The people wanted someone to do something about it and to do it quickly.

Mussolini improved conditions for many of the people. He had many public buildings, roads, and power plants built. He had the Pontine (*pawn* teen) Marshes drained and made into good farmland. These marshes covered a large section of the coastal lowlands south of Rome. But people who disagreed with Mussolini were put in prison, sent out of the country, or killed.

Italy joined Germany and Japan against the Allied powers in World

213

Italy

War II. As a result of defeat in this war, Italy lost several African colonies. Soon after the war the people voted to form a republic. Since 1946, Italy has had a president who is elected to a 7-year term by the members of parliament. The members of parliament are elected by the people. The real head of the government, however, is the prime minister who is nominated by the president and elected by the parliament.

Land of Italy

The Alps form a natural northern boundary for Italy. They do not isolate the country like the Pyrenees on the Iberian Peninsula, however. The Alps have several good passes that have been used for 2,000 years or more. Several of these passes link Italy with France and with Central Europe. Good roads have been built through the passes. Railroads run through tunnels many feet below the summits of these passes. A highway tunnel 12 kilometers (7 miles) long under Mont Blanc was opened in 1964. Traffic flows through it to France.

Another mountain range forms the backbone for the long Apennine Peninsula. This range, as you might expect, is called the Apennines. The Apennines are old mountains and much less rugged than the Alps. Some of the mountains in the Apennine chain are volcanic. Near Naples (*nay* pl'z) is the famous active volcano, Mount Vesuvius (veh *soo* vih us). Mount Etna (*eht* nuh), on the island of Sicily, is another active volcano in Italy.

This home in Pompeii was buried for almost 1,800 years. What was this room used for? What evidence can you find in the picture to support your opinion? The foods pictured on the next page have been excavated from the city. Can you identify them?

The Buried City of Pompeii

In A.D. 79, near where the city of Naples, Italy, now is located, there was a small city named Pompeii (pahm *pay*). The city was only about 1.6 kilometers (1 mile) from the volcano named Mount Vesuvius.

Pompeii had a wall that surrounded it. There were eight gates in the wall. Streets in Pompeii were paved with blocks of **lava**—rock formed from volcanic eruptions—from nearby Mount Vesuvius. No one knows for sure how many people lived in Pompeii in A.D. 79. It is estimated, however, that about 20,000 probably lived there.

Suddenly, Mount Vesuvius erupted. For days on end—perhaps for weeks or months—ashes and dust fell on Pompeii. Most of the people fled to areas farther from the volcano. Some, perhaps as many as 2,000 people, lost their lives. They probably died because they could not get their breath. Ashes and dust in the air made it difficult for them to get enough oxygen. Their bodies were covered with ashes—as was the whole town.

Pompeii lay there, covered with ashes, for almost 1,700 years. About 1750 some people started to dig where Pompeii had been buried. Some of them did so to get objects for museums, and some of them hoped to get rich. The careful **excavation** (the digging out) of Pompeii did not start for another hundred years. But since about 1860, work has been done by the Italian government to uncover Pompeii and restore it. More than half of Pompeii is now open to visitors.

We know a great deal more about how people lived in southern Italy in A.D. 79 because of the excavations. Near the center of town was an open square. It probably was used for celebrations and town meetings. Around the square were large, important buildings. Some of them were probably temples. Some were public bathhouses. Many of the homes in Pompeii must have been built by rich Romans. They were quite large.

Mount Vesuvius still erupts at times. Most of the eruptions are not as great as the one that buried Pompeii, of course. You might wonder why people live so near volcanoes. The reasons are fairly simple. For years and years, volcanoes may be **dormant**—that is, quiet or inactive. Then there may be a period of activity. The mountain may belch smoke, some ashes, and perhaps some lava. The mountain may then become dormant again. It may remain quiet for many more years. Moreover, the land where the ashes fall becomes more fertile. It soon becomes excellent farmland. In a country where there is not much good farmland, it seems a waste not to use the area. Even when volcanoes do erupt, there is usually time to move.

Not all the land of Italy is mountainous. Between the Apennines and Alps in northern Italy is a large plain. Known as the Po Valley, this plain is drained by the Po River and its branches. There are also narrow coastal plains along both coasts. And there is a large plain at the "heel" of the boot. Hilly land joins the plains and the mountains all along the peninsula.

Climate of Italy

Have you ever heard the expression "Sunny Italy"? Records show that Italy deserves this name. It has about twice as much sunshine each year as the countries to the north.

Southern Italy has a true Mediterranean climate with warm dry summers and mild rainy winters. Some snow falls in the highest parts of the Apennines. Some rain falls on the Apennines during the summer months, but there is little rain on the lowlands. Winds usually blow from the west during the winter rainy season. For this reason, western Italy gets more rain than the eastern part.

Northern Italy has a climate much like that in the southeastern part of the United States. Winters in northern Italy are colder than in southern Italy. Some snow also falls. Summers are warm, and rain falls throughout the year. Snowfall is heavy on the mountains of northern Italy in the winter.

REMEMBER, THINK, DO

1. Why is Rome known as the "City of Seven Hills"?
2. What was probably the Roman Empire's greatest gift to the world?
3. What kind of government does Italy now have?
4. Compare the amount of rainfall in Italy with that in Spain (See map, page 167.) Where would you prefer to be a farmer? Why?
5. How does the climate differ in northern and southern Italy?

Agriculture in Italy

About one-fifth (⅕) of Italy's people are farmers. Not enough food is raised to feed the people, however. Some food has to be imported.

Three regions contain the most productive farming areas. The most important area is the Po Valley in northern Italy. For ages swift mountain streams have brought soil from the mountains to the plain. Soil that has been carried by rivers or streams from higher land is known as **alluvial soil**. As a result, the Po Valley is very fertile. Water is available, moreover, to irrigate crops. The second most important region is the plain along the western coast north of Rome. The third important agricultural region is the country around

The farmer of this land is practicing three-story farming. Vegetables have been planted beneath grapevines. Citrus trees are scattered among the vines.

Naples where volcanic ash has weathered into rich soil.

Agriculture in Italy has been rapidly mechanized in recent years. At the same time millions of people have left farms and moved to the cities.

Among the many crops of Italy, several stand out because of the large amounts produced and consumed by the people. Of first importance is winter wheat. Thousands of hectares (or acres) are planted each year; yet some wheat has to be imported. Wheat is used in making bread, macaroni, and spaghetti.

Corn, rice, and other grains are raised in the Po Valley. A crop of corn can be grown there after a crop of hemp has been harvested. **Polenta** (poh *lehn* tah), a kind of cornmeal mush, is the chief food of many of the farmers. Sugar beets, tomatoes, and potatoes are also grown in the Po Valley. Fruits, such as apples, pears, and peaches, are now raised in such quantities that some can be exported. In southern Italy citrus fruits and figs are grown and exported.

Two major crops in Italy, as in other parts of Southern Europe, are olives and grapes. Most Italian foods are cooked in olive oil. Wine made from grapes is one of Italy's main exports. Much of it is consumed in Italy, too.

Some Italians try to make full use of their land by practicing what is sometimes called **two-story** or **three-story farming**. In this type of farming, two or three crops are grown at the same time. Grapevines are planted between fruit or nut trees. Between the grapevines, garden vegetables are grown. Thus, the vegetables grown on the ground make up the "first story," the vines the "second story," and the trees the "third story."

Terracing of the hillsides is also practiced in order to make use of as much of the land as possible. **Terracing** is the making of "steps" up a hillside. Usually the soil is held on the terrace, or level area, by making a wall of rock that rises from one level to the next. Some terraces, where the hillsides are not very steep, can be quite wide. When the hillside is steep, the terrace has to be narrow. Crops are planted, of course, only on the top of each terrace.

Mining

Italy has little mineral wealth, yet the nation has industrialized rapidly in recent years. Some iron ore is mined on the tiny island of Elba, east of Corsica. It is of high quality, but only small quantities can be obtained. Only a little coal, most of it **lignite**—a low grade of soft coal—is mined in Italy. As a result, Italy's big iron and steel industry depends almost entirely upon imported ore and coal. The ore and coal are obtained mainly from other members of the European Coal and Steel Community. Nevertheless, Italy has become the world's eighth largest steel producer.

Italy has few useful minerals, but it does have some sulfur, borax, and

Marble is mined in many parts of Italy. This quarry is located in Tivoli, east of Rome.

pumice. Italy also has plenty of building stone, a great asset for a country which lacks a good supply of timber. A fine grade of marble, which is used for stone sculpture and beautiful buildings, is mined at Carrara (kuh *rahr* ruh). It is shipped all over the world.

A natural gas field in the Po Valley supplies fuel for Milan (mih *lahn*) and other nearby cities. Some petroleum is obtained on the island of Sicily. Not nearly enough oil or natural gas is obtained to meet Italy's needs, however. A great deal of fuel has to be imported.

Manufacturing

For many years the lack of power slowed the development of manufacturing in Italy. Much use is now made of hydroelectric power. The Alps and the central Apennines both provide good sources of water for producing electricity. Additional power sources also have been developed by using natural gas, atomic energy, and coal. Because Italy is a member of the European Common Market, it can import coal from France and Germany without paying a tariff on it.

In the past 20 years Italy has become a major manufacturing country. Most of the large manufacturing centers are in northern Italy. They are close to the sources of power. Milan is the main manufacturing center of this area.

The government now is requiring that many new state-owned plants be built in southern Italy. This part of the country is much poorer than northern Italy. A new steel mill, for instance, has been built at Taranto

Italy

(tah *rahn* toh). A growing chemical industry is developing in Sicily near the oil fields. Food processing industries are also expanding rapidly in the south.

Many Italian products are well known in the United States. Among the best-known manufactured articles are sewing machines, typewriters, machine tools, electric motors, automobiles, tractors, and motor scooters. Many other products, including chemicals, plastics, fertilizers, drugs, furniture, and accordions, are manufactured.

In the textile industries cotton, woolen, linen, silk, and hemp goods are all manufactured. Rayon and other synthetic fibers also are made in large quantities. Handwoven textiles and laces, for which Italy was once well known, now have mostly given way to machine-made products.

The growth of manufacturing in Italy has helped raise the standard of living for the people. But it has also created problems as it has in other countries. Pollution is one of the main problems. Some of the famous buildings in Italian cities are being damaged by polluted air. And Italy's famous sunlight often "shines" through smog so thick that it seems like a cloudy day. In one city, workers were told to wear gas masks while at work! Italians are now beginning to worry about pollution.

Tourism

Italy exports many manufactured products. Even so, the country regularly imports more than it sells to other nations. If Italy had no way of making up the difference, the country would become very poor. In

Pollution is a major problem in the industrial areas of Italy. In Florence, smog often covers the city for days at a time.

order to pay for imports, Italy earns much income by taking care of millions of tourists each year. Most of them come from other European countries, but many thousands come from the United States.

Most of the tourists want to visit the famous Italian cities of Rome, Naples, Florence, and Venice (*vehn uhs*). Many also visit Milan and Turin (*toor* uhn), Italy's major manufacturing centers. Some visit Genoa (*jehn* uh wuh), Italy's busiest port city. Many go to visit the famous Leaning Tower of Pisa (*pee* zuh) and the Isle of Capri (kah *pree*).

Rome, the capital city, has about three million people. It is the largest city in Italy and one of the most famous cities in the world. Rome, as you know, served as the center for the Roman Empire. For more than 2,000 years it has been an important city. It is fondly called the "Eternal City."

Rome contains ruins from the time of the Roman Empire, and even before then. The most famous are the Roman Forum and the Colosseum. The Roman Forum was a wide place where returning victorious generals were welcomed home and political assemblies were held. Here, too, markets were set up, courts of law were established, and religious festivals were celebrated. The Colosseum, built much like a modern football stadium, was a place for holding huge public events. Sometimes wild animals battled each other in the arena of the Colosseum. More often wild animals such as lions or tigers battled with *gladiators*. **Gladiators** were people who were taught to fight to the death. Sometimes the gladiators fought each other until one was killed. And, sometimes Christians were put in the Colosseum to fight with lions.

Much of Rome today, of course, is like any modern city. Streets are crowded with buses, trucks, and automobiles. Office workers hurry to and from work. Factories turn out products needed by people in Rome and elsewhere.

The Smallest Independent State in the World

What is the smallest country you can think of? You remember tiny Liechtenstein between Switzerland and Austria. Do you remember even smaller Monaco in southern France? They both are independent states. They are not part of any other country. But are they the *smallest* independent states? Table V of the Appendix shows that Monaco is much smaller than Liechtenstein. Surely it is the smallest independent state in the whole world.

Look at the map on pages 200-201 at about 44° N. near the east coast of Italy. You will see an independent state, a

Vatican City is so small that almost the entire country can be shown in the photograph. Land outside the borders of Vatican City is shaded in blue.

republic, named San Marino. San Marino is much smaller than Liechtenstein—but much larger than Monaco. San Marino has about 61 km² of land (24 sq. mi.). It has about 19,000 people. Is, then, Monaco the smallest independent state anywhere?

No. There is a smaller one. And it, too, is in Italy. It is a city within a city as well as an independent state within an independent state. Do you know what its name is? It is Vatican City, and it is the smallest independent state in the world. Monaco has about three and one-half times more land and about 28 times more people than Vatican City.

What is Vatican City? It is the headquarters for the Roman Catholic church. It is where the pope and his staff live. Since 1929 it has been an independent state. When citizens of Rome go to the great square in front of St. Peter's Church, they are standing on "foreign" soil. They have left Italy. They are in Vatican City.

Within Vatican City is one of the world's most magnificent buildings, St. Peter's Church. It is the largest Christian church in the world. St. Peter's can hold about 50,000 people. Catholics from all over the world come to Vatican City to worship in St. Peter's. On special days such as Easter, the pope conducts the Mass. He is the supreme ruler of Vatican City.

Another famous building in Vatican City is the Sistine Chapel.

It is famous mainly because of the paintings on the ceiling. These were done by Michelangelo, the great artist and sculptor, between 1508 and 1512. They are considered to be among the most marvelous works of art in the world. The paintings show nine scenes from the Old Testament.

Vatican City, like most countries, has its own mail system, telephone system, water supply, bank, and printing plant. It does not have an army or navy—but it does have its own police force. Vatican City sends ambassadors to many countries.

The smallest independent state in the world is within an independent state—and it is a city within a city.

Italy's Large Islands

Italy's two large islands, Sicily and Sardinia, are both mountainous lands with many bare, rocky hills. Forests which once covered the highlands were cut down long ago. The hillsides are used mainly for grazing sheep and goats.

Most of the people on Sicily live near the shore or around settlements where there are irrigated orchards. Sicily is famous for its citrus fruits. About as many lemons are grown in Italy each year as are grown in the United States. Grapes, olives, and almonds are other important crops. Mining sulfur, working in the oil fields or petrochemical plants, and fishing are other ways of earning a living on this crowded island.

Although Sardinia has some land suitable for farming, not much farming was done on the island. Former swamp land has been drained, so more farming is now possible. Sardinia has some mineral resources, including some coal, zinc, lead, and silver.

REMEMBER, THINK, DO

1. What is the main crop grown by Italian farmers? How is much of this crop used?
2. What major crops in Italy are also important crops in Spain and Portugal? Why?
3. What kind of fruit is grown in southern Italy, especially on the island of Sicily?
4. How does Italy get the coal and iron ore needed for its large steel industry?
5. Where are most of Italy's manufacturing industries located? Why are they located there?
6. Why are tourists of such importance to Italy?
7. Where and what is Vatican City?

Italy

——————————WHAT HAVE YOU LEARNED?——————————

1. Is much of Italy mountainous or lowland? Where do most of the Italian people live?

2. Who was Benito Mussolini? What group backed him, enabling him to gain control over the government in 1922?

3. What mountains form a natural boundary between Italy and its neighbors to the north?

4. What—and where—is Italy's major plains region? Why does this plain have fertile farmland?

5. Why do Italians practice two- and three-story farming? Terracing?

6. Where is natural gas produced in Italy? Hydroelectric power?

7. Why have Italians become concerned about air pollution?

——————————BE A GEOGRAPHER——————————

You will need to use a globe or an atlas to answer the following questions:

1. Which is farther north?
   a. Rome, Italy or New York, New York
   b. Palermo, Sicily or Atlanta, Georgia
   c. Milan, Italy or San Francisco, California.

2. Which is farther from the equator?
   a. Rome, Italy or Sydney, Australia
   b. Venice, Italy or Capetown, South Africa

   c. Naples, Italy or Buenos Aires, Argentina

3. If you flew directly south from Naples to the South Pole, what nations in Africa would you fly over?

4. If you flew directly north from Naples to the North Pole, what nations in Europe would you fly over?

5. If it is noon in Rome on Tuesday, what time and day is it in:
   a. Washington, D.C.?
   b. London, United Kingdom?
   c. Honolulu, Hawaii?

QUESTIONS TO THINK ABOUT

1. If you could choose to live in either Spain or Italy, which would you choose? Why? Where in the country you have chosen would you prefer to live? Why?

2. What advantages and disadvantages does Italy have in terms of resources for modern industry? How does Italy compare with the state where you live?

3. Compare the latitude where you live with Naples, Rome, or Milan. Then compare the climate of both places, summer and winter. Are there differences? If so, why?

4. Italy, West Germany, France, and Great Britain all have about the same number of people. All of these nations hold free elections, as well. In recent years governments in Italy have not remained in power as long as the governments of the other three nations. Why do you think that is so?

5. Is having the same government for a long time always good? Spain had the same government for 36 years. Governments in Great Britain may change rather frequently, or may last for years. Is there a period of time that is the best length for a government to remain in power? Does it depend on the government, or what?

6. Why do countries have much influence at one period of time in history and less at another time? The Roman Empire was very powerful for a long period. Then Spain built a large empire. Later the British did. Today none of these three empires exist and the three nations have much less power. Why did it happen? Will the United States, the Soviet Union, and China lose their influence and power in the future? What reasons can you give for these changes?

Chapter Review

13

Chapter

──────YUGOSLAVIA──────

The broad peninsula east of Italy contains the countries of Yugoslavia, Bulgaria, Albania, and Greece, and a small part of Turkey. As you have learned, this peninsula is called the *Balkan* Peninsula. **Balkan** is a Turkish word meaning "mountain." Find the countries on the map on pages 200-201. Note that the Adriatic Sea and the Ionian (eye *oh* nee uhn) Sea are located west of the Balkan Peninsula. East of the Balkan Peninsula is the Black Sea. The small Sea of Marmara (*mahr* muh ruh) and the Aegean (ih *jee* uhn) Sea are south and east of the peninsula.

Land of the Southern Slavs

Yugoslavia means "land of the southern Slavs." About 1400 years ago Slavic tribes moved onto the Balkan Peninsula. They came from land farther east. At about the same time, other Slavs moved into Poland and Czechoslovakia.

Before World War I the southern Slavs were living under six different governments. After the war they became citizens of one country—Yugoslavia. Some of the land of Yugoslavia had been part of the Austro-Hungarian Empire. Some of it had been part of Italy, Bulgaria, and Turkey. Two formerly independent countries also became part of Yugoslavia. They were known as Serbia (*suhr* bih uh) and Montenegro (*mohn* tuh *nay* groh). The new country of Yugoslavia has about as much land as the state of Wyoming. It has about 22 million people. Since soon after World War II a Socialist government that permits only one political party has been in power in Yugoslavia.

The Regions

There are six main regions in Yugoslavia. Each is called a republic. The people of each of these republics have their own ways of living. In one of the republics, Serbia, there are two *autonomous provinces*. An **autonomous province** is a self-governing area, something like a state in our country. Find the six republics of Yugoslavia on the map on pages 200-201. The republic of

The traditional dress worn by these people is not usually seen in Yugoslavia today. These clothes are worn on special occasions, such as fairs and celebrations.

Slovenia (sloh *vee* nee uh) is in north-western Yugoslavia. It has about one-tenth (1/10) of the people. *Croatia* (kroh *ay* shuh) is southeast of Slovenia. Almost one-fourth (1/4) of the Yugoslavs live along the Sava (*sahv* uh) River that flows through the republic of Croatia. *Bosnia* (*bahs* nee uh) and *Hercegovina* (*hurt* suh goh *vee* nuh) is the republic southeast of Croatia. It has about one-sixteenth (1/16) of the people. And, southeast of Bosnia and Hercegovina is *Montenegro*. Montenegro has a border with Albania. It has about half as many people as the republic of Bosnia and Hercegovina. Most of eastern Yugoslavia, except in the far south, makes up the republic of *Serbia*. More than four-tenths (4/10) of the people of Yugoslavia live in Serbia. The capital city, Belgrade (*bell* grade), is on the Danube and Sava rivers. *Macedonia* (*mass* uh *doh* nyuh) is the republic farthest south and east. It has a western border with Albania and a southern border with Greece. Bulgaria is across the eastern border of both Serbia and Macedonia. Only about one-sixteenth (1/16) of the people live in Macedonia.

The People

Nowhere else in Europe (except for the U.S.S.R.) are there as many different peoples in one country. There are really five southern Slavic groups in Yugoslavia, and there are 17 minority groups. Quite a few Albanians, Hungarians, and Turks live in Yugoslavia, for instance. There are three main religious groups and three main languages. There are about 7 million members of the Orthodox church in Yugoslavia. Most of the members of the Orthodox church live in Serbia, Montenegro, and Macedonia. Also, most of them use a different alphabet than the one we use. (See page 229.) There are about 5 million members of the Roman Catholic church in Yugoslavia. The Croats, Slovenes, and Hungarians mainly belong to that religious group. And there are about 2 million Muslims (followers of Islam) in the country. They live mainly in Bosnia-Hercegovina and near the Albanian border.

Two Alphabets

Growing up in the United States, some youngsters learn to speak more than one language. Almost all boys and girls probably have learned only one alphabet, however. We are so used to our alphabet that we don't think much about it. Most of the letters we use were developed by the Romans many years ago.

Imagine living in a country where two quite different alphabets are used! In Yugoslavia, a country about the size of Wyoming, three different languages are used. One of these, called Serbo-Croatian, is spoken by most Yugoslavians. The Croats write their language using the **Roman** (our) **alphabet** of 26 letters. The Serbs and Montenegrins use the **Cyrillic** (suh *rihl ihk*) **alphabet** to write the same language! The Cyrillic alphabet was developed many years ago in Greece. It also is used in Bulgaria and in the U.S.S.R.

On the next page are the letters of the Cyrillic alphabet. The sounds are shown in parentheses. In school in Yugoslavia you would need to learn both alphabets. The sounds of the three languages are not very different, and you would be able to understand what other people are saying. But, you would need to know two alphabets in order to read books and newspapers printed in the country.

Which letters in the Cyrillic alphabet look like our letters? Do all the letters that look like ours have the same sound as our letters?

Try writing your name using the Cyrillic alphabet.

CYRILLIC ALPHABET

Cyrillic letter	Roman letter(s)	Cyrillic letter	Roman letter(s)
А а	a	Р р	r
Б б	b	С с	s
В в	v	Т т	t
Г г	g	У у	u
Д д	d	Ф ф	f
Е е	e	Х х	kh
Ж ж	zh	Ц ц	ts
З з	z	Ч ч	ch
Ии Йй	i, ĭ	Ш ш	sh
I i	i	Щ щ	shch
К к	k	Ъ ъ	*
Л л	l	Ы ы	y
М м	m	Ь ь	e
Н н	n	Э э	e
О о	o	Ю ю	yu
П п	p	Я я	ya

*This letter has no sound, it affects the sound of the letter which comes before it.

The Land of Yugoslavia

The land of Yugoslavia is about as varied as the people who live in the country. Along the Adriatic coast is a narrow coastal strip called *Dalmatia* (dal *may* shih uh). This area is the only part of Yugoslavia that has a Mediterranean climate. The mountains extend to the sea in many places, so the coastal strip is not continuous. Dalmatia is Yugoslavia's most famous tourist attraction. It has beautiful scenery and a warm climate. Adding to the beauty of this region are many long islands just off the coast.

Inland from Dalmatia are moun-tains and high rugged plateaus. These extend from the northwest to the southeast. These mountains make communication and transportation difficult between the lowland in the north and the sea. The mountains extend eastward into southern Bulgaria and southward into northern Greece. Rivers flow into three different seas from this high area. See if you can figure out from the map on pages 200-201 what the seas are.

Much of the western part of the mountain area has limestone under the surface. Water sinks very rapidly into limestone and dissolves some of the rock. Limestone has been called a "petrified sponge" because water sinks through it easily. Many underground streams and caves have been formed in the stone.

Some regions within Yugoslavia have frequent earthquakes. The city of Skopje (*skawp* yay) in Macedonia was almost completely destroyed in 1963. The city has since been rebuilt. It is much more modern than before the earthquake.

The best farmland in Yugoslavia and the area where most of the people live is the northern plains area. These plains along the Sava and the Danube rivers are covered with fine windblown soil called **loess** (less). Much rich soil has also been deposited by the rivers. What is such soil called? At one time this area was the bottom of a huge lake. There are some marshy and sandy sections in northern Yugoslavia. Most of the land is rich and fertile, however.

There are three types of climate in Yugoslavia. Along the Adriatic coast in Dalmatia, as you already know, there is a Mediterranean climate. Most rain falls during fall and spring months as in Italy and Spain. It is never extremely hot or extremely cold in Dalmatia.

In northern Yugoslavia, along the Danube, the climate is humid continental. Summers are hot and humid. However, in some years there are long periods without rain during summer months. Usually enough rain falls throughout the summer for crops to grow well. Winters are cold and snowy.

In southeastern Yugoslavia the mountain valleys are cold in the winter. They are hot and dry in the summer. Rain frequently falls in the

Yugoslavia's mountainous coastline has forced people to build their homes in places that are not always easy to get to.

mountains, but not often in the valleys during summer months. High in the mountains the temperatures are much cooler than in the valleys nearby.

Agriculture in Yugoslavia

About two-fifths (⅖) of the workers in Yugoslavia are farmers. The crops grown are much like those planted in the United States. In Dalmatia, crops suited to a Mediterranean climate are grown. These include citrus fruits, grapes, figs, and olives. On the northern plains the main crops are corn, wheat, potatoes, sugar beets, barley, oats, hemp, fodder crops, and other vegetables. Cotton and tobacco are grown in the Vardar Valley in the south.

Farming methods are somewhat different in different parts of Yugoslavia. On the river plains in the north, crops are rotated. Farm machinery and fertilizers are more and more being used. In the south, where the land is harder to cultivate, most farmers still use less scientific farming methods.

As in most of the other countries of East Central Europe, Yugoslavia tried to develop large state farms. Crop yields did not improve, however. Now, farmers may own up to 10 hectares (25 acres) of land. Cooperatives much like those in France have been developed. That way the farmers own their land and take pride in farming it. At the same time it is easier to try out new methods and new machines. The rugged nature of much of the land has made it difficult for farmers to use scientific agricultural practices in many parts of Yugoslavia. Much progress has been made, nevertheless. Grains that can grow during droughts are now used. Many farmers have improved their cattle herds through better breeding practices.

Minerals and Manufacturing in Yugoslavia

Yugoslavia has good supplies of several different mineral ores. Good possibilities exist for the development of additional hydroelectric power. As a result mining and manufacturing industries are important and are growing. The country has deposits of coal (although most of it is lignite), bauxite, chromium, copper, manganese, and petroleum. It also has some uranium.

Before World War II, the most important manufacturing industries were textiles and food processing. Tanneries used hides and skins to make leather and manufacture footwear. Only small amounts of metal and chemical products were made. Since the war the government has made strong efforts to increase manufacturing. The manufacture of steel and steel products, including machinery and tools, has increased. Gains have been made in copper production, cement making, and oil refining. Nevertheless, the standard of living is low in Yugoslavia compared to that in Western Europe. Unemployment is a problem. Many Yugoslavs have gone to western European countries to find work.

Yugoslavia

While working and living there, they send money home to their families.

Belgrade

Belgrade is the largest city in Yugoslavia. It is built where the Sava River flows into the Danube. Belgrade is about as large as Toledo, Ohio, or Birmingham, Alabama. It is the transportation "hub" of Yugoslavia. Rivers, railroads, and roads from all directions meet there. It is, therefore, a commercial center. Belgrade is also the center of government.

Much of Belgrade was destroyed during World War II. It has been rebuilt with many new sections. Because of its location in the Danube Valley, Belgrade has been occupied by armies from many countries at different times.

Government

Yugoslavia has had a government controlled by the Communist party since World War II. During that time, President Josip Broz Tito (*tee toh*) has been in charge. He was born in 1892. (In 1979, he was still an active leader. Figure out how old he was in that year.)

President Tito helped organize a Communist party in Yugoslavia. During World War II he was the leader of an army fighting the German occupation troops. Tito's forces were finally successful and after the war he took control of the Yugoslav government. Tito outlawed all other political parties and jailed Yugoslavs who criticized his way of governing the country.

Soon after World War II the Soviet Union tried to influence and control the government in Yugoslavia. Tito did not let that happen. He insisted that Yugoslavia would have a government that was controlled by himself and other Yugoslavians. Ever since, Yugoslavia has been quite independent of the U.S.S.R. From 1948 to 1955, Yugoslavia had almost no relations with the Soviet Union. It received aid and carried on trade with countries of Western Europe and with the United States. In 1955 the Soviet leaders decided to become more friendly.

For many years now Yugoslavia has been a leader among the Third World nations. As you know, these are countries that refuse to take sides when the Soviet Union and the United States differ on an issue. Instead, Third World countries try to be friendly to both sides. They try to get their own position on the issues understood.

Many political changes have taken place in Yugoslavia in the last ten years. Criticism of the government is now permitted in parliament. Many political prisoners have been released from jail and sometimes more than one candidate runs for a political office, giving the people a choice. Tito has, in other words, gradually developed a government that permits more freedom for individuals.

A presidential council was formed for the first time in 1971. That body has nine elected members. They

Josip Tito (left) is shown meeting with President Chadli of Algeria. Tito is an important leader among Third World countries.

represent the six republics and the two autonomous provinces. President Tito, the ninth member, chairs the council. The presidential council does most of the decision-making for the executive branch of government.

The **Federal Assembly** has 220 members. They are elected by various organizations in each republic. The Federal Assembly has two chambers, The **Federal Chamber** and the **Chamber of Republics and Provinces**. They discuss and then pass all laws.

REMEMBER, THINK, DO

1. What kind of land makes up most of the Balkan Peninsula?
2. Where is Dalmatia? Why do many tourists go there?
3. In which part of Yugoslavia do a majority of the people live? Why?
4. What changes are taking place in Yugoslavia's government?
5. Compare the governments of Yugoslavia and Romania. How are they alike? How are they different?
6. What have Romania and Yugoslavia built on the Danube River at the Iron Gate?

Yugoslavia

——— WHAT HAVE YOU LEARNED? ———

1. What does the word *Balkan* mean? "Yugoslavia"?

2. Into how many republics is Yugoslavia divided? Which republic has the most people?

3. What are the three main religious groups in Yugoslavia? What are the three main languages spoken in Yugoslavia?

4. Where is the best farmland in Yugoslavia? What are the main crops grown there?

5. What three climate types are found in Yugoslavia?

6. Why are farming methods more advanced in Yugoslavia's northern plains region than in the south?

7. What kinds of manufacturing have been increased in Yugoslavia since World War II?

8. Who is Josip Broz Tito? What position did he take regarding the U.S.S.R. after World War II?

——— BE A GEOGRAPHER ———

1. Two states in the United States have capital cities just about as far from the equator as Belgrade, Yugoslavia. What are the states and what are the cities? Do you think of these states as northern or southern states? Do you think of Yugoslavia as in northern or southern Europe? Why the difference?

2. Locate Belgrade by latitude and longitude, using the map on pages 200-201. Then check your work by looking for its location in an atlas. In some atlases the name Belgrade may be shown as "Beograd." That is the way Yugoslavs spell the name of their capital city.

3. Which of the following cities are farther north than Belgrade?
   a. Halifax, Nova Scotia
   b. Augusta, Maine
   c. Ottawa, Canada
   d. Duluth, Minnesota
   e. Bismarck, North Dakota
   f. Boise, Idaho
   g. Seattle, Washington

4. Which of the following cities are farther west than Belgrade?
   a. Stockholm, Sweden
   b. Warsaw, Poland
   c. Bucharest, Romania
   d. Budapest, Hungary
   e. Cairo, Egypt

5. According to the map on page 167, which has the most rainfall: Yugoslavia, Italy, or Spain? How can you tell?

6. Yugoslavia and the United Kingdom have about the same amount of land area, yet London has about ten times more people than Belgrade. Why do some cities grow very large while others do not?

QUESTIONS TO THINK ABOUT
---

1. Which of these three countries in Southern Europe has the most land: Spain, Italy, or Yugoslavia? Which has the most people?

2. What similarities and what differences exist in the governments of these three countries?

3. How are the governments of Yugoslavia, Romania, and Poland alike? In what ways are the governments different?

4. Which of these countries in Southern Europe—Spain, Italy, Yugoslavia—is the most industrialized? Can you explain why?

5. In which of the three countries listed above in Southern Europe would you prefer to live? Why?

14

Chapter

GREECE

The country farthest south on the Balkan Peninsula is Greece. It is the only nation on the peninsula without a government controlled by Communists. Greece is a small, mountainous country with many peninsulas. As the map on pages 200-201 shows, long inlets from the

Designed centuries before the automobile, streets in Greece are often narrow.

sea extend into the heart of the country. Mountain ranges at many places reach right to the water's edge. The valleys between the mountain ranges generally open into small lowlands along the coast. These lowlands usually are not connected. To go from one valley to another, the Greeks often use the sea rather than climb the steep mountain passes.

Greece has a population of about 9 million and about 69 people per km² (179 per sq. mi.). Since much of the land is mountainous, the lowland areas are crowded. Greece is not as crowded today as it might otherwise be. In recent years many Greeks have moved to other countries in Europe looking for better jobs. Most have moved to West Germany.

Islands and Seas

As the map on pages 200-201 shows, there are many islands off the Greek coast. These islands are peaks of the same mountain ranges that cover the country. Many of the Greek islands are very close to the

Turkish mainland. There are so many islands that they are like stepping stones across the Aegean Sea to Turkey. (On a large wall map of Southern Europe, count the islands near Greece. Did you count more than 400? If you counted 437, that is how many an encyclopedia says there are!) It is possible to go across the Aegean Sea without ever being out of sight of land. Notice which islands are part of Greece.

One of the largest of the Greek islands is Crete (kreet). Find it on the map on pages 200-201. Crete has high mountains running east and west. Peaks in these mountains reach more than 2450 meters (8,000 feet) above sea level.

A Brief History

In the early years the sea helped to protect the Greeks. Their ancient cities generally were built away from the coast on or near high land. Enemy ships approaching the land could be easily seen. Armies could rush to the shore to drive off the invaders. Athens, the capital and largest city of Greece, is an example of this kind of city. Athens was built a short distance from the coast. In the heart of the city stands a high hill, the Acropolis (uh *krahp* uh luhs). From the Acropolis, a person can see far out on the Aegean Sea.

The Athenians began to build ships to explore the seas nearby. It became necessary for them to have a seaport. This seaport is known as Piraeus (py *ree* uhs). In ancient times, Piraeus was connected with

High above the Mediterranean Sea, the acropolis at Lindos provided ancient inhabitants with protection against approaching enemies.

Athens by a road about 10 kilometers (six miles) long. High walls were built along both sides of the road. The walls made it difficult for enemy armies to cut Athens off from the sea. Today Piraeus is really a part of metropolitan Athens. It is a busy harbor for international trade and is the site of many factories.

"The Glory That Was Greece"

The ancient Greeks developed one of the finest civilizations in history. People from all over the world have studied the ancient Greek culture. Architects and artists regard ancient Greek buildings and statues as among the finest ever created.

The most famous Greek temple is

the Parthenon (*pahr* thuh *nahn*). The ruins of the Parthenon, which was built on the Acropolis in Athens, may still be seen and admired. Although much of the building was destroyed in wars, some of the temple today looks as it once did. Many people have contributed money to restore and preserve the temple. Other ancient buildings in Athens are also being rebuilt.

The ancient Greeks left us much besides temples and statues. They developed one of the first easily written languages. The word "alphabet" comes from the first two letters in the Greek language: *alpha* and *beta*. Some ancient Greek literature is still regarded by scholars as among the best ever written. Even more important to us, they developed the ideas of justice and freedom. These are among the basic ideas on which our government has been developed.

The Ancient Greeks as Traders and Fishers

The soils of mountainous Greece are poor except in the narrow valleys and on the small plains. As you would expect, the climate is Mediterranean. That means, as you know, that most of the rain falls during the winter. Only certain kinds of plants thrive well during summer months without irrigation. As the people increased in numbers, it became difficult to produce enough food. So the Greeks built ships and started to trade with their neighbors who had grain to spare. They established trading stations on the shores of other lands. In this way the Greeks became a nation of traders.

Quite naturally, also, the Greeks became fishers. First, they fished in nearby waters. Later, fishers went as far away as the Black Sea.

Greek Myths

Long ago people invented stories to explain things that they did not understand. Such stories are called **myths**. Most myths were told about people who could do super-human things. These people were thought of as gods and goddesses. Some were good and some were evil. The Greeks created many gods and goddesses. The stories helped explain how the world started and why things happened. Can you think of any television characters who can do super-human things today?

Zeus (zoos) was the most powerful of all the Greek gods. Some of the best-known gods and goddesses in Greek mythology are the following:

Apollo—god of music and poetry
Ares (*ehr* eez)—god of war
Hermes (*her* meez)—messenger of the gods
Poseidon (poh *sy* d'n)—god of earthquakes and the ocean

Athena—goddess of wisdom and war

Aphrodite (af ruh *dyt* ee)—goddess of love

Hades (*hay* deez)—ruler of the underworld

In addition to the gods and goddesses, there were many other godlike characters in Greek myths. And real Greek heroes sometimes got into the myths, too.

After the Romans conquered Greece, they took over most of the Greek myths. Some of the gods and goddesses were given different names by the Romans. Aphrodite, for instance, became Venus; Ares became Mars. Hermes became Mercury, and Poseidon became Neptune. Zeus became Jupiter, and Hades became Pluto. But Apollo remained Apollo. Do you see where we got the names for many of the planets in our solar system?

The myth telling the story of the creation of the world is a complicated one. Uranus (yoo *ray* nuhs), king of the sky, married Gaea (*jee* uh), goddess of the earth. Their children were called Titans (*tyt'*nz). The youngest of the Titans, named Cronus (*kroh* nus), was the father of six children—three boys and three girls. Cronus had taken control over the heavens from his father, Uranus. Cronus was afraid that his children would do

The myths of ancient Greece were a common subject for the artists of that time. The figures on the pitcher show part of the creation myth. Cronus is being handed a stone wrapped as a child. Who does Cronus think is in the cloth? Why is this important to the story?

the same thing to him. So, to prevent this, he swallowed five of his six children. He meant to swallow the sixth, too, but he was tricked. Instead he swallowed a stone wrapped in baby clothes. The sixth child, who was not swallowed, was Zeus. He got Cronus to drink something that made him vomit the other children. Then Zeus and his brothers and sisters fought Cronus for control of the heavens. Zeus won the war and became ruler of the gods. He and the other gods and goddesses decided to live on Mount Olympus.

Perhaps you'd like to read more about Greek mythology. Ask your librarian for help in finding some books that contain myths. Or perhaps you would like to write some myths of your own.

Modern Greece

At the close of World War I, a number of new countries were created in Eastern Europe. At that time the boundaries of Greece were set where they now are. Greece was also granted a little land on the mainland of Turkey near Izmir (ihz *mihr*). Find it on the map on page 311. Many Greeks had settled there in ancient times. The Turks objected to the Greeks having any land on the Turkish mainland. Their armies drove the Greeks out. As a result of this war, more than a million Greeks living in Turkey returned to Greece. A large number of Turks living in Greece also moved back to Turkey.

In World War II, Greece was a bloody battleground because of its location. Look again at the map of Southern Europe on pages 200-201. You will see how a country controlling Greece could control shipping lanes to the Black Sea. Submarines operating from Greek ports also could sink ships moving to or from the Suez Canal.

In World War II, Greece was first invaded by the Italian army. Greek soldiers fought fiercely for their homeland. They drove the Italian armies back into Albania. Later, German armies invaded Greece from Yugoslavia. The Germans had better equipment and greater numbers. They defeated the Greeks. The Germans occupied the country from 1941 to 1944. During this time the people suffered terrible hardships. Much of the country was ruined.

Greece Since World War II

When Greece was finally freed, trouble arose among the Greeks themselves. As in other Balkan countries, Communists attempted to seize control of the government. Some of the Soviet Union's best ports are on the Black Sea. The Soviets were naturally eager to control the entrances to the Black Sea.

War broke out between small bands of Greek Communists and non-Communist Greeks. British troops helped the non-Communist Greeks. Later the United States helped, too. The Communists were finally defeated in 1949. Greece is the only Balkan country today that does not have a government controlled by Communists.

Until 1967 the government of Greece was like Great Britain's. The king of Greece had much influence, but little real power. The head of the government was the prime minister. He had the most power.

Then, because different parties were struggling for power, King Constantine (*kon* stan teen) dissolved or ended Parliament. He called for new elections. Before the elections could be held, the army took over the government. King Constantine tried to lead the people in overthrowing the military. The king's forces did not succeed, however, and he had to flee Greece. The military controlled the government of Greece until 1974. They put many people in jail. No criticism of the government was permitted. At the same time, although freedom was limited, they tried to improve economic conditions. In 1974 the military invited the former prime minister to return to Greece and form a government. They admitted that their attempts at governing had failed. Since the prime minister's return, political prisoners have been freed. The people voted to become a republic, so Greece no longer has a king. And the military officers who seized power have been sent to a small island to live in exile.

Greek Traders and Fishers Today

Today many Greek traders live along the shores of other Mediterranean countries. Moreover, Greece has a large merchant fleet and carries on much foreign trade. Many freighters and tankers owned by Greeks move goods between port cities located around the world.

Many Greeks of today are also fishers. The Mediterranean and Aegean seas have been overfished for years, however. Most catches in

Sponge merchants sell their wares along the streets of many Greek towns.

Over the centuries, overuse and poor land management have made much of the land in Greece poor farmland.

nearby waters are, therefore, small.

The Greeks are noted for their sponge fisheries. The warm waters of the Mediterranean favor the growth of sponges. Thousands of pounds of natural sponges are prepared for market each year.

Coral, which is used in making jewelry, is another product which the Greeks obtain from the surrounding seas. Coral is the skeleton of tiny sea animals which live only in warm salt water, such as in the Mediterranean Sea.

REMEMBER, THINK, DO

1. Why do the Greeks sometimes use boats to travel from one town to another?
2. Why did the ancient Greeks build walls on each side of the road from Athens to the sea?
3. Why did the Greeks become traders and fishers?
4. What two products, other than fish, do the Greeks get from nearby seas?
5. Read about ancient Greece in an encyclopedia.
6. How do you suppose it felt to live in Greece when no criticism of the government was permitted?

Climate and Crops

Greece's Mediterranean climate means that the summers are dry and hot. The winters are mild with frequent rains. Winds which bring the rainfall blow mainly from the west. The western slopes of the mountains, therefore, receive the most rain. Highlands have cooler weather in both summer and winter than lowlands. It is also colder toward the north, with some snowstorms in winter.

Only those trees and plants that have long roots or those which ripen quickly grow well in Greece without irrigation. Olive trees and grapevines are examples of plants with long roots. Beans and cereal grains are examples of crops which ripen quickly. Wheat, barley, and corn are the leading grains raised. In recent years some wheat has been exported. In two parts of Greece, known as Macedonia and Thrace (thrays), tobacco and cotton are important money crops. Many Greek farmers raise enough vegetables for their families in gardens near their homes.

Olives, figs, citrus fruits, apples, grapes, and currants are leading fruit crops. Part of the fruit crop is exported. Olive oil, olives, raisins, and *currants* are also important exports. The **currants** that are shipped from Greece are small seedless grapes which have been dried.

Livestock

Few cattle are raised in Greece because of the lack of good pasture-land. Many sheep and goats, which are more easily cared for in a mountainous country, are raised. Some farmers in Greece, as in other mountainous areas of Europe, still lead a somewhat nomadic life. During the summer season they take their flocks of sheep and goats to graze in places high in the mountains. There, some rainfall makes the grass grow and the springs do not dry up. While in the mountains, farmers and their families live in tents or crude stone huts.

Farmers and Farming Methods

About half of the Greek people are farmers, but only about one-fourth (¼) of the land can be cultivated. Land suitable for farming is found only in the valleys, among the mountains, and on the small lowland plains. Farms and fields usually are small. Only in a few areas can farm machinery be used. Many farmers still use old farming methods. These farmers use oxen or horses to pull plows. They sow the seed by hand. They harvest their grain by cutting it with a sickle. Often oxen or horses are used to trample the grain from the stalks. The **chaff**—or husk covering kernels of grain—is then separated from the grain by tossing it up in the air from a tray. The wind blows the lighter chaff away.

Farming methods on the more level lands have improved. Modern farm equipment—such as tractors, cultivators, and threshing machines—is now used in such areas.

Separating the grain from the stalk has been achieved in the same way for hundreds of years.

Forests in Greece

Not many forests now grow on the mountains of Greece. But that situation will soon be changed. Due to poor cutting and grazing practices in the past, erosion had become a serious problem. Swift streams fed by the heavy rains cut deep gullies in the bare mountainsides. As coal is generally lacking, wood was the common fuel in mountain villages. Small trees had been cut for many years for fuel. Goats and sheep grazing on the mountainsides also ate young seedlings and kept them from growing. Such grazing practices prevented the growth of new forests on the cut-over areas. The Greeks saw their mistake. Millions

of young trees have been planted on the mountainsides. Farmers, moreover, are now raising more dairy cattle and fewer sheep and goats. Most of the trees being planted are fast-growing poplar trees. They will, within a few years, provide needed lumber and help prevent erosion. In time the Greeks hope they will be able to restore their forests.

Minerals and Manufacturing in Greece

Greece has a variety of useful minerals. But it lacks a sufficient supply of most of them except for bauxite. Greece lacks good coal, but has some lignite. The lack of good coal and abundant ores means that Greece is not likely to become a leading manufacturing nation.

Greece has already built dams on all the sites that can be used for producing hydroelectric power. Until new sources of power can be found, industry in Greece will be limited. A little petroleum is now produced on one island off the western coast. Although of some value, this petroleum supply is so limited that it does little to solve Greece's power needs. Exploration goes on, but no truly promising sources have been found as yet.

Leading products now made in Greece include textiles, chemicals, food products, wines, pottery, carpets, and leather. Athens is the main manufacturing center, and Salonika (suh *lahn* ih kuh) also has numerous factories.

An important mineral export from

Greece is marble. Many of the ancient temples and statues were made of marble. Greek marble is now shipped all over the world for buildings and sculpture.

Athens, Center of Greek Life

Today, as in ancient Greece, Athens is the main center for education, art, music, literature, and government. It is about as large as Baltimore, Maryland. Athens is a modern city with wide and busy streets, fine stores, and excellent hotels. Thousands of people visit the city each year to see the Parthenon and other famous buildings.

Greece and Turkey

For many years relationships between Greece and Turkey have been strained. The biggest problem has not been on their border in the area called Thrace. And it has not been on the islands near the Turkish mainland. The greatest problem has been on the island of Cyprus in the eastern Mediterranean. There, about three-fourths (¾) of the people are of Greek ancestry and about one-fifth (⅕) of them are Turkish. The Greeks on Cyprus want the island to be part of Greece. The Turks want it to be part of Turkey. Fighting broke out in 1974. Both Greece and Turkey were, at that time, members of NATO— the North Atlantic Treaty Organization. Greece pulled its military forces out of NATO because NATO did not stop Turkey from putting troops on Cyprus. Now the fighting has stopped and the island has a Greek area and a Turkish area.

Corinth Canal

Look again at the map showing Greece on pages 200-201. You will notice that the southern part of Greece is a large peninsula. It is entirely cut off from the northern section of Greece except for the narrow *Isthmus* (*ihs* muhs) *of Corinth* (*kor* ihnth). An **isthmus** is a narrow neck of land with water on both sides which joins two larger areas of land. In the last century a canal was dug through this narrow neck of land. Now small vessels can avoid the trip around the peninsula when sailing from Piraeus to the west.

Cut through rock, the Canal of Corinth has made the passage of ships between the Aegean Sea and the Gulf of Corinth safer and quicker.

You may have seen on television some of the world's best athletes taking part in the Olympic Games. The Games are held every four years. In 1976 the winter games (skiing, bobsledding, skating, etc.) were held in Innsbruck (*ihns* brook), Austria. The summer games (track, swimming, boxing, etc.) were held in Montreal, Canada. In 1980 the winter games will take place in Lake Placid, New York. The summer games will occur in Moscow, U.S.S.R.

The Olympic Games are a tradition started by the Greeks more than two thousand years ago. At first the games were partly religious. They were held to honor Zeus, the king of the gods. A huge stadium was built at Olympia. (See map pages 200-201.) At first only footraces were part of the games. Then other sports were added, such as throwing the discus, jumping, boxing, and wrestling. Only men participated in these ancient games.

When Rome conquered Greece, the religious part of the games was stopped. Then contestants took part to win money. In time the quality of the games became so bad that they were stopped. No Olympics were held for almost 1500 years!

Then, in the 1890s, a French teacher thought the games should be started again. He thought they should be international games. He thought they would help build friendship among peoples and lead to world peace. The first of the modern Olympic Games was held in Athens, Greece, in 1896. Women began to compete in the games four years later in Paris, France. The winter games were started in 1924. Except for 1916, 1940, and 1944—when world wars were being fought—the Olympic Games have been held every four years since 1896.

The opening and closing ceremonies of the Olympic Games are beautiful to watch. Athletes are there from all over the world. Their costumes are very colorful as are the flags of all the countries. At the opening ceremony a large flame is lit. It is lit from a torch that has been carried by runners from the valley of Olympia, Greece. The torch is taken onto airplanes to cross seas and oceans. The person chosen to carry the torch into the main stadium is watched by millions of people. Thousands of them are in the stadium—the rest are watching on television throughout the world.

The runner comes into the stadium holding the torch high. She or he runs around the track and then up a long series of steps. At the top of the stadium is a huge torch. The runner lights the flame of the large torch from the little torch. The flame burns until the end of the games. Then it is put out for another four years.

Maybe you will someday take part in the Olympic Games. There are now 21 different sports in the summer games. Many of them are individual sports. But some, like basketball and volleyball, are team sports. There are 7 winter sports, too. The last time the Olympic Games were held, more than 1,000 athletes took part in the winter games. And more than 8,000 took part in the summer games. What started in Greece about 776 B.C. is much bigger now. Athletes from more than 100 nations now take part. And the athletes are much better. Why not plan to take part in 1984, or 1988, or 1992?

A. R. Wilton of the United States finished fourth in the marathon run at the 1908 Olympic Games held in London, United Kingdom.

Greece

LEARN MORE ABOUT SOUTHERN EUROPE

You may wish to learn some more about the other countries in Southern Europe. There are two other countries on the Balkan Peninsula: Albania and Bulgaria. You may wish to learn more about the little republic of San Marino on the Apennine Peninsula. On the Iberian Peninsula are Portugal and the tiny nation of Andorra. In addition to these mainland countries, the island of Malta in the Mediterranean is also considered to be part of Southern Europe.

The Study Guide on page 31 may be of use to you as you learn about these countries. The following paragraphs provide a little information about each of them.

Andorra

The little country of Andorra is located in the Pyrenees Mountains on the border between France and Spain. It has only about 453 km² (175 sq. mi.) of land and only about 27,000 people. Spaniards make up more than half of the population. Most of the people in Andorra raise sheep or cattle for a living. A few crops are raised. Tourism, especially skiing, is a growing industry. Andorra is ruled jointly by a bishop from a nearby city in Spain and by the president of France. The two "rulers" have to agree before any changes are made. Find Andorra on the map on pages 200-201.

Portugal

West of Spain on the Iberian Peninsula is the interesting country of Portugal. It has about as much land as the state of Indiana and about twice as many people. Fishing and making wine are important ways of earning a living in Portugal. The Portuguese language and customs are much like those of Spain. This nation at one time had a number of colonies. Most of them are now independent countries. For many years in this century, Portugal was ruled by a dictator. The dictator died, and power has now been returned to the people. Free elections were held in 1976. A new constitution was approved, making Portugal a republic with a prime minister and a parliament.

San Marino

San Marino is even smaller than Andorra. It has only 61 km² (24 sq. mi.) of land and about 19,000 people. Most of the people are of Italian ancestry. San Marino is best known for its postage stamps. If you are a stamp collector, you probably have some stamps from San Marino in your collection. The main way of making a living in San Marino is caring for tourists.

Albania

The small nation of Albania is on the west coast of the Balkan Peninsula. It is one of the poorest countries in Europe. Albania is ruled by the Albanian Labour Party which is controlled by Communists. But the Albanian Communists do not like the Soviet Union. Instead, until recently, they cooperated with the Chinese Communist party. Albania is about the size of the state of Maryland and has about 2.7 million people. Much of the country is mountainous. About three-fourths

Much of the fresh food in Southern Europe is sold in open-air markets like this one in southern Portugal.

Students in Bulgaria learn a second language in elementary school. These students are listening to Russian as they read in their books.

(¾) of the people make their living by farming. Albania has few industries of any size.

Bulgaria

East of Yugoslavia and north of Greece is the People's Republic of Bulgaria. The country is governed by the Bulgarian Communist party. Bulgaria has a little more land than the state of Ohio and almost as many people. Much of Bulgaria is mountainous. Northern Bulgaria has very cold winters. About one-third (⅓) of the land is forested. Main crops include grains, fruits, corn, and potatoes. An interesting and unusual crop is rose petals from which perfume is made. Manufacturing industries include chemicals, steel, machinery, textiles, and leather goods.

Malta

The small island nation of Malta, with 320,000 people, is one of the most crowded lands in the world. Almost 1031 persons per km² (2,624 per sq. mile) live on Malta. The nation actually is made up of five small islands—the largest of which is Malta. For many years before 1964, Malta was a British possession. Now it is an independent nation. Until 1979 the British Mediterranean fleet had its headquarters there. NATO had its headquarters there, as well. Many people work in the shipyards and offices. Malta attracts many British tourists during the winter months. A major problem on Malta is the water supply. The people have to rely on rainfall and wells for their water.

A SUMMARY OF SOUTHERN EUROPE

Southern Europe consists of three main peninsulas that extend into the Mediterranean Sea. The western one, known as the Iberian Peninsula, contains the countries of Spain and Portugal. It also contains the small nation of Andorra. The central one, known as the Apennine Peninsula, has the country of Italy, Vatican City, and the small republic of San Marino. The eastern one, known as the Balkan Peninsula, contains Yugoslavia, Albania, Greece, and Bulgaria plus a small part of Turkey. In the Mediterranean Sea, south of Sicily, is the small, crowded island-nation of Malta.

Throughout Southern Europe much of the land is mountainous. Most rain falls during the fall and spring months. More rain falls in mountainous areas than at lower elevations. Summers are generally long, quite dry, and in many places, hot.

Greece, Italy, Spain, Portugal, and Malta have freely elected governments. Until recently Spain and Portugal had governments controlled by dictators. Albania, Bulgaria, and Yugoslavia have governments controlled by a Communist party. No other parties are permitted to nominate candidates for important offices in those countries. Both Albania and Yugoslavia have been able to resist being controlled by the Soviet Union.

Italy, partly because it is a member of the European Community, is the most industrialized of the southern European countries. Other countries such as Yugoslavia and Spain are making rapid progress toward industrialization. Nevertheless, the standard of living in southern European countries is generally lower than in Central or Northwestern Europe.

The countries of Southern Europe have had a great influence on the culture and history of our country. Greece gave us our alphabet and, along with Italy, many of our ideas about government. Italy and Spain had much to do with our religious history and early exploration. All three countries have influenced our art, architecture, literature, and music.

1. Why was Greece invaded by Italian and German armies in World War II?

2. What crops grown in Greece are also grown in Italy and Spain? Why?

3. Why did Greece pull out of the military part of NATO?

4. What is the name of one of the most crowded islands of the world that is located in the Mediterranean Sea?

5. What is a peninsula? Do you live on a peninsula? Have you ever been on a peninsula?

────────BE A GEOGRAPHER────────

1. On the left hand side of a piece of paper make a list of all countries in Europe, alphabetically. In two columns to the right, arrange the countries by placing a numeral opposite each name as follows:
   a. In the first column, number according to land area, number one being the largest.
   b. In the second column, number according to population. Again, use number one for the largest.

2. How do major cities of the three main regions in Europe compare? For instance, how do Athens, Rome, and Madrid compare with London, Paris, and Stockholm? How do those cities compare with Berlin, Warsaw, and Bucharest?

3. Is any part of Europe north of the northern coast of Alaska?

4. Is any part of Europe south of the southern tip of Florida?

5. How would you describe the location of Europe, generally?
   a. Is it farther north than most of the United States?
   b. Is it about the same latitude as most of the United States?
   c. Is it farther south than most of the United States?

6. How would you describe the size of Europe, in comparison with the land area of the Eastern Hemisphere?

QUESTIONS TO THINK ABOUT

1. Why do you think so much space has been given in this book to the countries of Europe?

2. What should be done with places on the Earth such as Gibraltar? Should Gibraltar be internationalized and placed under the United Nations? Should Great Britain do what the people who live on Gibraltar want? Should Spain be given control of Gibraltar? Should things remain as they are? Whatever position you take, be prepared to give reasons to support your position.

3. Is there any reason geographically why some nations, such as Italy, have a very rich history in music and art; while other nations, such as Greece, have a rich history in architecture and drama? What causes such differences? Why do some nations not have a rich artistic history?

4. Would you like to participate in the Olympic Games some day? In what sport or activity?

5. Do any buildings in your city have marble floors? Walls? Columns? Where did the marble come from? Does the building remind you at all of Greek or Roman buildings? Why or why not?

6. Where would you prefer to live in Europe if you could take your pick of all the countries and places? Why? Where in Southern Europe would you choose? Why?

7. Where did all the teachers in your school, or their parents, or their grandparents come from? Do they know why their ancestors came to the United States? Do they now keep alive any of the customs their ancestors brought with them to the United States?

8. What differences and what similarities are there in living in Catholic Poland and living in Catholic Spain? (Be sure to remember that some differences will be related to climate, resources, and government—not only to religion!)

Unit Review

Much of the fish caught by the fishers of the Soviet Union is processed in canning factories similar to this one.

Unit

THE SOVIET UNION

Unit Preview

● Why are rivers important to the people of the Soviet Union?

● What do you know about the important sights in these two pictures?

DO YOU KNOW?

- Why the Soviet Union is often just called "Russia"?

- Why the Soviets have always wanted ice-free ports?

- How large the U.S.S.R. is?

- Why Siberia and the Soviet Far East are so important to the U.S.S.R.?

- How many people in the Soviet Union are members of the Communist party?

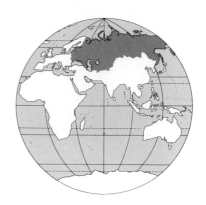

Unit Introduction

The Soviet Union

The Union of Soviet Socialist Republics has more land area than any other country in the world. It is about two and one-half times larger than the United States. From west to east, the land extends from the Baltic Sea to the Pacific Ocean and the Bering Sea. This distance is more than 11 250 kilometers (7,000 miles). From north to south, this vast land reaches from the Arctic Ocean southward to the Black and Caspian seas. This distance is about 3850 kilometers (2,400 miles). Almost one-seventh (⅐) of all the land on the Earth is part of this country. Locate this nation on the map on pages 260-261.

The Union of Soviet Socialist Republics is made up of 15 republics. The country is commonly called either the "U.S.S.R." or the "Soviet Union." Some people call this large country "Russia." However, Russia in only one of the nation's 15 republics.

Icebreaker Reaches North Pole

In August, 1977, for the first time ever, a surface ship reached the North Pole. Look at a globe or the map on pages 260-261 to see where the North Pole is located. It is located in the Arctic Ocean—an area that is almost constantly covered with ice. During summer months, however, some stretches of open water appear between ice **floes** (pieces of floating ice).

For many years the Soviet Union has been trying to keep the northern shipping lanes open longer. They developed ships strong enough to break ice. The engines of some of these are

The icebreaker Arktika *at the North Pole.*

driven by nuclear power. The *Arktika* (*ahrk* tee kuh) is such a ship. It is strong enough to break its way through ice four meters (about 12 feet) thick. The *Arktika* is used mainly to clear shipping lanes of ice, especially lanes east of Murmansk.

In the summer of 1977, the Soviet Union decided to try to reach the North Pole by surface ship. The *Arktika* was chosen to make the trip. On Tuesday, August 16, 1977, the *Arktika* reached the North Pole.

Who had been there before? How had they arrived? The first people to reach the North Pole were two Americans, Robert E. Peary and Matthew Henson. They used dog sleds and went north from Ellesmere Island, Canada. Peary and Henson reached the North Pole on April 6, 1909. In 1926 Admiral Richard E. Byrd flew over the North Pole in an airplane. And in 1958 the U.S.N.S. *Nautilus*, a nuclear-powered submarine, reached the pole under ice. Later that same year the U.S.N.S. *Skate*, another nuclear-powered submarine, surfaced at the North Pole. It broke through the ice from below.

But no surface ship had made it to the North Pole until 1977. Now that has been done. The *Arktika*, an icebreaker from the Soviet Union, did it.

Things You Might Like To Do As You Study About the Soviet Union

As you begin this study of the U.S.S.R., it would be helpful to begin a variety of other activities. These will help add to your information about this large and rapidly changing nation. Some suggestions are given below. You and your classmates might have other suggestions about things you might like to do.

1. Find articles in current magazines and newspapers about the Soviet Union. Because the U.S.S.R. is a large and powerful country, many articles are written and published about it. Bring articles to school which you think will be of interest to members of the class. You may wish to share the information by showing some of the pictures with an opaque projector.

2. On a large outline map of the Soviet Union, write the names of major rivers and seas. Locate all cities with more than 500,000 people. Outline major industrial areas and show different ways the land is used.

3. Learn about the lives of famous Russian musicians, dancers, artists, and writers. Use almanacs and encyclopedias as reference tools.

4. Many beautiful folk dances have been created and handed down from generation to generation in the Soviet Union. The ballet dancers of this nation are world famous. Find out all you can about the Russian dances. You might enjoy learning how to do some of the folk dances.

5. Form a group to prepare and give a report on one of the following subjects:

    a. Government in the Soviet Union

    b. The U.S.S.R. in the Second World War

    c. The Soviet Union and the United Nations

    d. Growth of Heavy Industry in the U.S.S.R.

    e. Agriculture in the U.S.S.R.

    f. Education in the U.S.S.R.

    g. Scientific Achievements of the U.S.S.R. since 1970

These boys live in a village east of Lake Baikal in the Soviet Far East. Locate this area on the map on the following pages.

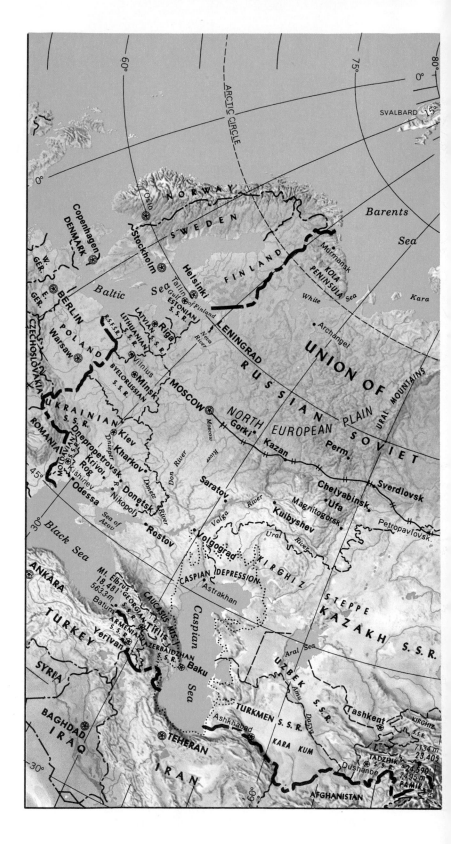

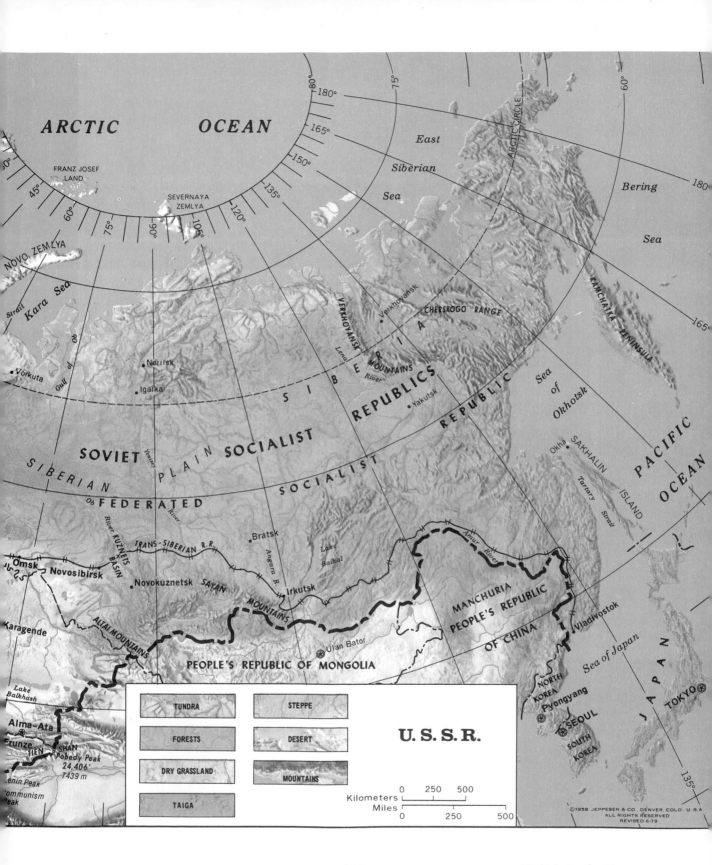

ARCTIC OCEAN

FRANZ JOSEF LAND

SEVERNAYA ZEMLYA

East Siberian Sea

ARCTIC CIRCLE

Bering Sea

NOVO ZEMLYA

Kara Sea

Strait

KAMCHATKA PENINSULA

Gulf of Ob

Vorkuta

Norilsk

Igarka

VERKHOYANSK

Verkhoyansk

CHERSKOGO RANGE

VERKHOYANSK MOUNTAINS

Lena River

S I B E R I A

SOCIALIST REPUBLICS

Yakutsk

Sea of Okhotsk

REPUBLIC

SOVIET

PLAIN

SOCIALIST

Yenisey

Okha

SAKHALIN ISLAND

SIBERIAN

FEDERATED

Ob River

SOCIALIST

Tartary Strait

PACIFIC OCEAN

Bratsk

Lake Baikal

Amur River

TRANS-SIBERIAN R.R.

KUZNETS BASIN

River

Omsk

Novosibirsk

Novokuznetsk

SAYAN MOUNTAINS

Angara R.

Irkutsk

MANCHURIA

PEOPLE'S REPUBLIC

Vladivostok

Karagende

ALTAI MOUNTAINS

Ulan Bator

OF CHINA

Sea of Japan

PEOPLE'S REPUBLIC OF MONGOLIA

Lake Balkhash

NORTH KOREA

JAPAN

Alma-Ata

Pyongyang

SEOUL

TOKYO

Frunze

TIEN SHAN

Pobedy Peak
24,406'
7439 m

SOUTH KOREA

Lenin Peak

Communism Peak

TUNDRA	STEPPE
FORESTS	DESERT
DRY GRASSLAND	MOUNTAINS
TAIGA	

U.S.S.R.

Kilometers 0   250   500

Miles 0   250   500

©1958 JEPPESEN & CO. DENVER, COLO., U.S.A.
ALL RIGHTS RESERVED
REVISED 6-79

Unit Introduction

15

Chapter

THE HISTORY, LAND, AND CLIMATE —————OF THE SOVIET UNION—————

A Brief History

The history of the U.S.S.R. began in the region later called Russia. Settlements of Slavic tribes were scattered about this area. They came into the central part of western Russia between the fifth and eighth centuries A.D. People from Finland were also found on this plains region.

Scandinavian traders, called Varangians (vah *ran* jih ans), or the Rus (roos), entered Russia by rivers that flow into the Baltic Sea. The Rus organized the Slavs and Finns into a political unit governed by a Scandinavian prince. In time these people intermarried. Their descendants were the original Russian people. Kiev (*kee* yehf) was established as the capital of this principality which included a number of neighboring states. Kiev is one of the oldest cities in the Soviet Union. It is still known to the Russians as the "Mother of Cities."

The Mongol Invasion

In the 13th century a group of people known as the Mongols (*mawng* gohls) crossed into Europe. The Mongols, from the area now known as the Mongolian People's Republic, had conquered China and other areas of Asia. (See the map of Eurasia on pages 8-9.) The invasion by the Mongols ruined the trade of Kiev. Some of the Russians near Kiev moved westward. Others went to the northeast and settled in the mixed forest region. There they were safer than in the open spaces of the grasslands. The modern Russian nation had its beginning in this forested area. Here the group known as the Great Russians started the city of Moscow (*maws* koh). From Moscow

Under the czars most peasants were able to provide only for the basic needs of food, clothing, and shelter.

the Great Russians slowly spread in all directions, colonizing or conquering neighboring territories.

Russia Under the Czars

From their earliest beginnings, the various tribes that lived in what is now Russia were governed by local chiefs. From the 16th to the 20th century, these tribes were brought under the rule of one ruler called a *czar* (zahr). **Czar** is the Russian word for king or emperor. Under the czars the peasants, or common people, had little freedom and opportunity. They were treated almost as slaves. In time they became the property of the landholders.

Repeated revolts by the peasants finally led to their being declared free in 1861. However, they had no money to buy land. They were not allowed to leave their villages without permission from the govern-

ment. It was not until 1904 that they were allowed to leave their villages and seek work elsewhere.

The Revolutions of 1917

During most of the 19th century revolutionary groups had tried to bring about change in the government of Russia. From 1914 to 1917 Russia fought on the side of the Allies in World War I. Then, in March of 1917, a revolution spread throughout the country. The Czar, Nicholas II, was forced to step down and a new government was formed. It lasted only a short time. Another government took over but it, too, did not last. Then, in November of 1917, a group known as *Bolsheviks* (*bowl* sheh vihks) seized power. Czar Nicholas II, his wife Alexandra, and the children of the royal family were killed by the Bolsheviks. The Bolsheviks' new government was led by

The Land of the Soviet Union

Nikolai Lenin (*lehn* uhn). **Bolsheviks** believed that all land should be owned and controlled by the government. This form of government was called "communism." After a long civil war, the various lands of the Russian Empire were united under the Communist government. The nation changed its name to the Union of Soviet Socialist Republics.

Efforts To Obtain Ice-Free Ports

In the early history of Russia, its leaders were concerned about the nation's isolation in wintertime. The lands of Russia are located in regions with very harsh winters. Rivers and ocean ports freeze during the main part of winter. This prevented the Russians from moving people and goods in and out of their lands. Trade with other nations just about stopped during the winter months. This is one of the reasons why, for hundreds of years, Russian leaders tried to expand their territory by taking neighboring lands to the south, west, and east. They wanted ports that were free of ice during the winter.

Ivan IV was the first czar of Russia. He was known as "Ivan the Terrible." He ruled the nation during the last part of the 16th century. At that time Russia's only winter port was at Archangel (*ahr* kayn jehl) on the White Sea. However, this port was a long distance from European markets. Ivan tried to take control of land along the Baltic Sea. A 25-year war was fought with Poland, Sweden, and Lithuania over this land. But the Russians were not successful in getting a warm-water winter port on the Baltic. After Ivan died other czars tried to take land in the south and west.

Peter the Great was another famous czar who tried to open Russia to the west in the early part of the 18th century. Under his rule territory was gained on the Baltic Sea. A city and port were built on the Gulf of Finland. It was called St. Petersburg. Today the city is called Leningrad (*lehn* ihn grad). Canals were also built linking the Volga (*vawl* gah) River to the Don (dawn) River. This provided a route to the Black Sea in the south of Russia. However, Russian vessels still had to pass through the Dardanelles in order to reach the Mediterranean Sea. These straits were controlled by the Turks. (See map, page 297.)

Catherine the Great ruled from 1762 to 1796. Under her rule the Russian territory was extended southward to the Black Sea.

During the 19th century, Russia fought several wars to gain control of the Dardanelles. However, both France and Great Britain aided Turkey in its defense of this land.

Because of the North Atlantic Current, Murmansk is the Soviet Union's only ice-free port along the Arctic Sea coast.

In 1905 Russia tried expanding southward along its eastern, Pacific, coast. However, Japan defeated Russia in its attempts to take over eastern warm-water winter ports.

To this day the Soviet Union is still bothered by frozen waterways during winter months. Goods sometimes have to be shipped thousands of miles overland to reach the sea in winter. Despite the vast size of the Soviet Union, it has many problems caused by the lack of ice-free ports. In many ways it shares problems faced by landlocked nations.

REMEMBER, THINK, DO

1. How much of the Earth's land is within the U.S.S.R.?
2. Why is Kiev known as the "Mother of Cities"?
3. What form of government was started in Russia after the revolution in November, 1917?
4. What similar transportation problems do you think the Soviet Union and landlocked nations might share?
5. On an outline map of the Soviet Union, locate the following places:
   - a. Arctic Circle
   - b. Archangel
   - c. White Sea
   - d. Baltic Sea
   - e. Arctic Ocean
   - f. Black Sea
   - g. Gulf of Finland
   - h. Leningrad
   - i. Volga River
   - j. Don River
   - k. Caspian Sea
   - l. Moscow

The Land of the Soviet Union

Land and Climate Regions of the Soviet Union

In the Soviet Union there are many natural regions. The main land feature is the vast plain which extends from the western border far into Siberia (sy *beer* ih uh). Most of the people living in the Soviet Union live on this plain. The plain has two main parts: the North European Plain and the Siberian Plain. On the south and east, the plain is bordered by plateaus and high mountains. The plain is hilly in places and gently rolling or level in others. Find the Ural Mountains and the Ural River on the map on pages 260-261. The Ural River and Ural Mountains are considered as the boundary between Europe and Asia. They also divide the European Plain from the Siberian Plain. The mountains are low and easily crossed, so they do not form a real barrier. Around the northern and eastern sides of the Caspian Sea, the land slopes to below sea level. This low area is called the Caspian Depression. It is enclosed by a dotted line on the map.

From north to south, the lowlands of the Soviet Union can be divided into natural regions based mainly on the climate and vegetation. These regions are the *tundra*, the *subarctic region* which contains vast forests called the **taiga** (*ty* guh), the *mixed forest region*, the *black earth* or *humid steppe region*, and the *dry* or *semiarid steppe* and *desert region*.

The Tundra. The tundra extends all along the Arctic coast of the U.S.S.R. On the south it reaches about to the Arctic Circle. (See map, pages 260-261.) In the far east, tundra conditions extend south of the Arctic Circle because the land is higher there.

Winters are long and cold in the tundra region. Summers are short and cool, but the days are long. Frost may occur, even in midsummer. Many days are cloudy in this region, but rainfall and snowfall are slight. Strong winds often blow, adding to the feeling of coldness. The cold increases from west to east. The Kola (*koh* luh) Peninsula (see map, pages 260-261), for instance, has much milder winters than coastal areas farther east at the same latitude.

In general, the tundra is a treeless area. There are some stunted trees and low shrubs in sheltered places. Farther south, where the tundra merges with the forest, trees grow only in the valleys. And still farther south, trees are found in scattered clusters on the hills between streams.

The soil in the tundra region stays frozen except for a short time in the summer. Then the ground thaws to a depth ranging from several centimeters to several meters (several inches to several yards). The ground becomes swampy when it thaws, and mosses, lichens, and **sedges** (grasslike plants with hollow stems) grow. On higher areas, which are drier and warmer, flowers and berry bushes come to life. They grow quickly during the long, sunlit, cool days of the short summer season. The cool summers and the nature of the soil make it difficult to grow

SOVIET UNION

NORTH ATLANTIC CURRENT

ARCTIC CIRCLE

Chelyuskin

Leningrad

Archangel

Moscow

Astrakhan

Batumi

Tselinograd

Rainfall and Temperature

Rainfall in:

centimeters	inches
under 25	under 10
25-50	10-20
50-100	20-40
over 100	over 40

→ Cold currents
→ Warm currents

Scale

Kilometers 0   500   1000

Miles 0   500   1000

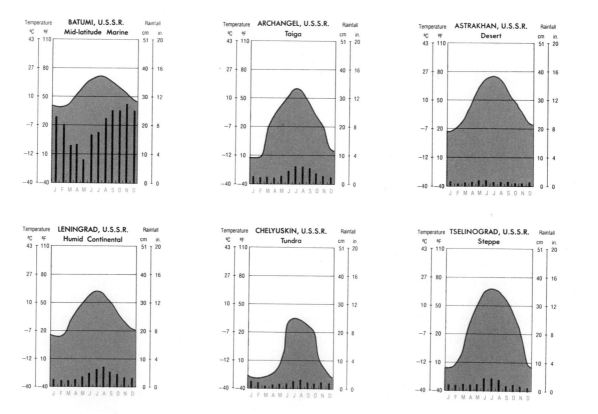

BATUMI, U.S.S.R.
Mid-latitude Marine

ARCHANGEL, U.S.S.R.
Taiga

ASTRAKHAN, U.S.S.R.
Desert

LENINGRAD, U.S.S.R.
Humid Continental

CHELYUSKIN, U.S.S.R.
Tundra

TSELINOGRAD, U.S.S.R.
Steppe

267   *The Land of the Soviet Union*

SOVIET UNION

vegetables out of doors. Some vegetables are grown near cities in greenhouses.

The tundra region is thinly settled. Much effort has been made in recent years to develop the area. The western part of this region, around Murmansk (moor *mansk*), has developed very rapidly. There the ports are kept open during the winter. Find this area on the map of the U.S.S.R. The North Atlantic Current, which flows northward along Norway, also warms this coast. Icebreakers sometimes have to be used, however, to clear a path for ships bound for Murmansk in the winter. Murmansk is a center for fishing fleets which catch herring, cod, and haddock in the Barents (*bar* ehnts) Sea. It is also the port through which coal is brought during the summer from the mines on the island of Svalbard

(*svahl* bahr). Lumber, fish, and phosphates are exported through the port of Murmansk.

During the summer months, from about May to October, ships can sail from Murmansk southeastward to ports along the White Sea. The Soviet Union has developed several atomic-powered icebreakers. Now ice-bound ports farther east along the Siberian coast may be kept open longer each year.

Other important cities in the tundra include Vorkuta (vawr *koo* tah) and Norilsk (naw *reelsk*). Find them on the map on pages 260 and 261. Vorkuta is a center for mining good coking coal. Norilsk is a center for mining nickel, platinum, and uranium. Most of the tundra region is almost uninhabited and probably will remain so. The few people who live there are mainly nomads. Like the Lapps in Norway and Sweden, much of their livelihood depends on reindeer. These animals furnish them food, clothing, tents, and a means of transportation. Fishing and hunting provide additional food.

During the summer the reindeer herders migrate with their herds from the interior to areas near the coast. They try to escape the swarms of mosquitoes which make life miserable for them and their animals. They feed fish to the dogs. The dogs help herd the reindeer and pull the sleds.

Wild animals found in the tundra include the lemming, Arctic fox, hare, wolf, and ermine. The ermine and the Arctic fox are valued for their fur. Polar bears are found on the offshore islands to the north. Walrus and seal are also hunted in this region.

Great numbers of birds, including ducks, geese, swans, snowy owls, and ptarmigans (*tar* mih ganz) are found on the tundra during the summer. They nest in the higher areas. Only the snowy owl and the ptarmigan remain for the winter.

The Subarctic Region and the Taiga. The subarctic region extends completely across the country. This region is just south of the tundra region. About half of the land area of the U.S.S.R. is in this region. Winters are long and cold. Summers are cool. The surface of the ground, as in the tundra region, is frozen for six or seven months each winter. In Siberia the ground is frozen to depths of more than 300 meters (1,000 feet).

The nomads of the tundra region depend on large herds of reindeer to support them during both the summer and winter months.

Such a condition is called **permafrost.** Most of the land in the subarctic region is covered by the northern coniferous forest called the taiga. Find this region on the map on pages 260-261.

The taiga is the largest continuous forested area in the world. It covers about as much land as all of Europe. The only similar forested area in the world is found in northern Canada. In the taiga, cone-bearing evergreen trees are the most numerous. Among these are fir, spruce, pine, cedar, and larch trees. There are also birch and aspen trees.

The taiga region provides the Soviet Union with a good supply of wood.

numerous in the Asiatic forests, as they have been largely killed off in Europe. For many years, furs were one of the Soviet Union's best-known commercial products. They were sold throughout the world. Among the most important furs obtained were squirrel, hare, fox, ermine, bear, sable, and marten. Many nations, however, now are trying to conserve wildlife species, and are limiting the import of furs.

The taiga, generally, has shallow, gray soils that are not very fertile. Much fertilizer has to be used in order for crops to grow. Only a few hardy vegetables, hay, and grains are grown in the southern part of this region during the summer.

This vast forest region is thinly settled. The largest city is Archangel, a port on the White Sea southeast of Murmansk. It is the main lumber and sawmill center in the Soviet Union. Another important saw-milling center is Igarka (ih *gahr* kah) on the Yenisey (yehn ih *say*) River. Both of these cities are important centers for exporting lumber. Industries which make wood and paper products are also important there.

Many bogs or marshes are found in the taiga. They are especially numerous in the low basin of the Ob (ahb) River east of the Ural Mountains.

The taiga is the home of many fur-bearing animals. They are most

REMEMBER, THINK, DO

1. What are the five natural regions of the Soviet Union?
2. In what way is the port of Murmansk different from most other northern ports in the U.S.S.R.?
3. What is permafrost?
4. Why is the subarctic region not a good place for agriculture?
5. Why is it possible to have ice-free ports in the area along the Barents Sea in the northwestern part of the Soviet Union?

The Mixed Forest Region. This region extends about 965 kilometers (600 miles) from north to south in the western part of the U.S.S.R. In Siberia, this region is only about 160 kilometers (100 miles) wide. Find it on the map on pages 260-261. Since it is south of the subarctic region, summers are longer and warmer. Winters are milder than farther north, although still very cold. More snow falls in winter, and more rain falls in summer. The climate in the mixed forest region is much like that in the Great Lakes region of the United States and Canada.

The mixed forest region contains broadleaf deciduous as well as coniferous trees. You will remember that deciduous trees are trees that shed their leaves in the fall. Common broadleaf trees in this region include oak, ash, elm, and maple. Although covering less area than the taiga, the mixed forest region has some of the best timber in the Soviet Union. Much of the best timber in the western part of this area has already been cut. Although the soils in this region are quite poor, cleared areas are used for farming. Flax and rye are two of the leading products. Dairy farming is fairly common.

Several of the major cities of the Soviet Union are located in the mixed forest region. First among these cities is Moscow, the capital and largest city. Second is Leningrad, the main port and second largest city. Another is Kiev, capital of the Ukraine (yoo *krayn*), a Soviet republic.

The large forests in this region and in the taiga provide the Soviet Union with a most valuable natural resource. Wood is plentiful and cheap in the U.S.S.R. Except in the larger cities, it is used for almost everything. Houses, public buildings, furniture, utensils, tools, wagons, and toys are made of wood. Much wood is also used for fuel.

Lumbering has long been one of the country's important industries. Machinery is used for cutting down the trees and handling the logs. Roads and railroads leading back into the forests have been built in many places. Rivers which run through the forests are also used in summer months for transporting logs.

The Black Earth Region. The black earth region is often called the humid *steppe*. **Steppe** is a word used to describe grassy plains with a climate of warmer, longer summers and less rain than in the mixed forest region. (See the Glossary.) Usually this region has less than 50 centimeters (20 inches) of rainfall each year. That is barely enough for growing crops without irrigating them. Some years only about 25 centimeters (10 inches) of rain falls. When that happens, the crops do not do well.

In spite of uncertain rainfall, this region is the best agricultural land in the U.S.S.R. Find it on the map on pages 260-261. The black earth region forms an irregular belt about 500 to 1000 kilometers (300 to 600 miles) wide and about 5000 kilometers (3,100 miles) long. Beginning at the western boundary of the Ukraine, the black earth belt extends

The western Ukraine, which is part of the black earth region, has many productive farms. Cattle as well as food crops are raised in this area.

eastward to beyond Novosibirsk (*noh* voh sih *bihrsk*). The soil of the black earth region is deep and rich in *humus*. **Humus** is fertile, dark soil formed as grass dies, decays, and is mixed with the earth year after year. The northwest section, where patches of forest grow, is often called the forest steppe.

The black earth region is the great wheat-producing area of the U.S.S.R. However, many other crops are also grown. For instance, sugar beets are a very important crop in the Ukraine. Raising wheat, rye, oats, and barley has long been important in the region west of the Urals. In recent years the steppes east of the mountains have also been farmed.

The Dry or Semiarid Steppes and Deserts. South of the black earth region are vast treeless plains called steppes. These plains get much less moisture than the black earth region. They are known as the dry or semiarid steppes. A **semiarid region** receives little rainfall. Here only enough rain falls to support a continuous cover of plant growth, such as grass. **Desert regions** are places where the land receives too little rainfall for a continuous cover of plant growth. For centuries the semiarid steppes in the U.S.S.R. were the home of nomadic herders. They made a living by caring for their sheep, horses, camels, and cattle.

The soils are fertile in some parts of this region. However, there is too little rainfall for crops to grow without irrigation. Some large farms are run by the government in the wetter parts of this region. Mostly, animals are raised on these farms.

The Caspian Sea, which is in this region, is about 26 meters (85 feet) below sea level. This body of water is the largest inland lake in the world. In the past 40 years the Caspian Sea has been shrinking in size very rapidly. The shrinking has been caused, in part, by the building of dams on the Volga River. The dams produce power and improve the inland waterway. Vast lakes have been created upstream. Water evaporates from these lakes before reaching the Caspian Sea. If present trends continue, the "sea" will soon be about half as large as it was in 1930. The area east of the Caspian Sea to Lake Balkash (bahl *kahsh*) and southward to Iran (ih *rahn*) and Afghanistan (af *gan* ih stan) is an inland basin and a desert region.

Fertile soils are found in the valleys of the rivers which flow into the Aral Sea. Water from these rivers and other streams is used for irrigating farmland. Cotton is the leading crop, with about three-fourths (¾) of the country's production coming from this area. Many textile mills have been built in this region. Tobacco, fruits, rice, nuts, and sugar beets are other valuable crops. In many ways this region is similar to the southern part of the Central Valley and the Imperial Valley in California. Tashkent (tash *kehnt*), the largest city east of the Ural Mountains, is the center of this agricultural area.

Mountains in the U.S.S.R. The Caucasus (*kahw* kuh suhs) Moun-

tains stretch from the Black Sea to the Caspian Sea. They are the highest mountains in the western part of the U.S.S.R. Peaks in these mountains are considerably higher than the Alps in the central part of Europe. The Caucasus Mountains form a great barrier in this area between the north and the south. It has always been possible to go around them by water, however. Railroads have been built around both ends of the range near the seashore. Several highways have also been built through high passes. The highest peak, Mount Elbrus (ehl *broos*), is more than 5600 meters (18,400 feet) high. In the higher parts of the Caucasus Mountains there are some glaciers.

On the southwestern slopes of the Caucasus Mountains, rainfall is heavy. Batum (bah *toom*), on the Black Sea, has about 230 centimeters (90 inches) of rain each year. Along the Caspian Sea, however, rainfall is light. Baku (bah *koo*), an oil center on the Caspian Sea, has about one-tenth (¹/₁₀) as much rainfall as Batum.

In the lower valleys and basins among these mountains, the climate is mild. Citrus and other subtropical fruits can be grown. Grapes, rice, tobacco, and cotton are produced in different sections. Tea is grown on the valley slopes where enough rain falls. Stock-raising is also important in the mountainous areas. North of the main range of the Caucasus, wheat, barley, cotton, and sunflowers are the leading farm products. The seeds of the sunflower are pressed to get an oil which is much like olive oil. Other important crops raised there are tobacco, flax, and sugar beets.

As the map of the U.S.S.R. shows, mountains extend eastward from the Caspian Sea most of the way to Lake Baikal (by *kahl*). This lake, one of the deepest in the world, is about 1740 meters (5,700 feet) deep. High mountains are located southeast of Tashkent. This high region averages more than 3950 meters (13,000 feet) above sea level. Communism Peak, 7495 meters (24,590 feet) high, is the highest mountain in the U.S.S.R. Nearby Lenin Peak is 7134 meters (23,405 feet) in altitude. Find these on the map on pages 260-261. The Tien Shan (tih *ehn shahn*) Mountains are also a high range. Peaks in the Altai (*al* ty) Mountains reach above 4500 meters (15,000 feet).

Northeastern Siberia is the most mountainous region of the U.S.S.R. Many ranges there are separated by long river valleys. Although the ranges are not nearly as high as those along the southern border, a few peaks are 3000 meters (nearly 10,000 feet) high. This region used to be almost uninhabited. However, many changes have occurred bringing new settlers to the area. Siberia is one of the coldest areas on the Earth. For many years it was known as the "cold pole." The lowest temperature ever recorded at Verkhoyansk (vehr koy *ansk*) was about −68° C (−90° F). In the Verkhoyansk region January temperatures average about −50° C (−58° F). Most people in eastern Siberia live along the Trans-Siberian Railroad. This railroad runs close to the northern border of Manchuria (man *choor* ee uh).

This town is located in the region of the Caucasus Mountains. Look at this picture. Why do you think this town developed here?

———————— WHAT HAVE YOU LEARNED? ————————

1. Which nation has more land area, the Soviet Union or the United States?

2. How would you describe the climate of the tundra region in the U.S.S.R.?

3. Why are wood products plentiful and cheap in the Soviet Union?

4. Why is the Caspian Sea getting smaller each year?

5. Where is the most mountainous region in the U.S.S.R.?

———————— BE A GEOGRAPHER ————————

1. Using the map on pages 260-261, trace the route of the Trans-Siberian Railway from Moscow to Vladivostok. About how many kilometers (miles) is it on this railroad from west to east?

2. Which of the following is farther north:
   a. Leningrad or Moscow?
   b. Leningrad or Murmansk?
   c. Leningrad or Archangel?
   d. Perm or Moscow?

3. Which of the following is farther east:
   a. Moscow or Perm?
   b. the Caspian Sea or the Aral Sea?
   c. the Altai Mountains or the Sayan Mountains?
   d. the Kara Sea or the Sea of Okhotski?

4. Use a map of the Soviet Union to identify three cities located in high latitudes.

———————— QUESTIONS TO THINK ABOUT ————————

1. Do you think that people in the Soviet Union should take steps to conserve their natural wildlife resources? If so, why?

2. What steps do you think could be taken to prevent the Caspian Sea from becoming smaller?

3. Why do you think that many of the people of eastern Siberia live along or near the Trans-Siberian Railway?

4. Can the "bigness" in land area of a country cause problems for a nation? Are there problems, for example, that the Soviet Union has to deal with because of its size? Are they different from those of Canada? The United States? China?

Chapter Review

Chapter

WORKING AND LIVING IN THE ——————SOVIET UNION——————

The western or European region of the Soviet Union has received most of the attention from the government in the past. It is in the European region of the Soviet Union where four-fifths (⅘) of the people live. Most of the cities and industries of the nation have been located in this region. Since the mid-1960s Soviet leaders have been looking more to Siberia and the Soviet Far East, however. Why do you think this harsh area has become important to the U.S.S.R.?

One reason has been the Soviet Union's concern over its southeastern borders with the People's Republic of China. Another major reason has been the discovery of needed natural resources in this vast area. About nine-tenths (⁹⁄₁₀) of the Soviet Union's known coal deposits are located in Siberia and the Soviet Far East. Oil and gas deposits have also been found. Three-fourths (¾) of the nation's timber is grown east of the Urals. More than half of the nation's water power resources are in Siberia and the Far East. Valuable gold and diamond deposits are also known to exist in the frozen lands of this region.

It is expensive to ship raw materials from the Far East and Siberia to industries in the European region. Therefore, new industries are being developed in the central and eastern regions of the U.S.S.R. Towns and cities are being built in Siberia and the Soviet Far East. This region is rapidly becoming another major industrial region. Higher wages, educational opportunities, and rapid job advancement are given to people who move to settlements in this remote area.

In 1956, 731,000 people lived in Novosibirsk. In 1977, 1,304,000 people lived in that Siberian city. These new apartment buildings have been built in the forest to house the new residents.

Help Wanted—in Siberia

Nikolai finished his day's work at the Ministry of Health. He left the office building and joined hundreds of other workers rushing to their bus stations in downtown Moscow. As he waited in line for his bus, he noticed a huge new sign on the Workers' News Bulletin Board. It read, "Come Work In The North!" The sign said that thousands of workers were needed in many regions of Siberia. The government wanted people to become pioneers and settle in many of the new towns and cities.

The lady behind Nikolai laughed as they both read the sign. "Who would want to leave a city like Moscow to go live in that wilderness?" she asked. The man next to them joined in the conversation. "Oh! some parts of Siberia are not that different from areas around Moscow," he noted. "The state pays very high wages to workers there," he added. "But what can a person buy with the money?" asked Nikolai. "Are there many stores, schools, or shops?" The sign said that housing would be provided and that many new settlements were growing rapidly. Nikolai had heard that vast deposits of oil had been discovered in regions of Siberia and the Soviet Far East. It was said that timber, gold, and diamonds were also to be found in this area.

Nikolai wondered what it would be like to become a pioneer in this region of the Soviet Union. Just then his bus pulled into the station. He jumped into the crowded bus, leaving the sign behind. But as he rode home he continued to think about Siberia. Maybe he would talk about it to his family. Maybe tomorrow he would try to learn more about pioneer life in Siberia. Maybe he would apply for a job in Siberia. He would think about it.

Mineral Wealth

The U.S.S.R. has a good supply of minerals and mineral fuels. Even so, mineral production there has generally not been as great as in the United States. However, the Soviet Union is developing rapidly. The U.S.S.R. leads the world in mining iron ore and manganese and in making cement. It has plenty of gold, chromium, titanium, platinum, and antimony. Known sources of bauxite, lead, copper, zinc, tin, and uranium are quite large. Production of these minerals is not sufficient to meet the U.S.S.R.'s own needs, however. Eastern European countries that have these minerals often send them to the U.S.S.R. Good supplies of phosphate, sulfur, salt, and mercury are now available.

The Soviet Union is reported to have the world's largest known deposits of coal. Much of it has been mined in the Donets (doh *nyets*) Basin and in the Urals. Vast deposits of coal exist in the Central Siberian Plateau and in the Arctic near Vorkuta. However, these areas have not been actively mined.

Petroleum and natural gas reserves in the U.S.S.R. may be the largest in the world. The Soviet reserves could last about 200 years at the present rate of use. The main field, from which more than three-fourths (¾) of the petroleum and natural gas comes, is called the Volga-Ural Field. It begins in the southwest along the Volga River near Saratov (suh *rah* tuhf) and extends northeastward for almost 800 kilometers (500 miles) to near Perm. Pipelines from this huge field take oil and natural gas to a number of major cities including Moscow and Gorki (*gawr* kih). As you have already learned, a pipeline carries oil to COMECON countries in Europe. Pipelines from northern Siberia connect to pipelines carrying fuel eastward to Irkutsk (ihr *kootsk*). In the Soviet Far East pipelines transport natural gas from Komsomolsk (kuhm suh *mulsk*) to Okha (*ohk* ha). Plans have been made to develop more gas pipelines from Yakutsk (yuh *kyootsk*) to the coast.

Although the Soviet Union has supplies of needed minerals and mineral fuels, their location causes special problems. Distances between major deposits of minerals are often great. Transporting ores or coal for long distances on land is costly and difficult. Because of this, the Soviet government is encouraging people to move to regions of the country nearer to the raw materials.

SOVIET UNION

Mineral and Fuel Resources

- Bx Bauxite
- Cr Chromite
- Cu Copper
- D Diamond
- Au Gold
- Fe Iron
- Pb Lead
- Mn Manganese
- Hg Mercury
- Ni Nickel
- Pt Platinum
- P Potash
- Ag Silver
- S Sulfur
- Sn Tin
- W Tungsten
- Zn Zinc
- Coal
- ☐ Natural Gas
- ○ Petroleum

Scale
Kilometers 0 500 1000
Miles 0 500 1000

△ Uranium
Major Industrial Regions

Manufacturing and Industrial Regions

Under the czars, Russia was mainly an agricultural country. Until the early 1900s, little attention was given to manufacturing. Since the Communists seized power in 1917, the country has been building new industries at a fast rate. The Soviet Union has, since 1928, made a series of "Five-Year Plans." Each Five-Year Plan set goals for building more factories and machines, increasing agricultural production, and improving living standards. Today only the United States leads the U.S.S.R. in industrial strength. Many Soviet-made goods, however, still lack the quality of goods made in Western Europe, Japan, and the United States.

Early Centers of Industry. Russia's first industrial center grew around the city of Moscow. That city is one of the world's largest cities today. **Lignite**, a soft coal used to produce electricity, furnished power for this region's early development. Lignite is still used to produce power. However, better grades of coal have to be brought to Moscow today from other areas. The Moscow industrial region still produces about one-fifth (⅕) of all goods manufactured in the U.S.S.R. Factories in Moscow itself produce almost one-tenth (1⁄10) of the nation's goods. Radios, television sets, refrigerators, and washing machines are made in factories near Moscow. More than one million persons work in industrial plants in Moscow alone.

After the revolution textiles were

Living in the Soviet Union

Since World War II the Soviet Union has become a major industrial nation. Many power plants, such as this one, have been built to supply the growing demand for energy.

the leading product in the Moscow region. The region is still an important center for textiles. Cotton, woolen, and linen fabrics are among the Soviet Union's leading manufactured products. Gorki, to the east, is a major center for automobile production. Trucks, farm equipment, and buses are also made there. Products such as electrical machinery, which require highly skilled workers, are also made in this region.

Leningrad, the former capital of Russia under the czars, is also an important manufacturing center. The volume of goods produced in the Leningrad region does not compare with the volume produced in the Moscow region. Shipbuilding is a major industry in Leningrad. Fac-

tories in Leningrad also make heavy electrical equipment, locomotives, textiles, precision instruments, and chemicals. Leather shoes, clocks, and watches are highly prized products made in this area. Most raw materials have to be brought to Leningrad from other parts of the Soviet Union.

The Ukraine. Find the Ukrainian S.S.R. in the southwestern part of the Soviet Union on the map on pages 260-261. The mineral deposits of the Ukraine have made it one of the U.S.S.R.'s major industrial regions. From the Donets Basin near Donetsk (duh *nehtsk*) comes about one-third (⅓) of the best coal mined in the Soviet Union. Krivoi Rog (kreh *voi rohgh*) has one of the largest deposits of iron ore in the U.S.S.R. But much of the best ore has been mined. Ways of using lower grade ore are being developed. Nikopol (nih *kaw* pawl), about 95 kilometers (60 miles) from Krivoi Rog, is the U.S.S.R.'s main source of manganese. You can see why the Ukraine has become a major center for steel production.

In addition to iron and steel, factories in the Ukraine produce chemicals from salt, another important mineral resource. Other factories make many types of machinery. The Ukraine is also one of the best farming areas in the nation, so there are many factories that process foods. Huge dams on the Dnieper (*nee* per) River provide needed power for these industries.

The Urals. The Urals region is a center for the production of iron and steel. Other metals, chemicals from potassium and salt, and petroleum are also found and refined there. The Urals region has become an important industrial region because of its rich mineral deposits. This region was developed by the Soviet government during World War II because it was far from the western border. The Ukraine had been captured by German armies. Leningrad was surrounded, and Moscow was threatened. The Urals region developed rapidly. About one-eighth (⅛) of the manufactured goods produced in the Soviet Union are made in this region. Magnitogorsk (mag *nee* toh gawrsk), in this region, is now the largest center for steel production in the Soviet Union. Much

of the steel is used in making tanks, guns, and other equipment for the country's large army.

The major problem of the Urals region is that it does not have any known deposits of high quality coal. The nearest deposit is in the Kuznetsk (*kooz* nehtsk) Basin near Novosibirsk. It is more than 1900 kilometers (1,200 miles) away. This basin contains vast deposits of coal. Some of the coal is carried by trains to the Urals. On their return trip the trains carry iron ore from the Urals to the Kuznetsk Basin. There the ore is used to produce steel.

Some New Industrial Centers. Novosibirsk, which not many years ago was a small city in the Kuznetsk Basin, now has more than a million people. It is a major industrial city and serves the developing areas in Siberia much as Chicago serves the Midwest in our country.

To house workers in new industries, entirely new cities are being built. Bratsk (brahtsk), once a small fishing village, is already a city of more than 300,000 people. Irkutsk has a population of more than 400,000. Its factories make airplanes, automobiles, and construction machinery. Long-range plans have been made to get more people to move from the western part of the Soviet Union to Siberia. However, settlers tend to stay there only a few years and then leave.

Important Industrial Developments

The U.S.S.R. is one of the leading nations in the production of aircraft.

Living in the Soviet Union

Factories make cargo and passenger planes as well as large numbers of planes for the country's air force. In recent years much effort has been given to the development of atomic energy and to the production of rockets.

Soviet achievements in space have been noteworthy. On October 4, 1957, the U.S.S.R. launched the world's first satellite into space. It was called *Sputnik I*. Since then a variety of space flights have contributed much to the world's knowledge of space. In 1975 United States astronauts joined Soviet cosmonauts in a "linkup" in space.

The development of electric power is being greatly increased in the U.S.S.R. Much use is made of it in the manufacturing industries. Much hydroelectric power has been developed in the Caucasus region. But the main hydroelectric developments to date have been on the Dnieper and Volga rivers. Six large dams have made the upper Volga River from Moscow to Kuibyshev (*kyoo* ih buh *shehf*) into what is almost a lake. Two more huge dams have been built at Saratov and Volgograd (*vawl* goh *grad*). The dams, like those constructed by the government of the United States, have more than one purpose. They produce hydroelectric power, prevent floods, make navigation easy, and provide water for irrigation. Many power plants have been developed throughout the Soviet Union. Hydroelectric power plants supply most of the nation's electrical energy needs. In 1964 the Soviets built their first nuclear power plant. **Thermal** power plants (ones using steam to generate electricity) have also been constructed in regions of central and eastern U.S.S.R. The U.S.S.R. has problems of uneven distribution of power resources. Plans are underway to construct a national network of power lines. Hopefully, this will provide more power throughout the entire nation.

United States astronaut Tom Stafford and Soviet cosmonaut Alexei Leonov meet in the hatchway that connects their two spacecrafts. Do you think that more joint Soviet-American space ventures should be undertaken?

Agriculture

Only about one-tenth (1/10) of the land in the Soviet Union can be farmed. The rest is too mountainous, too cold, too dry, or too poor to raise crops. Crop yields vary greatly from year to year depending on rainfall and harvesting practices.

More wheat and potatoes are grown in the Soviet Union than anywhere on Earth. Rye, corn, rice, barley, buckwheat, and millet are also grown. Cotton is the leading fiber crop. Farm animals are raised in practically all farming areas. As in the United States, the number of horses has decreased as the use of tractors and other agricultural machines has increased.

Farming Methods: Collective and State Farms. Since the Communists took control of the government, agricultural methods have changed a great deal in the Soviet Union. Almost all land is owned by the government. Individual families once had their own small farms. Now most of the cultivated land is part of *collective farms* or *state farms*. These farms are very large, averaging about 2000 hectares (5,000 acres) of cultivated land each. Some have as much as 12 000 hectares (30,000 acres) of land. As many as 1,000 families may live on a large farm.

On a **collective farm** everyone works together in the production of crops. Shares are credited to each family according to the number of hours worked and the amount and kind of work done. The farmers live in villages and each family has the use of a small plot of land for a garden. They also may keep a few chickens and farm animals for family use. Farm buildings are grouped near the village. Most of what is produced is sold to the government at prices set by the state. Collective farms grow mainly grain crops, cotton, and sugar beets. The government sets a **quota**—a fixed amount to be produced—as a goal for each collective farm. However, not all collective farms produce enough crops to meet their quotas. The government of the Soviet Union has often been disappointed in the collective farms.

On the **state farms**, workers are paid by the government. These farms specialize in raising certain crops or animals. For example, there are state grain farms, fruit farms, cattle, sheep, and pig farms, and even reindeer farms. Some of the farms are experimental stations where scientists study plants and animals and develop improved methods of farming.

Even state farms have disappointed the government. Often farm machinery on the collective and state farms was poorly maintained. Often, when machines were needed in the fields, they were not available. When machines needed repairs, spare parts were often lacking.

Workers received rather low wages on the farms also. Fertilizer and cattle fodder were often in short supply. This often resulted in low crop and cattle yields. During the government's 1971-1975 plan, extra help was given to agriculture. More machinery was sent to the farms. Workers were given extra money if their farm produced more than their quota. Farmers were also given more land for their private use. It was hoped that these benefits would encourage workers to produce more.

Occasionally, poor weather affects crop yields. For example, in 1972 the grain crop was ruined because of weather conditions. The U.S.S.R. had to spend $1 billion for wheat from the United States that year. Canada and Australia also sell grain to the Soviet Union from time to time.

SOVIET UNION

ARCTIC CIRCLE

Landuse

Livestock and food crops

Wheat, corn, or rice

Livestock

Special crops

Sheepherding
Mediterranean agriculture

Dairy farming

Non-agricultural, mainly forests

Nomadic herding

Non productive

Scale

Kilometers 0 500 1000

Miles 0 500 1000

Fisheries

The many rivers, lakes, and inland seas, as well as the bordering seas and oceans, give the U.S.S.R. rich fishery resources. The fishing industry, also, is run entirely by the government. The government owns and operates the fishing fleet, canneries, and refrigeration and storage plants.

Many kinds of fish are caught in the Baltic Sea and along the northern coast. Crabs and salmon are an important part of the catch in the Pacific region. **Caviar** (*kav* ee ahr), prepared from the eggs of the sturgeon (*stur* juhn), is a famous U.S.S.R. commercial product. Sturgeon is caught in the Caspian Sea. Some whalers from the Soviet Union travel to the Antarctic to hunt whales. The Soviet Union ranks next to Peru and Japan in the amount of fish caught.

Transportation

Because of its tremendous size, transportation within the Soviet Union has long been a problem. The country now has about 138 000 kilometers (86,000 miles) of railroad. Most of the railroads are west of the Urals. As the population map on page 286 shows, that area has the most people. The main railroad line is the Trans-Siberian Railroad. It goes from Moscow eastward to Vladivostok (*vlad* ih vuhs *tawk*) on the Pacific coast. Recently a number of branch lines from the Trans-Siberian Railroad were built. These branch lines connect various cities in Siberia with the main railroad.

Workers prepare caviar for shipment to all parts of the world.

Railroads carry most of the freight in the U.S.S.R. As in the United States, passenger travel by rail is small compared with the amount of freight moved by railroads. However, there are few highways and automobiles in the U.S.S.R. As a result, people in the Soviet Union do more of their traveling by train than people do in the United States.

Rivers in the Soviet Union are important transportation routes. During the summer months, rivers are used for carrying heavy, bulky products. These include coal, ores, petroleum, lumber, grains, and

Living in the Soviet Union

SOVIET UNION

Population

, People Per Square

kilometer		mile
uninhabited		uninhabited
under 1		under 3
1 to 10		3 to 25
10 to 51		25 to 130
51 to 102		130 to 260
over 102		over 260

• Cities over 1,000,000 population

Scale

Kilometers 0 500 1000

Miles 0 500 1000

cement. Even today, rivers are the only highways leading into many parts of Siberia. Rivers are used to transport goods from many areas in Siberia to the Trans-Siberian Railroad. One great drawback is that most rivers are frozen or clogged with ice several months each year.

A number of canals have been built to connect different rivers. Ships may pass through the canals and gain entry to other bodies of water. The White Sea and the Baltic Sea are connected this way. Canals connect Moscow with the Volga River and with Leningrad in the north. Using canals and Russia's famous waterway, the Volga River, it is possible to go by water from the White Sea to the Caspian Sea. A canal connects the Volga with the Don in the south. This canal makes it possible to ship goods by water between the Caspian and Black seas.

The long distances that have to be covered in the Soviet Union make air transportation desirable. Since many places cannot be reached by any other good means of transportation, the Soviet airline, Aeroflot (*air* oh floht), has grown rapidly. Transportation systems in the Soviet Union are so vast that one out of every ten workers is employed by some part of railway, highway, waterway, or airline departments.

Education

All children in the U.S.S.R. are required to attend school as soon as they are seven years old. They must continue their studies for eight years. Kindergartens are open for children between the ages of three and seven. Special military training is given to boys in school. Talented boys and girls are encouraged to study math and sciences.

Children are taught in the language of the republic or region in which the school is located. The Russian language must also be learned. In all, instruction is given in about 100 languages in the Soviet Union. English is a favorite foreign language. All schools are controlled by the government.

The U.S.S.R. has many colleges and universities. Special schools train scientists and workers needed in the various industries. The government decides who may go to such schools and what they study. Some of the college students are paid to go to school.

Religion

The Russian Orthodox church was the official church of Russia when the Communists came into power. The Communists are **atheists**—persons who believe there is no God. During the first few years the Communists held power, churches were destroyed and religious leaders were arrested or killed.

There still are many restrictions on religious worship in the U.S.S.R. The Communists are doing all they can to discourage religion through propaganda and education. They have been only partly successful, however. Many citizens of the Soviet Union risk punishment by worshipping in secret. Some attend the few churches still allowed to stay open in the country.

Moscow and Leningrad

Moscow is the capital and largest city in the Soviet Union. More than 7 million people live in the city area. Both modern skyscrapers and old historic buildings are found in the central section of the city. Red Square is the scene of many government activities, such as parades. Near it is the Kremlin (*krehm* lihn). The Kremlin consists of a group of old palaces, churches, and government offices. The Kremlin's huge walls surround many of these buildings. Lenin's tomb is located near the east wall of the Kremlin in Red Square. It is visited daily by thousands of people. There are many historic sights to be seen by tourists in Moscow.

Leningrad is the Soviet Union's second largest city. It is a beautiful city built on the Neva (*nee* vuh) Delta on the Gulf of Finland. It was originally called St. Petersburg. In 1917, it was named Petrograd (*peht* roh grad). Then, in 1924, it was renamed Leningrad. Many visitors marvel at the art collection in the Hermitage, a museum. Palaces, churches, museums, operas, and theatres add to the cultural richness of this great city.

Government

The Soviet Union is made up of fifteen republics. They are shown on the map on pages 260-261. Note that the largest of the republics is the Russian Soviet Federated Socialist Republic. The first part of this long name has survived as the common name for the country. This republic contains about three-fourths (¾) of the total area and more than half of the people of the Soviet Union.

The Soviet Union is ruled by the Communist party and no other political party is allowed. Only about 16 million people in the Soviet Union belong to the Communist party. This is about six percent of the population. Elections are held to choose government officials, but most voters really have no choice. Instead of two or more candidates being named for each office, only one is named. The candidates are selected by the members of the Communist party. Voters are not allowed to write in the name of any other person whom they would prefer. They may, however, vote against a candidate by drawing a line through the person's name.

The organization of the government is quite complicated. Each republic has its own government with officers to enforce the laws. At least every four years about 5,000 delegates of the Communist party from all the republics meet in Moscow to make laws and plans. The actual rule of the country, though, is in the hands of a small group known as the **Presidium** (pre *sihd* ih uhm). All of the members of the Presidium are leaders in the Communist party. They are elected by the delegates to the Party Congress. A person is selected to head the Presidium. That person is known as the Chairman of the Council of Ministers, and is sometimes called premier. The person who holds this position is the real head of the government. The headquarters of the national government is the Kremlin in Moscow.

The Soviet Union has seen great changes since the Communists came to power. The political influence of the Soviet Union on the rest of the world is very great. You will want to keep close watch on events that may shape the future of the U.S.S.R. and the rest of the world.

REMEMBER, THINK, DO

1. In what ways are the state farms and collective farms different? In what ways are they alike?
2. What do you think the government of the U.S.S.R. should do to improve production on the farms?
3. Take an imaginary trip on the Trans-Siberian Railroad. See if travel agents in your community might have information about the sights you would see on such a journey. Prepare a presentation about the trip to the class.
4. How does education in the Soviet Union differ from education in the United States?
5. Invite a visitor to class who has traveled in the U.S.S.R. to tell you about the country.
6. What group of people really rules the Soviet Union?

A SUMMARY OF THE SOVIET UNION

The Soviet Union contains more land than any other country in the world. At least five kinds of land are found within its boundaries. They are the tundra, taiga, mixed forest region, "black earth" region, and the dry steppe region. Almost every kind of agricultural product can be grown somewhere in the Soviet Union. The nation's mineral resources are of great amount and variety. Throughout its history farming has been the chief occupation of its people. However, since the Revolution of 1917 the Soviet Union has transformed itself into one of the world's leading industrial powers.

For more than 300 years the government of Russia was an absolute monarchy headed by a czar. However, in 1917, a revolution headed by Lenin and the Bolsheviks established a new form of government, communism. The new government has attempted to eliminate religion, provide an education for all the people, and create an advanced industrial society.

The Soviet Union has experienced much in the 20th century. Its people have fought in two world wars and in a revolution. Many changes have taken place in social, economic, and political conditions. Today the Soviet Union is one of the major nations of the world. Its Communist government influences the activities of most nations of East Central Europe. And, its influence is felt in nations in Africa, Asia, and South America, as well.

1. What are "Five-Year Plans" used for in the U.S.S.R.?

2. What are some of the important natural resources found in Siberia and the Soviet Far East?

3. Why was the Urals region developed into an industrial region during World War II?

4. What are some of the new industrial cities that have grown rapidly in Siberia?

5. What similarities and differences are there between the agricultural and mineral resources of the Soviet Union and those of:
   a. Northwestern Europe?
   b. Southern Europe?
   c. Central Europe?

———————— BE A GEOGRAPHER ————————

1. What nations touch
   a. the U.S.S.R.'s southern border?
   b. the U.S.S.R.'s western border in Europe?

2. Between what parallels of latitude and meridians of longitude is the mainland of the Soviet Union located? (Figure to the nearest degree.)

3. What cities in the Soviet Union have more than 2,000,000 people?

4. How many cities in the Soviet Union with more than 500,000 people are west of the Ural Mountains? How many are east of the Urals?

5. What two large rivers flow into the Caspian Sea? Which is the longer river?

6. How is it possible for ships to sail between the Caspian Sea and the Black Sea?

7. What large rivers flow northward to the Arctic Ocean?

8. On an outline map of the Soviet Union, shade in areas that you think are: a. too cold for growing crops; b. too dry for growing crops (except through irrigation). The maps on pages 267 and 284 will be useful in locating such areas.

9. Does any nation have land (that is not an island) closer to the North Pole than the Soviet Union's northernmost tip?

QUESTIONS TO THINK ABOUT

1. The Soviet Union is not only building a much larger merchant fleet, but is also increasing the size of its navy. Why do you think that this is happening? Should the United States be concerned about this? Why or why not?

2. The United States and the U.S.S.R., as the two most powerful nations on Earth, have made some important agreements. They have agreed, for instance, to do all testing of nuclear weapons underground. They have agreed to try to keep other countries from building nuclear weapons. They also have agreed to fly passenger planes between New York City and Moscow. Do you think these two countries should try to find other areas of agreement? What kinds of agreements might be wise for them to make?

3. Compare Siberia and its problems with the western areas of the United States in the years after the Civil War. Are there similar problems? Are there differences? May Siberia become a much more important part of the Soviet Union than it now is? What handicaps will be hard to overcome?

4. Almost every day in Moscow, thousands of people line up to visit Lenin's tomb. Why would they do so? Are there any places in the United States at which many people also line up to visit? Why?

5. The United States and the Soviet Union frequently take opposing sides in debates in the Security Council of the United Nations. Do they ever vote together? The U.S.S.R. frequently vetoes a vote in the Security Council. Does the United States ever do so? Why are these two nations often on different sides of issues?

6. Do you think that individual nations should (or should not) be able to claim land areas (as on the moon) explored in outer space? Why?

7. Do you feel that it is the proper role of government to decide if a person should continue in school? For example, should a government decide whether a student should go on to college after completing high school? Why?

Tiles often are used to decorate mosques in the Middle East and North Africa.

VII
Unit

THE MIDDLE EAST
AND
NORTH AFRICA

Unit Preview

- Can you identify these national flags of the Middle East and North Africa?

- Can you name the nations of the Middle East and North Africa shown on this outline map?

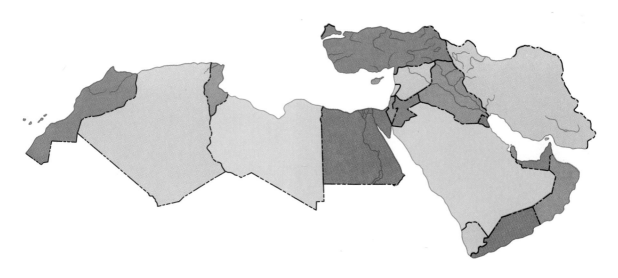

DO YOU KNOW?

- Why Christians, Jews, and Muslims all consider parts of the Middle East to be their Holy Lands?

- Why the nations of the Middle East and North Africa are thought to be alike?

- What the oil-rich nations of the Middle East and North Africa are doing with their oil profits to improve conditions within their lands?

- Why the meetings of President Sadat of Egypt and Prime Minister Begin of Israel in 1978 and 1979 were such important international events?

- Which nation in the world has the highest per capita income?

Unit Introduction

The Lands of the Middle East and North Africa

The area of land known as the Middle East includes parts of Asia, Africa, and Europe. Look at the map of the Middle East on page 311. As you can see, the Black and Caspian seas are north of this area. The Mediterranean Sea is on the west, and the Arabian Sea is to the south.

Many countries make up this area known as the Middle East. Israel (*ihz reh ehl*), Jordan (*jawr d'n*), Syria (*sihr ih uh*), and Lebanon (*lehb uh nuhn*) are located in the central coastal part of this area on the Mediterranean Sea. North and east of these nations are Turkey, Iraq (ih *rahk*), and Iran. West of them is the island country of Cyprus (*sy* pruhs). To the south is the Arabian Peninsula. Saudi Arabia

(sah *oo* dih a *ray* bih uh) takes up most of this peninsula. On the eastern edge of the peninsula is the Persian Gulf. Here Kuwait (koo *wayt*), Bahrain (bah *rain*), Qatar (*kah* tahr), and the United Arab Emirates (ih *mihr* uhtz) are located. Oman (oh *mahn*) and the People's Democratic Republic of Yemen (*yehm* ehn), or South Yemen, are located along the southern edge of the peninsula. The Yemen Arab Republic, or Yemen, on the Red Sea, completes the nations of the Arabian Peninsula. Egypt, in North Africa, is considered to be part of the Middle East, also. Locate these Middle East nations on the map on page 297. Notice the bodies of water near these land areas.

Unit Introduction

The cities of the Middle East and North Africa are often a blend of the traditional and the modern. Amman, Jordan, is such a city.

Throughout history the Middle East has been important because of its location. It seems almost as if the Middle East holds the continents of Europe, Asia, and Africa together. In the past many civilizations and traders had used the lands of the Middle East as trade routes. In more recent times the nations of the Middle East Have controlled the waterways of the Dardanelles and the Suez (soo *ehz*) Canal. Find these on the map on the next page.

This area is also very important as a major religious center. The holy lands of three major religions are located in the Middle East. *Judaism* and *Christianity* started in areas now controlled by Israel and Jordan. **Judaism** is the way of life developed by the Jewish people as a result of historic and religious experiences. The basic guide for the Jewish way of life is the Torah, which dates from the time of Moses. **Christianity** is the religion of the people who believe in the teachings of Jesus Christ. Followers of **Islam** also consider some of these areas as holy. The Islamic religion had its beginnings in Saudi Arabia. The main prophet of the Islamic faith is Mohammed (moh *ham* ehd). The followers of Islam are known as Muslims. People from all over the world come to the holy lands to worship and pay respect to these sacred places of Judaism, Christianity, and Islam.

Ownership of these holy lands, however, has caused conflict many times throughout history. The Bible describes some ancient battles over these lands. History books tell of the Crusades and of the Islamic Holy Wars fought to win these sacred lands. During the middle of the twentieth century, several wars have occurred between Israel and Arab nations over parts of these holy lands.

Also included in this section are the nations of North Africa. These nations are the Arab Republic of Egypt, Libya (*lihb* ee uh), Tunisia (too *nee* zhuh), Algeria (al *jeer* ih uh), and Morocco (muh *rahk* oh). These nations can also be located on the map on the next page.

MIDDLE EAST AND NORTH AFRICA

Scale

Kilometers 0   300   600

Miles 0   300   600

DRY GRASSLAND

DESERT

FARMLAND

FORESTS

MOUNTAINS

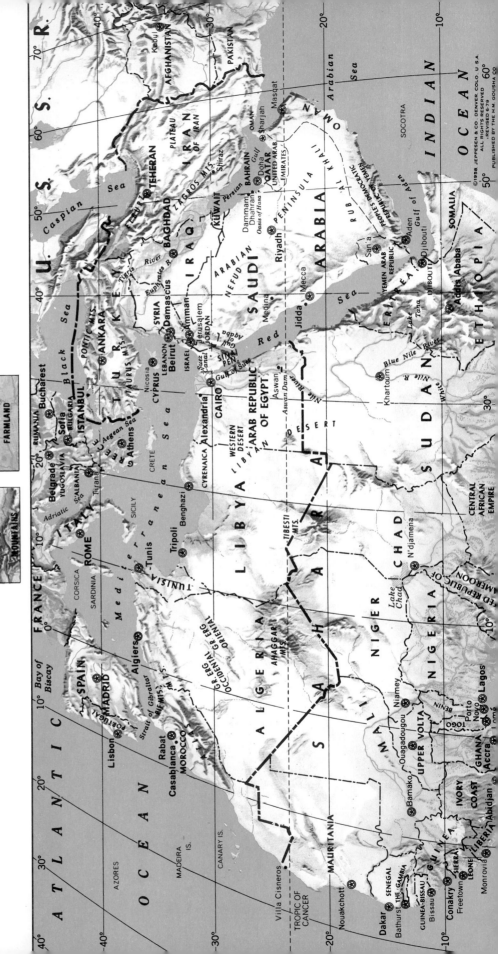

MIDDLE EAST AND NORTH AFRICA

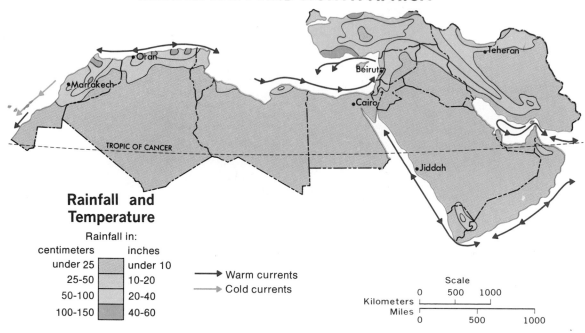

Rainfall and Temperature

Rainfall in:

centimeters	inches
under 25	under 10
25-50	10-20
50-100	20-40
100-150	40-60

→ Warm currents
→ Cold currents

Scale

Kilometers 0 500 1000
Miles 0 500 1000

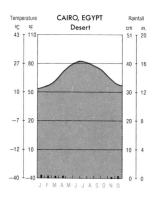

CAIRO, EGYPT
Desert

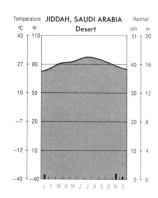

JIDDAH, SAUDI ARABIA
Desert

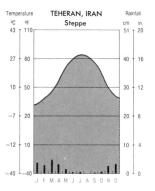

TEHERAN, IRAN
Steppe

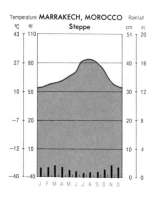

MARRAKECH, MOROCCO
Steppe

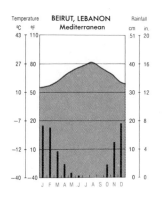

BEIRUT, LEBANON
Mediterranean

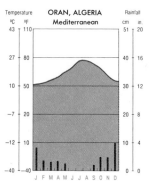

ORAN, ALGERIA
Mediterranean

How Are the Lands of the Middle East and North Africa Alike?

Look again at the map of the Middle East and North Africa on page 297. In what ways do you think the lands of the Middle East and North Africa might be alike?

• Most of the countries of the Middle East and North Africa have a similar climate. Coastal areas along the Mediterranean Sea have, for the most part, a Mediterranean climate. Inland, however, much of the land is desert. Look carefully at the rainfall map. Notice that most of the land of the Middle East and North Africa receives very little rain. More moisture falls in Turkey and in the mountainous areas of Iran, Yemen, and Morocco than elsewhere in this region. Snow falls in these mountains during the winter months. However, throughout most of the lands, a desert climate is common.

In the desert rain falls sometimes. Now and then **cloudbursts** (storms which bring heavy amounts of rain in a short period of time) occur. These storms often cause great damage. They cause erosion and flash floods. Long periods of drought are common in the desert. Day after day the skies are clear and sunny. In the summer season temperatures go as high as 49° C (120° F) at noon. In the evenings the temperatures drop rapidly. Nights are cool compared to the days. In the winter months daytime temperatures can reach 27° C (80° F). Nighttime temperatures close to freezing are common.

• A majority of the people of the Middle East and North Africa have the same written language and believe in the same religion. Except for Turkey, Iran, and Israel, most of the people use Arabic as their written language. Although the people of the Middle East and North Africa speak different languages they read and write in Arabic. The majority of these people practice the Islamic religion. Sometimes these nations are called the Arab nations or the Muslim nations. However, most of the people of Israel and some people in Lebanon follow other religions.

• Ways of living and working are alike in the Middle East and North Africa. Many people live around the fertile river valleys or coastal regions in farming communities. Some have settled on *oasis* lands in the deserts. An **oasis** is a fertile area in the desert where water is found. Others lead a nomadic life as herders of sheep, goats, donkeys, horses, and camels. Most people in the Middle East and North Africa are poor. They usually have just enough food or money to feed their own families. Not much is left for other needs. Only about one out of five people live in the cities. Many city dwellers are also poor.

• Land resources of the Middle East and North Africa are alike. Most of the farmland of this region requires irrigation in order for crops to grow well. Forest resources are usually found only in the higher mountains. There are not many forested areas in the Middle East and North Africa. Few mineral ores have been discovered. However, huge quantities of

Unit Introduction

Throughout the Middle East and North Africa there are many people skilled in working with gold, silver, and copper. In front of his shop in Baghdad, this man proudly displays what he has made.

• A new spirit of *nationalism* and *cultural pride* has developed in nations of the Middle East and North Africa. During the first half of this century, much of the land in the Middle East and North Africa was controlled by Western nations. After the end of World War II in 1945, these areas became independent nations. New governments were formed as new nations were created. It became important for the people to be loyal and helpful to their new government and nation. A spirit of *nationalism* grew. **Nationalism** is a feeling of interest in and devotion to one's country.

The new governments of the Middle East wanted people to be proud of their culture and history. They did this in many ways. Some governments tried to bring back local languages and traditions. Teachers taught in the local languages in school. Religious teachings and local customs were often taught, too. All of these actions helped people develop *cultural pride*. **Cultural pride** may be defined as a good feeling one has about the deeds and ways of life of one's people and ancestors. People had to learn to live with governments that were often quite different from what they had known.

Many Islamic nations have tried to work together to improve both their culture and economy. However, it is often difficult to work together with neighboring countries. These nations have different forms of government. While nations may agree on religious beliefs, they may differ on political issues.

petroleum are located in parts of the Middle East and North Africa. The region contains perhaps half (½) of the world's known oil reserves. The Middle East is the leading producer of petroleum in the world. The valuable petroleum deposits have turned some of these nations from poor countries to rich countries in a short period of time.

OPEC and the Middle East and North Africa

The Organization of Petroleum Exporting Countries (OPEC) was formed on November 14, 1960, at a conference held in Baghdad (*bag dad*), Iraq. Representatives from the Middle Eastern nations of Iran, Iraq, Kuwait, and Saudi Arabia joined with Venezuela to form this organization. Later other petroleum-producing countries of Asia, Africa, and Latin America joined OPEC.

Before the formation of OPEC, much of the oil in the Middle East and North Africa was controlled by large companies from Western nations. These companies had spent huge sums of money exploring for oil. Once oil was discovered, these companies built oil refineries. They also built pipelines to move the oil to

Two of these pie charts show the percentage of energy each region of the world produces and consumes. Compare these two charts with the one showing population. What do these charts tell you about the world's energy situation?

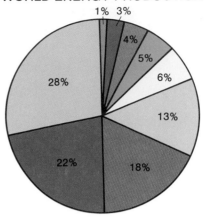

WORLD ENERGY PRODUCTION*

*Energy includes: petroleum, natural gas, nuclear, hydroelectric, coal, solar

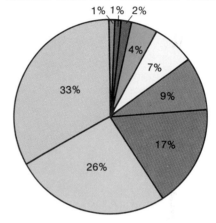

WORLD ENERGY CONSUMPTION*

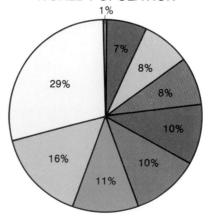

WORLD POPULATION

United States and Canada
Middle East and North Africa
Europe*
South and East Asia**
Soviet Union
China
Latin America
Africa South of the Sahara
Australia and the Islands of the Pacific

*does not include the Soviet Union
**does not include China

Unit Introduction

markets. The oil companies sold the oil and then kept most of the profits for themselves.

With the creation of the Organization of Petroleum Exporting Countries, this was changed. By joining together the OPEC nations controlled much of the world's known oil reserves. They decided to take over control of all the oil companies within their lands. They also decided to regulate the price of oil themselves. OPEC members wanted the majority of oil profits to stay within the oil-producing nations. They wanted this money to be used for improving the conditions of their countries.

To do this many OPEC nations **nationalized**—took over the ownership and control of—their oil resources. The OPEC nations obtained about two-thirds (⅔) ownership in the oil companies. OPEC also decided to control the price at which oil would be sold. They could raise or lower the price. Nations in need

of oil had little choice except to pay the prices demanded. This price fixing rapidly increased the profits from petroleum to these oil-rich nations.

OPEC nations also decided that industries could be started within their borders. Some decided to build oil refineries and chemical plants. Steel mills and other industries were also built. Some nations demanded that oil companies train local citizens for jobs that had been held by foreigners.

OPEC nations believe that developing their countries is important. They believe that these developments will make their nations more self-sufficient. They are looking ahead to the day when oil supplies will not be so plentiful. If this happens, other industries will have been developed to provide jobs and income.

The OPEC nations' demand for high oil prices, however, hurt many poor nations. Therefore, the OPEC

The map below shows the three regions that will be studied in this unit.

REGIONS OF THE MIDDLE EAST AND NORTH AFRICA

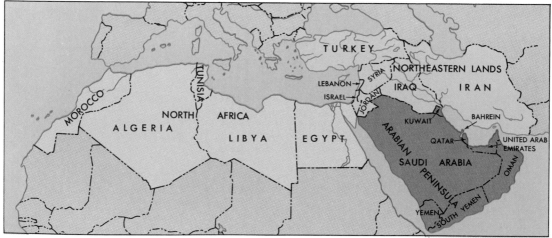

nations agreed to establish a $5 billion fund for Third-World developing nations. This money came from the profits earned from oil. Developing nations could borrow from this fund to help improve their local economies.

Many OPEC nations of the Middle East and North Africa were poor nations themselves before the discovery of oil. These nations have been using some of their oil profits to improve the health, living conditions, and education of their people. Improvements in transportation and industry have also been made.

An Oil Crisis? It Happened Before! Will It Happen Again?

In 1973 there had been talk in the United States of a possible fuel shortage. Nonsense!, thought many people. There had always been plenty of fuel in the United States. Then, in early winter some gas stations began to run out of gasoline. Motorists often had to search for gas. Fuel prices increased rapidly. Sometimes people had to wait in line for several hours to buy gasoline. Drivers sometimes lost their patience. Not everyone was always lucky enough to buy all the fuel needed. The same thing happened again in 1979.

Home heating fuels were also scarce in 1973. Within the first few months of the oil crisis, fuel prices almost doubled. Some families could not afford to pay for their heating fuel. Gas stations and some heating fuel companies sometimes were forced to close their business. They had run out of fuel. The United States had not experienced such a severe fuel shortage since the days of World War II in the early 1940s. What happened to cause the sudden oil shortage? How could a nation as rich as the United States suddenly find itself in such a crisis? Part of the answer to these questions is "OPEC." Part was caused by events in the Middle East and North Africa.

OPEC is the Organization of Petroleum Exporting Countries. More than half of the members of OPEC are from the oil-rich Arab nations of the Middle East and North Africa. Some of the Arab nations went to war with Israel in October, 1973. The United States was supplying Israel with military equipment and money. This angered many of the Arab nations. Therefore, the Arab members of OPEC decided that they would not ship any petroleum to the United States or any other country that was helping Israel. The result of this action was to cause an oil

Unit Introduction

In June 1979 the members of OPEC met in Geneva, Switzerland. They were meeting to decide what price to charge for their oil.

shortage in the United States. This action also helped to raise the price of oil.

When the Arabs began to ship oil to the United States again, after the war, the price did not go down. This was because all the members of OPEC had decided that they were not getting a fair price for their oil. The nations of OPEC decided in 1973 that the price that they were getting for the oil was too low. Also, they thought that the people who were using the oil should conserve it better. They felt that a higher price for the oil would make people more careful about using the oil.

REMEMBER, THINK, DO

1. Make a jigsaw puzzle of nations of the Middle East and North Africa. See if you and your classmates can correctly reassemble the pieces.
2. Why were the lands of the Middle East important in the past? Why are they important today?
3. How are the lands of the Middle East and North Africa alike?
4. What is OPEC?
5. What have OPEC nations done with some of their profits from oil?

A LOOK AHEAD

In this unit several nations of the Eastern Mediterranean will be presented first. These nations include Israel, Jordan, Lebanon, and Syria. Then the northeastern lands, with an emphasis on Iran, will follow. Saudi Arabia and the rest of the Arabian Peninsula will complete the lands of the Middle East. Lastly, the nations of North Africa will be presented with emphasis given to Egypt and Morocco. You may wish to begin some individual and group activities about the Middle East and North Africa.

Things You Might Like To Do As You Study About the Middle East and North Africa

1. Make a large outline map of the Middle East and North Africa.

   a. Mark the major bodies of water in and near the Middle East and North Africa.

   b. Label each nation of the Middle East and North Africa.

   c. Indicate on your map the early trade routes to and from Europe, Asia, and Africa. (See articles in encyclopedias and maps in atlases to get the data you need.)

2. Form four groups to collect current events items on the lands of the Middle East and North Africa. Group #1 should concentrate on news items of the Eastern Mediterranean lands; Group #2, the Northeastern nations; Group #3, the Arabian Peninsula; and Group #4, North Africa. News items should be shared with the class. Attach these articles to your class outline map in their proper locations.

3. Use the same groups as listed in Activity #2. Have each group find out the date of independence for each nation in their region. Also note what nation controlled this land before independence. Mark this information on your outline map.

4. Develop a chart of oil prices for the years 1965, 1970, 1975, and 1980. Your local auto association might help you get the facts you need.

5. During the study of the Eastern Mediterranean countries, have four groups make product maps for Israel, Jordan, Syria, and Lebanon. Then compare the maps.

6. Before studying the Arabian Peninsula, write to the Washington, D.C., embassy offices of Saudi Arabia, Kuwait, Bahrain, Qatar, United Arab Emirates, Oman, South Yemen, and Yemen for up-to-date information on their countries.

7. Write to the United States branch office of the Arab League. Request up-to-date information on the activities of the League. Prepare a display of your materials to share with other students.

Unit Introduction

17

Chapter

THE NORTHEASTERN LANDS ——— OF THE MIDDLE EAST———

In geographic features, the nations of Israel, Jordan, Syria, and Lebanon are much alike. All have an outlet to the sea. Israel, Syria, and Lebanon have coastlines along the Mediterraenan Sea. Jordan has a small coast along the Gulf of Aqaba (*ahk* uh buh). This gulf is part of the Red Sea. Jordan has one of the shortest coasts of any nation.

Each nation has much desert land. All of these nations also have a section of land with a Mediterranean climate. This is most obvious in the northern and western parts of this region. Inland from the sea, most of the land is desert.

Most of the rainfall in these countries occurs in the winter. Summers are hot and dry. The skies are usually sunny and clear. In the desert areas winter temperatures sometimes drop below freezing. This is very common in Syria. Little rain or snow ever falls in the desert regions. The rain tends to fall in the mountain areas inland from the Mediterranean Sea. In Lebanon and Syria, the mountains are often covered with snow in winter. Of course, slight differences exist in both climate and landform throughout these four nations. However, the greatest difference that exists within these lands of the Middle East is more cultural than geographical.

Israel

Israel was created as a national state for the Jewish people in 1948. The area had been known as Palestine. Israel is located at the southeastern end of the Mediterranean Sea. Egypt is on the southern border of Israel. To the east are Jordan and Syria. Lebanon is north of Israel. Locate Israel on your map on page 311. Some of the boundary lines have changed at times because of wars. Through military victories Israel now has control of four times as much land as it had when it became independent. This Israeli-occupied land is a major stumbling block to a lasting peace in the Middle East. Solving the problem of what to do with this land is taking a long time. Check the local news reports for the latest news concerning the conflict between Israel and its Arab neighbors.

When Israel was created there were about 873,000 people living in the country. Most of them were Jews. Many were **Zionists** (*zy* unn ihsts) who believed in the establishment of a national homeland for the Jews in Palestine. As a result of the Zionist Movement, thousands of Jewish immigrants came from Europe. Many more came from North America, Asia, and Africa. More than a million people moved to

Israel during its first ten years as a nation. By the late 1970s almost four million people were living in Israel. About two million Muslim-Arabs live in Israel as well.

Crisis in the Middle East

Within hours of its 1948 independence celebrations, the country of Israel was attacked by military forces from Egypt, Jordan, Syria, Lebanon, and Iraq. Israeli troops were able to defend their lands. Later in 1949, an uneasy **truce** (agreement to stop fighting) with Arab leaders was agreed upon. However, not all remained peaceful. Periodic **guerilla** (hit and run) raids and counterattacks threatened both Israeli and Arab lands. In 1956 fighting again broke out between Israel and Egypt. The United Nations was able to

President Sadat, President Jimmy Carter, and Prime Minister Menachem Begin sign the Egyptian-Israeli peace treaty March 26, 1979.

work out a truce agreement which stopped the fighting. But in 1967 the Six-Day War occurred between Israel and Egypt. As a result of this, Israel gained control of large land areas that were part of Egypt, Jordan, and Syria. This angered many Arab nations that sided with Jordan, Egypt, and Syria.

In 1973 bitter fighting again broke out between Israel and the Arab nations of Egypt, Syria, and Iraq. Once again the United Nations tried to establish peace in the Middle East. Leaders from the United States and other nations also tried to influence the leaders in the Middle East to try and work out a lasting peace agreement.

In 1977 Egyptian President Anwar Sadat accepted an invitation to visit Israel. This was the first time that an Arab leader had visited the nation of Israel. Later Prime Minister Begin of Israel visited Egypt. The leaders of Israel and Egypt tried to plan a peaceful solution to the problems between these two nations. The peace talks continued for a long time.

At last in 1979, Egypt and Israel

Since 1948 wars and treaties have altered the borders of many countries in the Middle East. The maps below show the border changes from 1948 to 1979. Have there been any changes since 1979?

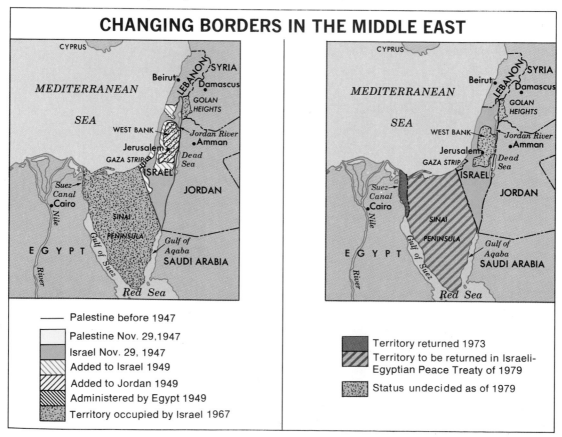

CHANGING BORDERS IN THE MIDDLE EAST

signed a peace treaty. A number of Arab nations then stopped all trade with Egypt. They also stopped all aid that they had been furnishing Egypt. Will peace really come at last to the Middle East? One hopes so, but you will need to check the newspapers and newscasts to see.

The Land and Climate of Israel

The land of Israel (excluding occupied territories) has four main regions. There is a low coastal plain along the Mediterranean Sea. Citrus fruit is raised in this sandy soil. The deep-water port of Haifa (*hy* fah) is one of Israel's main ocean ports in this region. Tel Aviv (*tehl* uh *veev*), a shallow-water port, is located nearby. Tel Aviv is Israel's largest city. Major manufacturing plants have been built in this coastal region.

Inland from the coastal plain is the hill country. This low range of hills extends from the Egyptian border northward. On the western side the hills slope gently toward the Mediterranean Sea. At their highest, these hills are only about 600 meters (2,000 feet) above sea level. The hills are much steeper on the east toward the Jordan River and the Dead Sea. Jerusalem (jeh *roo* seh lehm), Israel's capital, is located in this hill country.

A third region of Israel is the Jordan Valley and the Sea of Galilee (*gal* uh lee) area. Parts of these lands have been made into fertile farmlands. This region is the best area for growing wheat and vegetables in all of Israel. Much of this area was swampland in 1948. Many swamps were drained. Then new crops were grown on the land.

From its mountain source the Jordan River descends rapidly southward. In the 16 kilometers (10 miles) between Lake Hule (*hoo* leh) and the Sea of Galilee, the river drops from about 2 meters (6 feet) above sea level to almost 200 meters (650 feet) below sea level. Then the river flows into the Dead Sea. The shoreline of the Dead Sea is the lowest land area on Earth. It is about 400 meters (1,300 feet) below sea level. The waters of the Dead Sea are also the saltiest of any sea water on Earth.

The last region is the Negev (*nehg* ehv), a desert in southern Israel. The desert is hilly in many places. South of the Dead Sea, however, is a low, flat land which extends to the Gulf of Aqaba. Find this important body of water on the map on page 311. The shallow-water port of Elath (*ee* lath) is Israel's main city on the Gulf of Aqaba. This port provides Israel with a sea route to Asia and Africa. Egypt has not always let Israel use the Suez Canal. Therefore, this port has been very important to Israel for its trade to and from Asia and southern Africa.

The greatest amount of rainfall in Israel occurs in the northern part of the nation. About 100 centimeters (40 inches) of rain falls there annually. Most of this rain comes during the winter. Near Beersheba (beer *shee* bah) in the Negev, only about 20 centimeters (8 inches) of rain falls each year. At Elath in the far south, about 5 centimeters (2 inches) of rain falls during most years.

Agriculture in Israel

The people of Israel have worked hard to increase the amount of usable farmland in their country. Some of the land along the Jordan River was once marshland. Certain sections of the fertile land had been overworked and were no longer useful. Vast areas of land were dry, barren desert land.

Farmers began land reclamation projects. They improved their farming techniques and irrigation systems. As a result, Israel has more than doubled the amount of land for farming. Now, four times more land is irrigated than in 1948. Pipelines carry water from a river near Tel Aviv to Beersheba in the Negev. Many dams and canals have been built to use the water from winter rains and summer cloudbursts.

There have been many disputes with Jordan, Syria, and Lebanon over the water rights to the Jordan River. These disputes prevented full use of that water supply during the early years of Israel's existence. By 1975 most of the dry land that could be farmed was already under cultivation. However, only one-third (⅓) of the possible irrigated farmland was in use. At that time Israel was using almost all of its water to irrigate this land. Additional sources of water are needed to increase Israel's irrigated farmland areas.

To improve agricultural production Israel uses scientific farming methods. Fertilizers, farm machinery, and improved farming methods have helped increase the crop yields per hectare. Agricultural researchers are continuing to discover new ways to use desert lands to grow crops.

Some farmers have also worked together with neighbors in cooperatives. Others have often joined a *kibbutz* (kihb *oots*). A **kibbutz** is a collective settlement. In such a settlement workers share the land, most of the property, the work, and the profits from the sale of products.

MIDDLE EAST

Kilometers
0   150   300
Miles
0   150   300

SEMIARID GRASSLAND AND FOREST
DESERT
MOUNTAINS
FORESTS
FARMLAND
DRY GRASSLAND

Islamabad

Kabul

AFGHANISTAN

PAKISTAN

KARACHI

Arabian
Sea

TROPIC OF CANCER

Meshed

U. S. S. R.

PLATEAU OF IRAN

I R A N

TEHRAN
Mt. Demavend
18,934
5771m
ELBURZ MTS.

Esfahan

Shiraz

Muscat

O M A N

UNITED ARAB EMIRATES

OMAN

R U B   A L   K H A L I

Caspian
Sea

Resht

Tabriz

ZAGROS   MOUNTAINS

Ras Tanura
BAHRAIN
Manama
Dhahran
QATAR
Doha

Dubai
Persian   Gulf

Gulf

Dammam

Oasis of Hasa

Riyadh

PEOPLE'S DEMOCRATIC
REPUBLIC OF YEMEN

SOCOTRA

Gulf of Aden

Mt. Ararat
16,946
5165m

Erzurum

PONTIC
MOUNTAINS

Black
Sea

Kuwait
KUWAIT

Shatt al
Arab
Basra
Abadan

Neutral
Zones

S A U D I

A R A B I A N

P E N I N S U L A

A R A B I A

10,561
3219m

YEMEN ARAB
REPUBLIC
San'a
12,366
3769m

Aden
Mocha

Mosul

Tigris

MESOPOTAMIA

BAGHDAD
River
River

I R A Q

SYRIAN   DESERT

N E F U D

Euphrates

ANKARA

T U R K E Y

TAURUS   MOUNTAINS

Iskenderun

Nicosia

CYPRUS

Tripoli
LEBANON
Beirut

Latakia

Aleppo

SYRIA

Damascus

Amman

JORDAN

Yarmuk R.
Lake
Tiberias
Nazareth
Haifa
ISRAEL
Tel Aviv
Gaza
Jerusalem
Beer-sheba
NEGEV
DESERT

Medina

Mecca
Jidda

Port Sudan

Asmara

ERITREA

ETHIOPIA

Red

Sea

Istanbul
Sea of Marmara
Dardanelles Strait

Izmir

Mediterranean
Sea

Alexandria

Port Said

CAIRO

Suez

Suez Canal

ARAB REPUBLIC
OF EGYPT

Dead
Sea
Elath
Aqaba
Gulf of Aqaba
SINAI PEN.
Gulf of Suez

Nile   River

Aswan
Aswan Dam

TROPIC OF CANCER

S U D A N

Atbara

Khartoum
Omdurman

Blue   Nile   R.
Sennar Dam
White   Nile   R.

Important crops raised in Israel include wheat, oats, barley, and cotton. Sugar beets and a variety of green vegetables are also grown. Peanuts, bananas, and olives have become important export crops. The most important export crops are citrus fruits. Chicken and dairy cattle provide enough eggs and milk for local use. Food products, however, have to be imported to satisfy the needs of the Israeli citizens.

Manufacturing in Israel

Israel is the only country in the Middle East and North Africa where more people work in manufacturing than in agriculture. Food processing and textiles are the major manufacturing activities of Israel. Many factories import raw materials. They are then processed into finished products for export. Raw cotton, for example, is brought into Israel. It is then processed into fabrics and clothing. Rough diamonds are imported, cut, and then exported as expensive jewelry. Parts of autos, radios, precision instruments, and other household appliances are imported. These parts are assembled, and then sold both locally and abroad. Tires, leather goods, drugs, and chemicals are also manufactured. Building supplies are another important item made in Israel. Aircraft and military equipment are also produced.

As you have learned, Israel has a great need for increasing its supply of fresh water. It has been experimenting with turning salt water from the Mediterranean Sea into fresh water. This process is called **desalination** (dee sal ih *nay* shun). By using nuclear energy, desalination plants are able to produce both electricity and fresh water. However, the cost of producing fresh water in this way is very high. Scientists are working hard to create more practical ways for making sea water usable. Specialists from the United States have helped Israeli scientists design and test new equipment to use in desalination plants. Once this procedure is less costly, Israel may well be exporting fresh water to other desert countries.

Mineral Production in Israel

Israel has limited mineral resources. Small oil deposits have been discovered near the Dead Sea and in the Negev near the Mediterranean Sea.

Copper mines, which were used more than 2,000 years ago, are again being worked for their valuable ore. Desert sand is also readily available. The sand is used in manufacturing glass. Clay, limestone, and marble are available for building materials and supplies. Salt from the Dead Sea is processed in factories on its southwestern shores. Potash, phosphates, and magnesium are also obtained in this same area.

Israel's supply of fuel and power is limited. Nuclear energy may provide this nation with another power source as its industries expand.

Transportation in Israel

Good highways in Israel connect major cities and rural towns and villages. The railway system has also been improved. Since Israel has been partly surrounded by some unfriendly Arab nations, access to the sea and air has been very important to Israel. Israel maintains one of the most up-to-date shipping fleets in the world. Its ports along the Mediterranean and the Red Sea receive and send goods to major cities all over the world. Its national airline, El Al, has flights to many countries in Africa, Europe, and North America.

Education and Government in Israel

Education in Israel is free. All children are required to attend school until they are 14 years old. Many adult immigrants have come from a variety of countries, speaking many different languages. To help them, adult classes have been started. Adults may study Hebrew (*he broo*), the national language. Trades and business skills can also be learned. Many high schools, technical institutes, and colleges have been started. Excellent university programs are also available. Free nursery schools have been opened for children of working parents. Schools for Arab children provide instruction in Arabic for the Muslim population of Israel.

The government of Israel is a parliamentary republic. The parliament is called the **Knesset** (kuh *ness eht*). Members of the Knesset are elected to serve a four-year term. The president of the country is elected for a five-year term of office. The position of president is more honorary than powerful. The real head of the government is the prime minister. The prime minister is usually selected from the majority party elected to the Knesset.

Israel has a strong military force. Its armed forces must always be alert to any possible threat from neighboring Arab nations.

An Observation

Israel's progress in its first thirty years of existence has often been considered a miracle of development in the Middle East. Its modern cities and rural developments are among the best in the world. The Israeli people took the unproductive, undeveloped land and created a prosperous nation. Indeed, the people have been a valuable resource to the nation. In addition, much money was brought into Israel to help aid in its development. Between 1952 and 1965, the West German government paid Israel almost a billion dollars in goods. These payments were made because of what Germany, under Hitler, had done to the Jews before and during World War II. A great amount of money was also contributed by Jewish people and friends from the United States and other nations of the world.

1. What are some of the major agricultural and industrial products of Israel?
2. Why do you think the government of Israel maintains an up-to-date transportation system on the land, sea, and air?
3. See if your school or community library has any stories about life on an Israeli kibbutz. If so, read one story and make an oral report to the class about the story.
4. Why do you think the government of Israel maintains a separate educational system for its Arab population? Can you think of reasons why a separate educational system might be unwise?
5. On an outline map of Israel, locate the following places:
   a. Current national boundary lines
   b. Dead Sea
   c. Sea of Galilee
   d. Jordan River
   e. Haifa
   f. Gulf of Aqaba
   g. The Negev
   h. Jerusalem
   i. Tel Aviv

Jordan

The Kingdom of Jordan is about the same size as the state of Indiana. Almost three million people live in Jordan. This includes over 700,000 Arab refugees. These people were moved from their homelands because of the battles fought during the Arab-Israeli conflicts. The entire western border of Jordan touches Israel.

The Land and Climate of Jordan

Jordan claims an area of about 97 700 square kilometers (37,700 square miles). This includes land west of the Jordan River. However, this area was occupied by Israel in the Six-Day War of 1967. As the map shows on page 311, the Jordan River flows southward to the Dead Sea. In the Jordan River Valley, the climate tends to be subtropical. The highlands on either side of the Jordan Valley have a Mediterranean climate. In summers it is warm to hot. The winters are often cold enough for some light snow. Rain also falls in the winter months. From about November to April, 40 to 80 centimeters (15 to 30 inches) of rain falls each year.

East of the Dead Sea are low mountains. These mountains extend north and south. Continuing east and south from these mountains are vast desert lands. In fact, about three-fourths (¾) of Jordan is desert.

Agriculture in Jordan

Most of Jordan's farming lands are located near the Jordan River. Its waters are important for irrigation. However, disputes between Israel and Jordan have stopped the talks on rights to the water of this river. A canal has been built from the Yarmuk (*yahr* muhk) River to bring extra water to farmlands on the eastern side of the Jordan River Valley. The main crops grown are vegetables, citrus fruits, and bananas. Corn, sesame, millet, and tobacco are also raised.

Winter wheat and barley are the main crops grown on nonirrigated lands. Olives, grapes, nuts, and figs are also raised. Almost all of the food grown in Jordan is used by the local people. At times some food must be imported.

A great number of people living in eastern Jordan are nomadic herders. Flocks of sheep, goats, camels, and fine Arabian horses are raised by these people.

Minerals and Manufacturing in Jordan

The land of Jordan lacks valuable minerals. Some limestone is found. Phosphates, used in fertilizers, are Jordan's main export. Small deposits of magnesium, salt, and potash are known to exist in the Dead Sea area. A lack of money has prevented some of these minerals from being mined.

Food processing is the main type of manufacturing. There are mills for processing grains. Factories also produce olive oil, cigarettes, shoes, fruit, and vegetable products. Oil pipelines from the Persian Gulf cross

The land in the Jordan River Valley has been used as productive farmland for many centuries.

The manufacturing of fine rugs is an important industry in Jordan. Today, machines help do the work that in the past was done completely by hand.

northern Jordan. An oil refinery has been built to process the crude oil.

Tourism used to bring income to Jordan. Many tourists flocked to visit the holy city, Jerusalem. It was once controlled by both Jordan and Israel. It is a holy city to Christians, Jews, and Muslims. Since the 1967 war, however, all of Jerusalem has been under the control of Israel. Jordan, therefore, lost much of the money it received caring for tourists.

Transportation in Jordan

Jordan has a major transportation problem. It has only one ocean port. It is located in the southwest on the Gulf of Aqaba, on the Red Sea. This is a long way from the main cities which are in the north. A good highway system, however, has been developed from Aqaba to Amman (a *mahn*) in the north. Amman is the capital and largest city of Jordan. Most goods being imported or exported move through the port of Aqaba. Trucks use highways to travel to and from this port.

Government in Jordan

The land of Jordan is ruled by a king. In 1977 King Hussein (hoo *sayn*) celebrated his Silver Jubilee as ruler of this kingdom. Do you remember who else had been on a throne for 25 years in 1977?

Syria

The Land and Resources

Syria is about the size of the state of North Dakota. It has a population of almost eight million people. Much of the land area of Syria lies in a region known as the Fertile Crescent. The Fertile Crescent has been an important farming area for thousands of years. Northeastern Syria is also part of a region known as *Mesopotamia* (*mehs* uh puh *tay* mee uh). **Mesopotamia** means land between the rivers. The area between the Tigris (*ty* grihs) and Euphrates (yoo *fray* teez) rivers was the home of one of the world's oldest civilizations. See if you can locate this region on the map on page 311.

The Euphrates River flows from Turkey across Syria. Irrigated crops are raised in its valley. Syria has two other fairly good farming regions. One is the desert oasis which surrounds its largest city, Damascus. The other is the northeastern corner of Syria. This is sometimes known as the country's granary. Most of the land in the southeast is desert.

The Anti-Lebanon Mountains are a low range located in central and southwest Syria. Near the Mediterranean coast is another low mountain range. The Orontes (oh *rohn* tez) River Valley between these ranges and the narrow coastal plain contain two of the nation's best farming areas.

Other than its fertile soil and water, the nation has few natural resources. Some petroleum has been found. Production from these fields is not great. Gypsum and basalt are also mined in small quantities. Timber from the hills was once a valuable resource. Much of it has been destroyed. The government is now trying to reforest these areas.

Agriculture in Syria

Farming is the main way of making a living in Syria. Ancient methods are still used by many farmers, however. The government has been trying to teach new methods to the farmers. Water pumps are used by some farmers to lift water from streams and wells to irrigate their land. Tractors and combines are being used in some of the wheat and barley fields. In addition, the government has tried to provide more land to poor farmers. Land was taken away from large landowners. It was then divided into smaller pieces and given to landless people. Very small farms were also made larger.

Winter wheat is the main crop grown in Syria. It is raised mainly in the northeast. Farmers use dry-farming methods. If the year's rainfall is adequate, enough wheat is usually harvested for the nation's people.

Cotton is the major export crop. Its production has increased rapidly. Barley, millet, corn, and rice are also grown. Where irrigation is possible, vegetables, tobacco, sugar beets, and hemp are raised. Some olives, nuts, and fruits are grown in the oasis areas.

In the semiarid and desert lands, herds of sheep, goats, and camels provide a living for the nomadic herders. The government has been trying to improve the standard of living for these people. Some land has been set aside for the nomads. Schools have been provided for their children. Animal doctors have been trained. One hopes these doctors will help improve the quality of the herders' livestock.

Manufacturing in Syria

Syria is proud of its industrial growth. Among Arab nations it claims to be second to Egypt in manufacturing. Since 1970 most of the industries, trade, and farms have been taken over by the government.

Textiles are one of the biggest industries in Syria. Cotton fabrics and yarns are made from the nation's fine cotton crops. Factories also make olive oil, cottonseed oil, sugar, canned fruits, and vegetables. Soap, glass, cement, tobacco products, and leather goods are also manufactured. Many fine handicrafts are still made in Syria. Silk brocades, furniture inlaid with mother-of-pearl, and mosaics are some of the fine handicraft items.

Much of the commerical trade of Syria is located in its modern capital city, Damascus (dah *mas* kuhs). The city is situated on a desert oasis inland from the sea. Damascus has both an old world charm as well as the buildings and traffic of an up-to-date urban center. It is one of the most interesting cities in the Middle East. Aleppo (uh *lehp* oh), in the northwest, is about as large as Damascus. It is a major industrial and trade center also. Latakia (*lat* uh *kee* uh), on the coast, has a new harbor, one designed and built by engineers. It is an important shipping port.

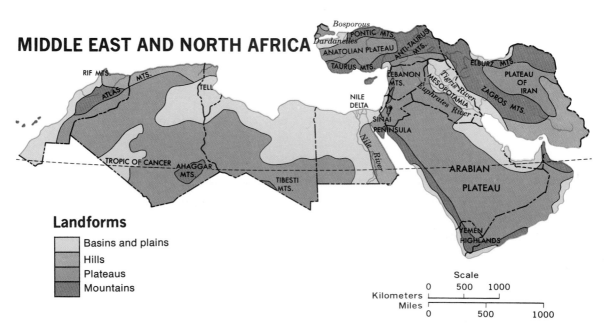

MIDDLE EAST AND NORTH AFRICA

Landforms

- Basins and plains
- Hills
- Plateaus
- Mountains

Scale

Kilometers 0 500 1000

Miles 0 500 1000

The manufacturing of television sets is one of the new industries that is being developed in Syria.

Transportation in Syria

Three pipelines carry petroleum products from Iraq across central Syria. A fourth pipeline from the Syrian oil fields leads to the Mediterranean. Another line from the Persian Gulf crosses the southwestern corner of the country. Syria gets a considerable amount of money for permitting oil in these pipelines to flow through its territory.

Syria's railroad system does not meet the nation's needs. Its railroads were built before present national boundaries were set. As a result, the railroad between Aleppo and Damascus was built through what is now the country of Lebanon. The Mediterranean port of Latakia lacks a connection to this rail line. However, the highway system is quite good. Trucks can move most goods to and from the major cities.

REMEMBER, THINK, DO

1. How does the climate of Jordan compare with your area's climate?
2. Why is the location of Jordan's main ocean port a problem for Jordan?
3. What has the government of Syria done to improve the working conditions for farmers and nomadic herders?
4. Why do you think the government of Syria took control of most industries and businesses within the country?
5. On an outline map of Jordan and Syria, locate the following places: Amman, Damascus, Latakia, Aleppo, Euphrates River.

Lebanon

Lebanon has been called "The Riviera of the Middle East." It was once one of the most economically developed countries in the region. Its capital city, Beirut (bay *root*), was like a European city in the heart of the Arab world. It served as a banking center for the Middle East. But a severe political crisis occurred in Lebanon in 1975. Political differences between the Christians and Muslims led to the outbreak of a civil war. Many people were killed. Homes and businesses were damaged or destroyed. Beirut was badly damaged in the fighting. By mid 1977, an Arab peace-keeping force was trying to maintain a truce between these two rival groups. The people of Lebanon then began the difficult job of rebuilding their nation. However, a year later fighting again broke out. Be sure to check to see what has happened in Lebanon since 1980.

A Brief History

A people known as the Phoenicians (fih *nihsh* unz) established communities and trading centers along the coastal lands of Lebanon thousands of years ago. The armies of Alexander the Great occupied this area about 320 B.C. The land became part of the Roman Empire in 64 B.C. During this period many people of the region became Christians. In A.D. 640 Muslims conquered the land. They introduced the Islamic religion and brought the land under Arab rule. For hundreds of years, Muslims and Christians lived together in Lebanon without having armed conflict.

The Land of Lebanon

Lebanon is located on the Mediterranean Sea. It is directly north of Israel. The small country is about 195 kilometers (120 miles) long and 55 kilometers (35 miles) wide. Its population is about half Muslim and half Christian.

Lebanon has four distinct geographic regions. Each extends from north to south in strips, starting from the coast. Along the Mediterranean Sea is a narrow coastal plain. This fertile area is very densely populated. Next to this are the Lebanon Mountains. The highest peak is about 3000 meters (10,000 feet) high. Among the mountains are great **canyons**—deep valleys with high, steep sides. Some drop as much as 300 meters (1,000 feet). This is a very beautiful part of Lebanon. The Bekaa (beh *kah*) Valley is just east of these mountains. In the valley are the Orontes and Litani (lih *tah* nih) rivers. The land in this valley is very fertile. Most of the grain crops of Lebanon are grown here. The fourth region is the Anti-Lebanon Mountains. They form the border with Syria. At 2814 meters (9,232 feet), Mount Hermon's snow-capped peak makes a beautiful boundary marker between the two countries.

The climate in the summer along the coast is hot and humid. The Mediterranean Sea breezes have

Their shops destroyed during the civil war, these merchants of Beirut, Lebanon, attempt to earn a living.

only a minor cooling effect. The higher elevations of the nearby hills and mountains have cooler temperatures. The winter season brings cooler temperatures and rain to the lowlands. Rain and some snow fall in the mountains in winter.

The Economy of Lebanon

Until 1975 Lebanon was the commercial and financial center for most of the Arab nations. The country had become an international banking center. Many oil-rich Arabs of the Middle East invested money in Lebanon. Many international business firms had main offices in Beirut, the capital of Lebanon. Beirut also served as an important transporta-

tion center for the Middle East. Its port served the neighboring countries of Syria and Jordan. Roads through the mountains connect Beirut to Damascus, Syria, and to Amman, Jordan. Trucks, buses, and autos moved people and goods to and from these three nations. Two oil pipelines from Middle East nations end in Lebanon. Oil refineries have been built at the port cities of Tripoli (*trihp* oh lih) and Sidon (*sy* d'n). Oil was shipped to European markets. Locate these places on the map on page 311.

The city of Beirut was also a major educational and cultural center for the Middle East. More books, magazines, and newspapers were printed and read in this country than in any other Arab nation of the Middle East. Until the civil war Beirut had many fine elementary and secondary schools. Several famous universities were also located in this city. One of the universities was started by Americans. Another was built by the French. The Lebanese had also established a university. Many students from Middle East countries came to study in these schools.

Until the civil war Lebanon had also been a favorite tourist center in the Middle East. Hot summer days can be very uncomfortable in many Middle East nations. Beirut's ocean resorts attracted many Middle East tourists. Hotels and lodges had also been built in the Lebanon Mountains

Snow covers the terraced hillsides of the Anti-Lebanon Mountains. Over the centuries the farmers of Lebanon have carved farmland from the mountains. It is a process that is still going on.

near the coastal region. The air is almost always cool in these high areas. There were many ski resorts open during the winter in the snow-covered mountains.

The Cedars of Lebanon. For many centuries Lebanon was known for its forests of cedar. Loggers cut most of the trees in order to sell them. Now most of the forests are gone. Many of the mountain slopes are bare. For many years thousands of goats have grazed upon these slopes. They have not only eaten the grasses, but also the bark of young trees. The government has started a reforestation program to try to return the cedars to the mountain slopes in Lebanon.

Agriculture in Lebanon

About one-third (⅓) of the land of Lebanon is cultivated. More than half the people make their living by farming. Most farming is done along the coastal plain and the inland Bekaa Valley. Fruits and vegetables are grown on the coastal plain. Citrus fruits and bananas are plentiful. On the hillsides much of the land has been terraced. This provides more land suitable for farming. In the fertile inland valleys farmers grow olives, grain crops, apricots, peaches, and grapes. Apples are grown on the eastern slopes of the Lebanon Mountains where irrigation is possible.

In 1953 the Litani National Office was created. This office has been responsible for developing irrigation and hydroelectric power from the

Litani River. So much water is used in summer that the Litani River is dry before it reaches the Mediterranean Sea.

Some Problems of the Future

The government of Lebanon had been concerned about the population in Beirut. This city was becoming overcrowded. People were flocking to the city in hopes of finding employment. Many thought that city life was more enjoyable than rural living. The government had also been concerned over the nation's heavy dependence upon international commerce and trade for its main revenues. Plans had been made to build new agricultural and industrial centers inland, away from Beirut. It was hoped this would have encouraged people to move from the overcrowded coastal region. It would also have provided new employment to many people in the inland regions of the country.

The civil war in 1975, however, changed all plans. It forced many international businesses to leave Lebanon. During the civil war many of these companies moved their international headquarters to other countries. As Beirut is rebuilt, one wonders how many companies will return. Will this nation regain the trade that was so profitable to Lebanon? Some of the answers to these problems may depend upon the ability of the people in Lebanon to achieve and maintain a lasting peace in their country.

REMEMBER, THINK, DO

1. Why was Lebanon once called "The Riviera of the Middle East"?
2. Prepare a current events update on Lebanon. What caused the crisis in Lebanon? Has the civil war really ended, and has the war-torn nation begun to rebuild Beirut?
3. On an outline map of the nations of the Eastern Mediterranean do the following:
   a. Locate Israel's three major ocean ports.
   b. Identify territory gained by Israel from Jordan, Syria, and Lebanon as a result of the Arab-Israeli conflicts.
   c. Locate and name each nation's capital.
4. How do you think the waters of the Jordan River should be controlled for use by Israel, Jordan, Syria, and Lebanon?
5. Why do you think some Arab oil-producing countries ship some of their oil by pipelines to ports on the Mediterranean Sea?

Middle East

Iran

Iran is the name for the country some people still call Persia. Persia was the nation's name until 1935. The Persian Empire had given this region its fame in history and legend. At one time the Persians ruled practically all of southwestern Asia from India to the Aegean Sea.

Now Iran is about as large as the states of Texas, Oklahoma, Arkansas, New Mexico, and Arizona combined. As the map shows on page 311, Iran is twice the size of Turkey. It has fewer people, however. Iran has only about 22 people per square kilometer (57 per square mile). The population is unevenly distributed throughout the land.

A devout Muslim faces the city of Mecca as he bows and prays to Allah. This man is saying his noontime prayers at a mosque in Isfahan, Iran.

No-Ruz: A Traditional New Year's Holiday in Iran

March 21st is usually the day to celebrate the New Year in the Islamic country of Iran. According to Islamic tradition, a solar calendar is used for identifying the exact times and dates of major holidays and ceremonies. No-Ruz (noh *rooz*) is the first day of the traditional New Year in the calendar of Iran. This day also marks the end of the cold winter and start of the warmer springtime season. No-Ruz is a very popular celebration in Iran.

Preparations for the New Year often begin a month earlier. Homes are usually given a thorough spring cleaning. New clothing is prepared for the children of most families. Special holiday sweets and cakes are made. Two weeks before No-Ruz, wheat seeds are placed in clay saucers of water. By New Year's Day the seeds usually have sprouted into small green seedlings. These are later used in ceremonies which end the holiday season.

No-Ruz, which means "New Year's Day," is a very happy holiday. Adults usually recite prayers of blessing from the holy book of Islam, the *Koran* (kuh *ran*). Presents are sometimes given to children. During then next twelve days families visit the homes of their relatives and friends. Holiday greetings and gifts are often exchanged. The special holiday cakes and sweets are shared among the many visitors. On the thirteenth and final day of the New Year's celebrations, the wheat seedlings are placed into a running stream to float away. According to ancient beliefs, this act means that families wish to share their happiness and good fortune with others. Families thank *Allah* (*ahl* uh), or God, for their past year's survival and look forward to a warm and prosperous spring and happy new year.

How does this holiday differ from the way your family celebrates the New Year? In what ways are the customs similar to your holiday activities?

The People of Iran

Most of the Iranian people are farmers who live in small villages and cities near their farms. About one-twelfth (1/12) of the nation's people live in Teheran (teh eh *rahn*). It is the capital of Iran. Few people live in the country's vast desert areas.

About one-fifth (1/5) of the people are nomadic herders. They spend their summer months leading their flocks of sheep and goats to highland pastures. Winters are spent in the lowlands where they do some gardening. Herders also raise horses, donkeys, and camels. The camels are usually used to carry the nomads' tents and possessions. The rugs used for the floors of their tents are often fine handmade carpets. Persian rugs, as they are still called, are highly valued all over the world. The diet of the nomads includes cheese, yogurt, dried fruits, and some rice and meat. Many nomadic people are very proud of their independence. They enjoy the freedom to travel when they want and where they want.

Until a revolution in 1979, women had more opportunities and freedom in Iran than in most other Islamic nations. The **purdah** system was not required. (Under that system women have to wear veils over their faces when in public.) After the revolution in 1979 some steps were taken to restore old Muslim laws. It is difficult to tell, as this book is written, what will happen. In 1978, women had the right to vote. They were also encouraged to take an active part in civic affairs. Whether that will continue is not clear. You will need to check in the library to learn what has happened in Iran since 1979.

Middle East

This couple is tilling the soil on the Iranian Plateau.

The Land of Iran

Most of Iran is a plateau that is almost surrounded by mountains. This plateau is from 300 to 1200 meters (1,400 to 4,000 feet) above sea level. Between the Persian Gulf and the plateau are the Zagros (*zag* ruhs) Mountains. South of the Caspian Sea are the Elburz (ehl *buhrz*) Mountains. There are also mountains along the eastern border.

As the map on page 311 shows, much of eastern Iran is desert. The edges of the plateau are higher than the land between. As a result, many of the rivers flow inland, draining into salt flats. During the dry summers, most of these salt flats have a solid crust of salt on the surface. They are sometimes solid enough for a person to walk upon. In the rainy seasons, however, a person might sink into the swamp underneath. A few forested areas are found on the Caspian Sea side of the Elburz Mountains. Some are also located in the Zagros Mountains. Good grasslands for grazing are found in the mountain areas during the summer.

The Climate of Iran

The climate in Iran varies from region to region. The lowlands along the Caspian Sea and the Persian Gulf tend to be hot and humid in the summer. On the plateau, the air is dry and generally cooler than in the lowlands. In the dry basin areas, strong winds often sweep over the land in summer. Summer temperatures in Teheran may reach from 38° to 43° C (100° to 110° F). Winter temperatures reach lows of −4° to −9° C (25° to 15° F). Cold rain and light snow also occur in this season.

Rainfall is generally light except in the mountains and along the Caspian seacoast. Much of northern and western Iran has about 30 centimeters (12 inches) of rain a year. Only about 10 to 15 centimeters (4 to 6 inches) fall annually in the central and southeastern regions, however. In the Elburz Mountains about 100 centimeters (40 inches) fall. Much of this moisture comes as winter snow. Rainfall is not as great in the Zagros Mountains.

The Economy of Iran

Agriculture. Most farmers still use hand methods of farming. Only a little more than one-tenth (1/10) of the total land area is cultivated. Much of the land is used for grazing animals.

Most of the farmers use either dry farming methods or irrigation. In **dry farming**, moisture is conserved, or "stored" in the soil over a period of time. Soil on idle land is broken up so that it will soak up rain during the year. After a year, this land is planted in a crop, usually wheat or barley. Crop yields, however, are low. Water for irrigation comes from mountain streams through canals and underground tunnels. When proper farming methods and irrigation are used, farmers have been able to double their yearly harvest. In spite of the hardships faced by many farmers, farming is still the chief occupation of most people.

Wheat is the nation's main crop. Much barley is also grown. Rice is cultivated on the lowlands along the Caspian Sea. Tobacco, citrus fruits, and tea are also grown in this area. Raw silk is produced, too. Mulberry trees are grown for their leaves which are fed to silkworms. The worms produce cocoons that are unwound to produce silk threads. A great variety of fruits, such as those grown in California, are produced in Iran. Melons, peaches, apricots, and cherries are among some of the favorite fruit crops grown. Dates are also an important export crop. Most dates are raised in villages near the

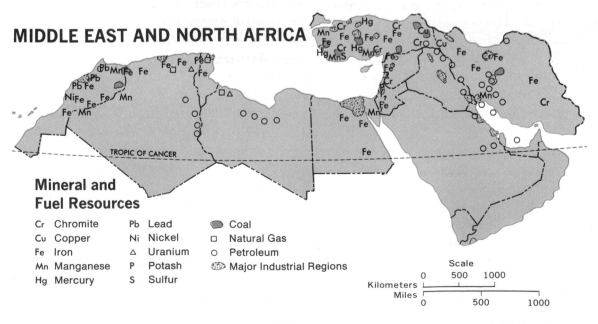

MIDDLE EAST AND NORTH AFRICA

Mineral and Fuel Resources

Cr	Chromite	Pb	Lead
Cu	Copper	Ni	Nickel
Fe	Iron	△	Uranium
Mn	Manganese	P	Potash
Hg	Mercury	S	Sulfur

- ⬤ Coal
- □ Natural Gas
- ○ Petroleum
- ⬤ Major Industrial Regions

TROPIC OF CANCER

Scale
Kilometers 0   500   1000
Miles 0   500   1000

Persian Gulf. Potatoes and sugar beets add to the nation's list of farm produce. Farmers usually raise enough food to feed all the people in Iran. Tea and sugar are the only main food items imported. Cotton and wheat are often exported as surplus cash crops. A large part of Iran's cotton crop is used in the local textile mills.

Until 1964 most good farmland in Iran was owned by a few wealthy landowners. *Tenant farmers* worked these lands. **Tenant farmers** usually pay a landowner cash or part of the crops grown for use of the land. In Iran, they had to give the owner four-fifths (⅘) of the crops grown. The *Land Reform Act of 1964* changed this system. It limited the amount of land that any one person could own in one village. Excess land was taken away from the large landowners. It was divided into smaller plots and given to the tenant farmers. Over one and a half million tenant farmers have received land since this Land Reform Act became law.

Mineral Wealth. Iran has a variety of mineral and fuel resources. It is one of the world's leading oil-producing nations. A major part of Iran's income comes from the sale of its oil. It also has vast quantities of natural gas. Huge deposits of copper were discovered in 1967. Zinc, chrome, lead, coal, sulfur, and salt are found in Iran. Turquoise, a very popular semiprecious stone used in jewelry, is mined here. Building materials are also made from Iran's deposits of marble, alabaster, limestone, and gypsum.

Manufacturing. Many industries are fairly new to Iran. Using income from its oil sales, Iran has taken steps to become an industrial nation. Cotton, wool, and silk fabrics are now manufactured in textile factories. Tires, glass, paper, and cement are also made locally. Imported raw tea and sugar are processed and refined in Iranian factories. Some chemical and steel products are also manufactured. Auto assembly plants have been built in Iran. Radios, refrigerators, and other electrical appliances are sometimes assembled in local factories as well.

As you know, Persian carpets are another famous product of Iran. The rugs are made from dyed wools. The handmade carpets are highly prized throughout the world as beautiful home decorations. They are also regarded as financial investments and as art objects.

Transportation. Transportation systems in Iran have been improved greatly during the last half of the 20th century. Thousands of kilometers (or miles) of roadways have been built. Many more are under construction or are being improved. A major highway is planned to link Iran with both Turkey and Pakistan (*pak* ih stan). Camels, horses, and donkeys are still used for transportation by some people. A railway, with several branch lines, extends from the Persian Gulf to the Caspian Sea. Airplanes fly to most major cities throughout the country: The international air terminal at Teheran links this nation with the rest of the world.

The Ayatollah Khomeini prays with a group of his followers. An important religious leader in Iran, the Ayatollah led a revolution in 1979 that overthrew the Shah.

International Importance of Iran

At times other nations have been interested in the land and oil resources of Iran. In the past the Soviet Union made several attempts to control the government of Iran. But the people of Iran have been independent for thousands of years. They strongly resisted the Soviet threat to take over their government. The Soviet Union has since improved its relations with Iran. It has contributed labor and equipment for the building of Iran's first steel mill, a machine tool plant, and a farm machinery factory. In turn, the Iranian government pays for some of this aid by supplying natural gas to the Soviet Union. The United States has also assisted the Iranian government with development programs.

The Government of Iran

Until 1979, the government of Iran was ruled by a powerful monarch, named Mohammed Reza Pahlavi. Although there was a parliament, as well as a king, the major power for ruling the nation was in the hands of the king. In Iran, the king was called "The Shah." During his rule, the Shah kept a tight control over the press. Newspapers were forbidden to print things against him. People who openly criticized his actions were often treated harshly.

During his years in power, he started many programs to "modernize" parts of the country. Because many poor people had no land on which to farm or live, the Shah brought about changes in the system of land ownership. This was done through the Land Reform Act of 1964. Under this law, some of the land was taken away from wealthy landowners, divided into smaller parcels, and given to poor, landless people. The Shah used millions of dollars of oil revenues to improve agricultural methods and to develop or improve industries. In addition, he started a Literacy Corps. Members of this Corps were school graduates who gave their services to the government. They went to remote rural areas and taught basic reading, writing, and counting skills to children and adults. By 1970 over half the nation's children were attending some type of school.

But these social reforms were not always wanted by some of the people. Many Islamic leaders in the country felt that actions taken to "modernize" the nation were against their religious beliefs. Many people felt that education should be left to religious Islamic leaders. Discontent began to grow among many classes of people in Iran. There were more and more demands that the people have a stronger voice in the government. By 1978 people in large numbers were demonstrating and marching to protest rule by the Shah of Iran. Finally, in 1979, the Shah was forced to leave the country. People the world over are wondering what will happen in Iran as the new government and rulers establish new rules. It appears that Iran will become an Islamic republic. That means that religious leaders will be in charge.

REMEMBER, THINK, DO

1. Why do you suppose some people still call Iran by its old name, Persia?
2. Use an outline map of Iran to locate the following places:
   a. Caspian Sea
   b. Persian Gulf
   c. Zagros Mountains
   d. Elburz Mountains
   e. Teheran
3. Why do you think the Iranian government has tried to be on friendly terms with both the Soviet Union and the United States?
4. What mineral resource is a major source of income for Iran?
5. Invite a local carpet dealer to your class to show you real Persian carpets. Your library might have books showing a variety of designs.
6. Do a mini-report on Iranian turquoise and turquoise jewelry. Try to find out how it compares with turquoise jewelry made in the United States.

LEARN MORE ABOUT NORTHEASTERN LANDS OF THE MIDDLE EAST

Vast differences in government and ways of living exist in the countries of Cyprus, Turkey, and Iraq. You may wish to add to your knowledge of this area of the world by learning more about these countries. The Study Guide provided on page 31 will help you plan your independent study of these countries.

Cyprus

Cyprus is the third largest island in the Mediterranean Sea. In the days of the Roman Empire, copper mined on this island provided metal for shields and weapons used by Roman soldiers. These mines are still worked for their valuable ore. For hundreds of years, people from both Greece and Turkey have settled on this island. In the mid-1950s disputes between the Greeks and Turks caused fighting on Cyprus. More fighting occurred in the 1960s and 1970s. United Nations peace forces have tried to keep peace on the island. Maybe you will have some idea about ways to get lasting peace for the citizens of Cyprus.

Turkey

Turkey is a nation divided between two continents. Thrace (thrays) is the part of Turkey that is in the continent of Europe. The much larger part of Turkey is in Asia. It is sometimes called Asia Minor.

The history of Turkey is quite interesting. Some scholars believe the ruins of Noah's Ark may be located on Mount Ararat (*air* uh rat) in eastern Turkey. This nation was once the site for the eastern capital of the Roman Empire. Constantinople (*kahn stan* tih *noh* puhl) was the center of government during the

Customers examine the daily catch displayed along the European shore of the Bosporus. Across the water is Asia.

These women are attending school in Baghdad, Iraq. In the past it was not possible for most Arab women to get an education. Today this situation is changing in many Arab nations.

Byzantine Empire as well as the later Ottoman Empire. Now called Istanbul (*ihs* tuhm *bool*), the city is one of the most interesting places to visit in the Middle East. Since 1923 the people of Turkey have been updating their way of life. Led by the progressive leader, Mustafa-Kemal (moos tah *fah* kuh *mah'l*), many changes took place in Turkish life. You may find it interesting to learn about the changes made in this Muslim nation.

Iraq

Iraq prides itself on its ancient historical background. This country contains a large part of the Fertile Crescent. The land between the Tigris and Euphrates rivers is rich farmland. Some legends state that this region was the site of the Garden of Eden. The civilizations of Mesopotamia existed in this area. Its developments equaled those of ancient Egypt.

With the discovery of oil in the mid-20th century, the land of Iraq began a new era in its development. Military leaders overthrew the king in 1958. They have since governed the country. The military leaders are trying to use the profits from the oil resources to give the citizens of Iraq a better way of life. You may wish to discover how this money is being spent to benefit the people of Iraq.

WHAT HAVE YOU LEARNED?

1. How has the population of Israel changed from 1950 to 1980?

2. What is desalination?

3. What difficulty does Syria have with its railway system?

4. What are the four major landform regions of Lebanon?

5. How did the Land Reform Act help the poor farmers of Iran?

BE A GEOGRAPHER

1. Which nation has land farther north:
   a. Lebanon or Syria?
   b. Lebanon or Israel?
   c. Lebanon or Jordan?
   d. Israel or Jordan?

2. When the central plateau in Turkey is covered with snow and mountain passes are blocked, what season is it in:
   a. Minneapolis, Minnesota?
   b. Edmonton, Alberta, Canada?
   c. Sydney, Australia?
   d. Athens, Greece?
   e. Cape Town, South Africa?

3. Make an agricultural product map of Iran.

4. Is it farther across Iraq from its northernmost point to its southernmost point, or from its easternmost point to its westernmost point?

QUESTIONS TO THINK ABOUT

1. How do you think the holy lands should be governed so that the followers of Christianity, Judaism, and Islam can worship peacefully?

2. What do you think could be done to encourage businesses back to the country of Lebanon?

3. Why do you think some people in the Middle East dislike attempts to modernize their way of life?

4. Why do you think you might like, or dislike, the purdah system?

5. Why do you think the United States government has tried to maintain friendly relations with both Israel and the Arab nations?

Chapter Review

·❧[18]❧·

Chapter

——THE ARABIAN PENINSULA——

The Arabian Peninsula is one of the world's unusual land areas. It is almost all desert. For centuries many people thought that this region was a poor wasteland. However, the people and natural resources of the peninsula have greatly influenced much of the world.

During the mid-20th century, vast deposits of oil were discovered in sections of the Arabian Peninsula. The oil deposits here are believed to be the world's richest. Many nations have been eager to buy this oil. Yet oil is not the peninsula's most famous contribution to the world. In A.D. 622 the Islamic religion was started in Medina (muh *dee* nuh), Saudi Arabia. By 1977 there were more than 500,000,000 followers of the Islamic religion in the world. As you have learned, the religion owes its beginning to an Arab named Mohammed. His followers helped spread the Islamic faith to the rest of the world.

Mohammed—the Prophet of Islam

Mohammed was born in Saudi Arabia near Mecca (*mehk* uh) about A.D. 570. His parents died when he was very young. His grandfather and uncle helped raise him. In his early adult years Mohammed worked as a shepherd and herder. Later he married a well-to-do widow and settled down with her to raise a family.

Mohammed often wondered about the way people lived in his homeland. He did not think that slaves were treated fairly. He believed that women were not treated properly. He was unhappy that people worshipped many gods. It is said that Mohammed would often seek a quiet cave in the nearby hills. Once alone, he would think about the ways people should treat one another. According to Islamic belief, an angel appeared to Mohammed in one of these caves. The angel was supposedly a messenger from God. The angel visited Mohammed many times.

334

وقال يا محمد ربك يقرءك السلام ويخصك بالتحية والكرامة
حق تعالى نكا سلام فلدى ايندى عايشته جبرائيل

سكا كوندددومكه سنوك امروكه مطبع اولاسنوك
دوشمنلروكى هلاك ايليه بن نكيه كوكلوك دنلو

Muslim law prohibits showing the likeness of Mohammed. Thus, when Muslim artists picture Mohammed his face is never seen. In this picture the veiled Mohammed is shown leading two followers through the desert.

During the visits the angel repeated messages to Mohammed from Allah (or God). The messages were instructions to people about how they should worship God. They also instructed people how they should treat one another. Mohammed began to teach these messages to his family and friends. Soon the number of his followers increased. The messages were finally written into a book. These messages became the **Koran**, the Islamic holy book.

Not everyone was interested in the teachings of Mohammed. Many of his townspeople did not like Mohammed's preachings. Some thought his ideas would destroy the religion already in practice. Mohammed was forced to leave Mecca. He went to the town of Medina. Many Islamic believers went with him to Medina. The number of converts to Islam increased rapidly. Later Mohammed returned to Mecca with thousands of faithful supporters. The Islamic religion had begun! It soon spread throughout the world. Mohammed died in Mecca in A.D. 632. He is remembered daily by Islamic followers everywhere when they recite their prayer: "There is no God but Allah, and Mohammed is his Prophet."

Saudi Arabia makes up most of the land of the Arabian Peninsula. Seven smaller nations occupy the rest of the peninsula. Kuwait, Bahrain, Qatar, and the United Arab Emirates are near the Persian Gulf. Oman and the People's Democratic Republic of Yemen (Democratic Yemen) are located at the southern end of the peninsula. The Yemen Arab Republic, near the Red Sea, completes the nations of the Arabian Peninsula. Find them on the map on page 311.

The Arabian Peninsula is about as large as the United States east of the Mississippi River. The population of the peninsula is estimated to be about 21 million people. Like the land of Turkey, this peninsula bridges the lands of two continents. People have moved between Asia and Africa across the Arabian Peninsula for hundreds of years.

The Land and the Climate

Much of the Arabian Peninsula is a high plateau. Along the Red Sea mountains rise quite steeply. Many peaks in the northwest are more than 1500 meters (about 5,000 feet) high. In the southwest the mountains are higher. Some peaks are between 3000 and 3700 meters (from about 10,000 to 12,000 feet) high. Mountains are also found along the southern and southeastern coasts. The surface of the main plateau area is hilly. The average elevation is about 1500 meters (about 5,000 feet) above sea level. The plateau decreases in elevation near the Persian Gulf.

Except in the mountain areas of the peninsula, there is very little rainfall. Nowhere in the Arabian Peninsula is there a river that flows all year long. Instead, there are many dry streams and river beds

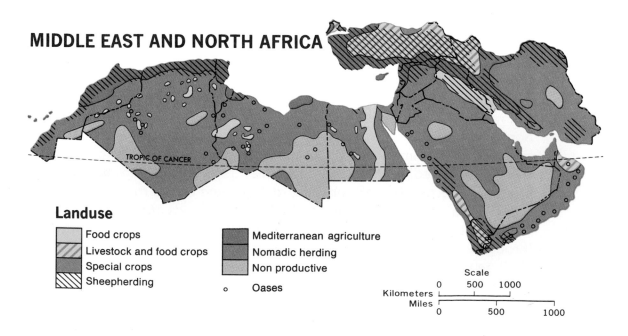

MIDDLE EAST AND NORTH AFRICA

TROPIC OF CANCER

Landuse

Food crops	Mediterranean agriculture
Livestock and food crops	Nomadic herding
Special crops	Non productive
Sheepherding	
	o Oases

Scale

0   500   1000
Kilometers

Miles

0   500   1000

called **wadis** (*wahd* ees). After rainstorms the wadis fill up with water and flow as streams. Most of the time, however, they are dry.

On the steep slopes leading down to the Red Sea, the wadis are deep, with steep sides. Little use has been made of them for irrigation. However, dams have been constructed along the wadis to aid irrigation. In the central Arabian Peninsula the wadis are shallow. Some have springs very near the surface. Wells have been built to bring this water to the surface. Many oases are also found along the wadis.

As the map on page 311 shows, most of the Arabian Peninsula is desert land. A large part of this desert is named the Rub 'al Khali (roob al *kahl* ee). This Arabic name means "empty quarter." As its name suggests, very little grows in this area. Few animals and people live there. The surface of the Rub 'al Khali is covered with gravel and sand dunes.

Much of the high central plateau is gently rolling dry steppe land. Some scattered vegetation and water holes may be found. This land is the home of many nomads. These people depend mostly on camels for their food and clothing. Their camels can survive on little water and the coarse vegetation of the desert.

Summer temperatures on the peninsula are very hot except in the high mountains. Near the coastal regions the air is very humid. Inland it is dry. At midday during the summer, the temperatures often reach 50° C (about 120° F) in the shade. Nights are cooler in the interior than along the coast. The southern coast is usually hot both winter and summer, day and night.

Farming and Grazing

Irrigation is necessary for farming in most areas of the peninsula. Only in the mountain areas is it possible to grow trees, vegetables, and grains without irrigation. Dates are also grown. They are the main food for people who live near the oases.

Animals provide milk, meat, wool, and transportation for the desert people. Most people have camels, goats, and sheep in their herds. Arabian horses are also raised. They are famous for their speed and endurance.

Governments and Oil

Most of the nations of the Arabian Peninsula have recently become independent. However, tribal *sheiks* (sheekz) or *sultans* have ruled these lands for centuries. **Sheiks** are Arab chiefs. **Sultans** are Muslim princes. Leadership was passed down from father to son. During the first half of the 20th century, many European nations took control of lands in the peninsula. After World War II, the Arab rulers regained their leadership. Boundaries between nations were sometimes changed during the colonial period. At times some Arab leaders fought with one another over sections of land. In certain cases nations were willing to share disputed territories.

Some of the new wealth of the Arabian Peninsula has been used for modernization. Cities, such as Riyadh, Saudi Arabia, have begun to look like cities in the United States.

The oil discoveries in the Arabian Peninsula have made this region one of the richest peninsulas in the world. It is thought that about half the world's known oil reserves are located in and around the peninsula. The oil has brought sudden wealth to some nations of the Arabian Peninsula. However, not all nations in this region have rich mineral deposits. It is possible to find some of the world's richest nations as well as some of the poorest nations on the Arabian Peninsula.

Oil profits increased greatly during the 1970s. At that time OPEC recommended that countries nationalize their oil companies. Most oil-rich nations demanded at least 60 percent ownership of the companies. How these nations use their new wealth and power is of great interest to people the world over.

REMEMBER, THINK, DO

1. How does the land of the Arabian Peninsula differ from that of the northeastern lands of the Middle East?
2. Prepare a chart noting the differences between a sheikdom, a kingdom, and a sultanate.
3. What are wadis, and why might they be dangerous at times?
4. Use an outline map of the Arabian Peninsula to do the following:
   a. Label the nations of the Arabian Peninsula.
   b. Put a red "X" on each nation of the Arabian Peninsula crossed by the Tropic of Cancer.
   c. Put a blue "X" on those countries of the Arabian Peninsula which are located along 40° E.
5. Visit a mosque, compare it with other religious buildings in your community. Possibly a Muslim member of your community could visit your class and explain some of the major characteristics of the Islamic religion.

Saudi Arabia

The largest country of the Middle East is Saudi Arabia. It is well known for two reasons. First, it is the holy land for Muslims. As you have read, the founder of Islam, Mohammed, was born and raised in Mecca. He successfully introduced the Islamic religion first to the people of Medina. Find Mecca and Medina on the map on page 311. They are situated near the Red Sea between 20° N. and 25° N.

Mecca is a very old and sacred city for Muslims. Every Muslim is supposed to make a **pilgrimage** (or religious trip) there during his or her lifetime. Thousands of Islamic pilgrims from all over the world visit Mecca each year. Many pilgrims also visit Medina. Some people still travel to these cities by camel caravan. People on the peninsula, however, mostly travel there by auto or bus. Many foreign visitors arrive by sea at the port of Jidda (*jihd* uh) on the Red Sea. They then travel overland to Mecca by car or bus. Others come by air. The holy shrines of Islam in Saudi Arabia are world famous. Only Muslims are allowed to enter both Mecca and Medina.

Oil is the second reason why Saudi Arabia is so well-known. Its oil reserves are the largest known in the world.

Most of the oil production takes place along the Persian Gulf near Dhahran (dah *rahn*). Some of this petroleum is refined in the area and then shipped by tankers through the Persian Gulf. There is also a pipeline which carries some of the oil to the Mediterranean port of Sidon, in Lebanon. This pipeline is 1600 kilometers (about 1,000 miles) long. The oil can reach European markets quicker through the pipeline. A trip by tanker from the Persian Gulf through the Suez Canal takes nine more days than it does through the pipeline. Plus, there are canal tolls that must be paid.

In 1974 the Saudi Arabian government took over all American petroleum companies in the country. Revenues from the oil royalties amount to billions of dollars each year. Some of this money has been used to improve transportation in the country. A railroad has been built from the Persian Gulf port of Dammam (duh *mam*) to the inland capital, Riyadh (rih *yahd*). Rail lines are also being extended to Jidda and Medina. Money has also been used to improve roadways to other cities.

Many new schools have been built. All citizens now receive free medical services. Large investments have been made by the government to improve housing facilities. Additional electric power and water supplies have been introduced to many cities. Wells are also being dug in remote regions. This water is useful to the nomadic herders.

Some students are being sent to foreign colleges and universities for training in a variety of professions. It is expected that these students will return one day to Saudi Arabia and aid in the nation's development. Some oil money is used to support the military forces of Saudi Arabia.

The Kaaba in Mecca, Saudi Arabia, is the most holy of all Muslim shrines. It is a large square building that houses the Black Stone. Each year thousands of Muslims make a pilgrimage to the site.

The rest of the revenue goes into the treasury of the royal family.

Gold, iron ore, and silver provide some additional mineral wealth to this land. The government has started a Five-Year Economic Plan. Under this plan it hopes to develop a number of additional industries.

The Agriculture of Saudi Arabia

There is some farming in western Saudi Arabia. The oases are also productive. The Oasis of Hasa (*hahs uh*) is the largest oasis in Saudi Arabia. It supports more than 200,000 people. More than sixty natural springs can be found in this oasis. The Oasis of Hasa is famous for its huge crops of dates. Dates are the main cash crop of Saudi Arabia. Dates are also the staple food for most local people. Coffee is grown in the highlands. Wheat, alfalfa, barley, cotton, grapes, and melons are also raised.

Swarms of locusts sometimes cause great damage to crops in the Middle East. These insects often breed in the hot desert lands of the Arabian Peninsula. At times locusts have been known to travel about

2400 kilometers (1,500 miles) in one flight. The swarms of locusts can ruin grain crops. Once a swarm of locusts has been sighted, scientists can sometimes stop their movements. Airplanes are often used to spray chemicals on the millions of insects. If this is done in time, the locusts are destroyed. If not, the locusts can destroy the crops!

The Government of Saudi Arabia

Saudi Arabia is ruled by a king. The king appoints a prime minister to help rule the nation. The prime minister has a Council of Ministers for assistance. The ministers actually run the government. They recommend rules and laws to the king. Once approved, they see that the laws are enforced. Islam is the official religion of Saudi Arabia. The holy book of Islam—the Koran—contains the main laws for the people in Saudi Arabia. This is true in most other Islamic nations as well. Many changes have occurred in a very short period of time for many people in Saudi Arabia. Check the news reports for current developments and progress in this nation.

REMEMBER, THINK, DO

1. Why does Saudi Arabia ship some of its petroleum by pipeline to Lebanon?
2. Write to major petroleum companies in the United States. Find out how oil is processed for use in automobiles. Sometimes companies will loan documentary films which explain the entire oil refining process. Possibly your classroom teacher can obtain such a film.
3. Find out what the oil-producing nations of the Arabian Peninsula are doing to stop pollution in the Persian Gulf waters. What are they doing to decrease air pollution from their many refineries?
4. What types of businesses and industries might be developed in neighboring, non-oil-producing nations in the Arabian Peninsula?
5. Use a globe or a map of the world to answer the following questions:
   a. Is Saudi Arabia farther north or farther south than the United States? Argentina? India?
   b. What nations of the Middle East are found along 40° E.?
   c. Which is farther east:
      (1) Moscow or Riyadh?
      (2) Teheran or Riyadh?
      (3) the Ural Mountains or the eastern part of Saudi Arabia?

The Arabian Peninsula

LEARN MORE ABOUT THE ARABIAN PENINSULA

Not every nation in the Arabian Peninsula has rich mineral resources. Not all of the people in the wealthy countries receive benefits from oil royalties. Some people still live as their ancestors did hundreds of years ago. There are many great differences in ways of living and types of governments in the countries of the Arabian Peninsula. Some of these countries are unknown to many people in the United States. It would be good for you and your class to learn more about these rapidly changing nations in the Middle East. Use the Study Guide on page 31 to help you plan your study of these countries. Be sure to include topics according to your needs and interests.

Kuwait

Kuwait is located in the northeastern corner of the Arabian Peninsula. It is smaller than the state of Massachusetts. In 1977 Kuwait ranked tenth in the world in the production of crude oil. As the map shows on page 311, some of the land between Kuwait and Saudi Arabia is neutral territory. Both countries have agreed to share this land. Its oil royalties are shared.

Kuwait's oil royalties are being

This desalination plant was built with the money Kuwait earned from the sale of its oil resources. It provides the people of Kuwait with much-needed fresh water.

used to improve housing, roads, and commercial facilities. Free schooling and medical care are also provided. Kuwait has almost no fresh water. Desalination plants turn sea water into fresh water. More than 19 million liters (5 million gallons) of fresh water are produced daily. Fortunately, Kuwait is wealthy enough to be able to afford this water.

Bahrain

Bahrain is an independent *sheikdom*. A **sheikdom** is a country ruled by an Arab chief called a sheik. This nation is composed of four small islands in the Persian Gulf. The largest island is about 50 kilometers (30 miles) long and 16 kilometers (10 miles) wide. Some dates and citrus fruits are raised on the islands. Fishing and pearl diving provide some income. Bahrain's major revenues, however, come from oil. Bahrain's international airport is served by major American airlines with weekly flights to and from New York, New York.

Qatar

Qatar is an independent *emirate*. An **emirate** is territory ruled by a local leader called an emir. This nation is located on a peninsula southeast of Bahrain. Almost all of this nation's income comes from petroleum. Commercial fishing is also quite important to some people in Qatar. There has been great con-

In many regions of the Middle East the deserts are growing. This oasis on the Arabian Peninsula gets smaller each year as the desert moves.

cern over pollution of the waters of the Persian Gulf. Many oil tankers are polluting its waters. Some of the fishing companies fear the pollution will destroy the valuable fish of these waters.

The United Arab Emirates

The United Arab Emirates was formed in 1971. It was previously known as the Trucial States. In 1958 the oil boom came to this land. Most of the oil comes from offshore drilling operations. Because of its oil resources, the United Arab Emirates has the highest income per person in the world. In addition to oil, fishing, agriculture, and commercial banking provide jobs and income for the people.

The Arabian Peninsula

Sultanate of Oman

The Sultanate of Oman is a narrow strip of land about 1600 kilometers (1,000 miles) along the southeastern end of the Arabian Peninsula. A **sultanate** is a country ruled by a king called a sultan. It has a few mountains where limited rainfall permits some agriculture. Dates, pomegranates, and vegetables are the main crops raised. Most of the people of Oman are Arabs. However, some traders from India, Pakistan, and Africa have settled there. Oil was discovered in Oman in 1964.

People's Democratic Republic of Yemen

The People's Democratic Republic of Yemen, or South Yemen, occupies the southwestern portion of the Arabian Peninsula. This area was a British crown colony for 128 years. British warships then used the port of Aden (*ah* den). From this port the British controlled ship traffic into the Red Sea and the Suez Canal. Finally, in 1967, the British were forced to give up the port of Aden and nearby lands. The closing of the Suez Canal in 1967 brought economic problems to this once busy seaport. However, the canal reopened in 1975.

Yemen Arab Republic

Yemen Arab Republic, or South Yemen, is about the size of the state of Nebraska. Its coastal plain is hot and sandy. Inland is a central plateau with high forested mountains. Enough rain falls in parts of Yemen for agriculture. Nuts, fruits, millet, barley, wheat, and some livestock are raised. The main export crop is coffee. It is known throughout the world as *mocha* (*moh* kuh) coffee. Mocha is the port from which this coffee is shipped. Yemen is also believed to be part of the ancient site of the empire of the Queen of Sheba.

Date palms are an important crop in the Middle East and North Africa. The meat of the fruit is a major food source.

———————— WHAT HAVE YOU LEARNED? ————————

1. Who was Mohammed?

2. What two reasons are given for Saudi Arabia's fame in the world?

3. What has Saudi Arabia's govern-ment done with the money that it has received from selling its oil resources?

4. What is the Koran?

———————— BE A GEOGRAPHER ————————

1. Use an outline map of the Arabi-an Peninsula and correctly locate each nation.

2. Which countries of the Arabian Peninsula are completely north of the Tropic of Cancer?

3. What body of water is located at 20° N., 40° E.?

4. What body of water is located at 30° N., 50° E.?

———————— QUESTIONS TO THINK ABOUT ————————

1. Why do you think the waters of the Persian Gulf might be pollut-ed easily?

2. Why do you think non-Muslims are not allowed within the holy city of Mecca?

3. Why do you think the govern-ment of Saudi Arabia wanted to take control of the American pe-troleum companies?

4. Do you think that the many changes which have occurred in Saudi Arabia in the last few years were welcomed by all the people? Why or why not?

5. If you controlled the oil revenues in a nation of the Arabian Penin-sula, how would you spend the money to improve the nation's welfare?

⊶❦〖 19 〗❦⊷
Chapter

── NORTH AFRICAN LANDS ──

The lands of North Africa have been linked to those of the Middle East for many hundreds of years. People have often moved back and forth across the Sinai (*sy* ny) Peninsula and Red Sea to the Middle East and to the lands of North Africa. At the beginning of this unit, similarities between these two regions were mentioned. You might wish to review those similarities on pages 299 and 300 before continuing. Look at the map on page 347 and identify the countries of North Africa.

One of the strongest links between the Middle East and North Africa is that of "pan-Arabism." The majority of people in the Middle East and North Africa are of Arab ancestry. They are mainly followers of the Islamic faith. Many Islamic-Arab nations have similar problems. During the 1960s and 1970s, Arab leaders worked together to solve many of their problems. One major example of this cooperation between Arab nations of the Middle East and North Africa is the growth of the Arab League.

As one studies about North Africa, one must also remember that these nations form an important part of the continent of Africa. "Pan-African" ties are quite strong. African nations have also joined together to help one another.

The Arab League

The Arab League was formed in 1945. It was formed to try to unite the Arab nations to help one another in common causes. In 1945 Egypt, Syria, Lebanon, Jordan, Iraq, Yemen, and South Yemen were members. Arab League members did not do very much during those early years. However, during the 1960s a number of Arab nations became independent. Oil discoveries made many of these nations very rich. New nations joined the Arab League. Wealthy nations began to contribute extra money to support projects of the Arab League. The league began to be more powerful as a result. By 1977 there were twenty nations in the Arab League. Eight member nations are from northern Africa.

How do Arab League nations help one another? They often vote together on issues brought before the United Nations. Arab League committees also try to help member-nations solve disputes among one another. The Arab League has also suggested simple rules for importing and exporting goods. It has been encouraging Arab people to retain their cultural traditions.

Arab Studies programs have been started in many villages, cities, and university campuses. Cultural exchange programs are often made between Arab League countries. Financial assistance is provided to needy member-nations. Arab League nations have joined together in a boycott of all trade and most communication with Israel. They also boycott any foreign business firm that has branch offices in Israel. As time goes on, the activities of the Arab League may change. It would do well to keep up-to-date about practices of the Arab League as you continue this section of the book.

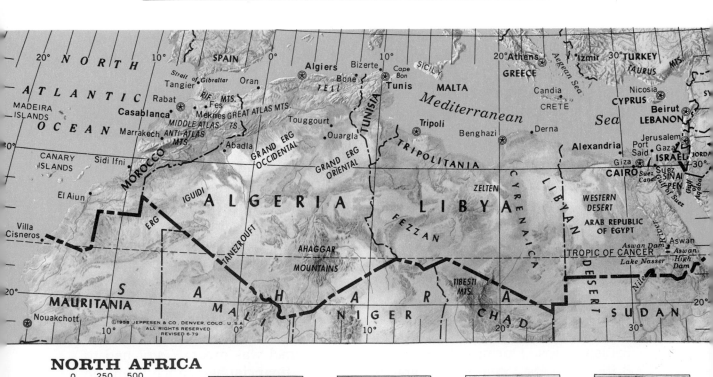

NORTH AFRICA

Kilometers 0 250 500

Miles 0 250 500

DRY GRASSLAND | SEMIARID GRASSLAND AND FORESTS | DESERT | MOUNTAINS

Arab Republic of Egypt

The Arab Republic of Egypt is more commonly known as "Egypt." Egypt occupies the extreme northeastern section of North Africa. A small part of Egypt is on the eastern side of the Gulf of Suez. This is the Sinai Peninsula. Much of this peninsula was occupied by Israel during the 1967 war. The coastal land of the Sinai Peninsula is mountainous. The ranges vary from about 1200 to 2600 meters (3,900 to 8,400 feet) high. Locate this region on your map on page 318. Notice that, aside from the Nile Valley, most of the land of Egypt is desert.

Egypt's History and Government

Ancient Egypt was one of the world's early centers of civilization. Early Egyptians contributed many ideas to the world. Some of them include the calendar, writing on paper, and using metal tools. Egyptians were among the first people to manufacture glass and fine linen. Egyptian mathematicians and scientists also discovered many facts and principles now taught to us in arithmetic, astronomy, and geometry.

For several centuries before World War I, Egypt was part of the Ottoman Empire. However, since 1882 the United Kingdom had been in control of the country. The United Kingdom wanted to make sure that the Suez Canal would be safe for British ships. During World War I British troops occupied Egypt in order to protect the canal. From 1921 until 1952 Egypt was an independent monarchy under the protection of the United Kingdom.

In 1952 some Egyptian military officers took control of the government and established a republic. The new Egyptian government wanted control over the Suez Canal. In 1956 British troops were finally forced to leave Egypt. Egyptians took over the canal.

The new government tried to improve the standard of living for the people. Large estates were broken up. No one was permitted to own more than 40 hectares (100 acres) of land in the Nile Valley. Land was distributed to the many needy farmers. The government also took over many businesses and industries. New factories, employing many workers, were built. Hospitals and clinics were built in many villages and cities. Schools were built. Educational programs for adults were also started.

During these years Egypt and Israel were not on friendly terms. In 1967 the Egyptian military fought bitterly with neighboring Israel. Israeli troops occupied the Sinai Peninsula and crossed the Suez Canal into the African part of Egypt. The United Nations then worked out a cease-fire and Israeli troops pulled back across the Red Sea. In 1979 Israel and Egypt signed a peace treaty. Egypt began to get back land on the Sinai Peninsula.

The cost of the military conflicts with Israel was great. Egypt had to receive financial assistance from the Arab League to pay its war debts.

The population had also increased more rapidly than the social services in the country. For many Egyptians, their way of life has changed very little.

The Nile

Throughout history, Egypt has been called the "Gift of the Nile." Without this river, most of the country would be barren desert. Most of the people of Egypt live in the valley and delta of the Nile. Twice as many people per square kilometer live in the Nile Valley as in the most densely populated part of western Europe.

In most places the Nile Valley is only about 3 to 16 kilometers (2 to 10 miles) wide. On both sides of the valley are high cliffs of limestone. Beyond them are the huge expanses of desert land. At Cairo (*ky* row) the Nile River divides into two parts. The river has formed a wide delta. It is about 240 kilometers (150 miles) wide. This delta extends about 160 kilometers (100 miles) to the Mediterranean Sea. This delta area is called "lower Egypt." It is the most populated part of all Egypt. Along the Mediterranean coast, much of the land is marshy.

The Nile is probably the longest river in the world. Except for a few **cataracts** (large waterfalls) and rapids, it is navigable for a great distance. This distance is equal to that between New York and San Francisco. For centuries the Nile flooded much of the valley floor every year. The flood season was from August to January. September was usually

the month when the water was at its highest level. From then on the water's volume decreased.

This picture taken from outer space shows the Nile River Delta. The reddish area is cropland, the dark blue areas are bodies of water, and the white areas are desert. The light blue spots within the reddish area are population centers. See if you can locate the following: the Nile River, Cairo, the Mediterranean Sea.

North Africa

Normally, floods can be very destructive. However, the Nile's annual floods were very useful. In its lower course, the river flows slowly. Here, the land is not high above sea level. The water, therefore, dropped **silt** (soil carried by water) from the higher lands of the south upon the floor of the Nile Valley. This action helped rebuild the soil each year. It was nature's way of fertilizing the soil. For thousands of years, this process has made the Nile Valley one of the best farming areas of the world.

As you can see on the map on pages 297 and 311 the Nile River has two main branches: the *White Nile* and the *Blue Nile*. The White Nile begins at Lake Victoria (vihk *towr* ih uh) near the Equator. There, rain falls throughout most of the year. Thus, the White Nile flows steadily northward. The Blue Nile originates at Lake Tana (*tahn* uh) in Ethiopia (*ee* thih *oh* pih uh). Rainfall there is heavy only in the summer months. Until dams were built, water from the Ethiopian highlands caused flooding in the lower Nile in the fall. Now the Nile rarely floods. Dams have been built to store the water.

Aswan Dam

The main dams on the Nile have been built near Aswan (as *wahn*). (See map on page 347.) The older and lower dam was built by the British in 1902. The upper dam, the *Aswan High Dam*, was built by the Egyptians with the help of the Soviet Union. Work on this project began in 1960. It was completed in 1971. The dam is almost 4 kilometers (2.5 miles) long. It rises 111 meters (365 feet) above the river. The lake that was made behind the dam is about three times as large as Lake Mead in Nevada-Arizona. Lake Nasser (*nah* sur) stretches 550 kilometers (340 miles) towards neighboring Sudan (soo *dan*). It is estimated that an additional 800,000 hectares (about 2,000,000 acres) of land can be irrigated because of this dam. A great deal of power can also be generated at this dam.

REMEMBER, THINK, DO

1. Why is Egypt often called the "Gift of the Nile"?
2. How do members of the Arab League help one another?
3. Use a globe to locate the nations of North Africa. Also, trace the Nile River from its sources to its mouth.
4. Why do you think the European countries near the Mediterranean Sea are different from the North African countries near the Mediterranean Sea?
5. Why was the Aswan High Dam built? Does it block boats from moving up the Nile River?

These Egyptian workers are setting up a watering system for this field. The water stored by the Aswan High Dam has been used to irrigate new farming areas in Egypt.

The Climate of Egypt

During the spring and summer, most of Egypt is warm and dry. Extremely hot temperatures are common in the Nile Delta. Alexandria (al eg *zan* drih uh), on the edge of the Delta, has recorded temperatures as high as 43° C (110° F). In the winter a small amount of rain falls. Inland the amount of rainfall decreases rapidly. Cairo has almost no rain at all. The winters near Cairo are mild. Occasionally, cool winds from the north make it necessary to have some heat in homes. The deserts of the interior have high daytime temperatures throughout the year. There is quite a difference between a desert's daytime and nighttime temperatures.

Agriculture in Egypt

The principal crops grown in Egypt are cotton, wheat, barley, corn, rice, sugar cane, beans, onions, and dates. Because there is so little rain in Egypt crops cannot be grown in the Nile Valley without irrigation.

Important crops. The chief food crop for livestock is Egyptian clover. The clover is grown during the coldest part of the year in order to enrich the soil. The bacteria which grows around the roots of the clover "fix" or transfer the element nitrogen from the air into the soil. Since the Nile no longer floods the land, artificial fertilizers must now be used.

Cotton is Egypt's most valuable money crop. Egyptian cotton's long,

North Africa

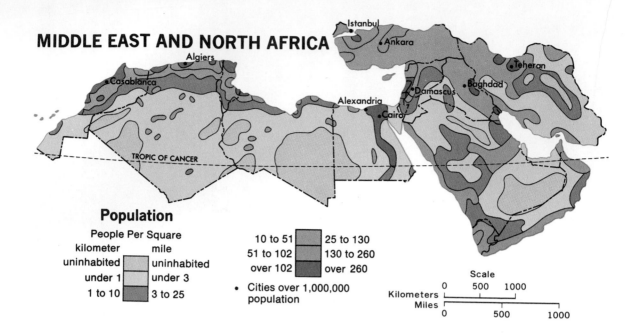

MIDDLE EAST AND NORTH AFRICA

Istanbul
Ankara
Algiers
Casablanca
Teheran
Damascus
Baghdad
Alexandria
Cairo

TROPIC OF CANCER

Population

People Per Square

kilometer	mile
uninhabited	uninhabited
under 1	under 3
1 to 10	3 to 25

10 to 51	25 to 130
51 to 102	130 to 260
over 102	over 260

• Cities over 1,000,000 population

Scale

Kilometers 0   500   1000

Miles 0   500   1000

silky, strong fibers are world famous. Textiles made of Egyptian cotton bring high prices. Because of the climate and year-long irrigation, farmers in the Nile Valley can now raise two or three crops a year on each plot of land.

Many farmers who live near cities also plant vegetables. Potatoes, tomatoes, cabbage, lettuce, cucumbers, peppers, green beans, and peas are grown. Many farmers sell these extra vegetables in the city markets.

The government has been encouraging farmers to plant citrus fruits on some of their land. Oranges, tangerines, and lemons are now being exported. Farmers also raise chickens, cattle, and water buffalo. Only about 2 out of every 40 hectares (5 out of 100 acres) of land in Egypt can be used for growing crops. Where farming is possible, the land is densely populated.

Farming Methods. On many farms, modern machinery is used to prepare the soil and sow seeds. Machines are also used in harvesting crops. Machines pick much of the cotton. Motor-driven pumps help lift water from irrigation canals onto the land. However, there are still many farmers in Egypt who own less than two hectares (five acres) of land. These small landowners cannot afford to buy modern equipment. Many farmers still use wooden plows drawn by oxen or buffalo. They cultivate their fields with hoes and harvest with hand sickles. Many farmers still use a long pole with a bucket at one end to lift water from irrigation ditches onto the land. Others use a water wheel turned by a buffalo or an ox.

Beyond the Nile Valley. The contrast between farmland and desert is very great in Egypt. Where water is used for irrigation, the land is green.

Where this is not possible, the brown sandy desert exists.

Scattered throughout the Western Desert, which is part of the Sahara, there are a few fertile oases. In these areas, farmers irrigate their land and grow many date palm trees. Manufactured products are still brought to some oases by camel caravans. Roads have been built to some of the larger oases. Trucks now move goods to and from these areas.

Major Cities of Egypt

Two world-famous old cities in Egypt are Cairo and Alexandria. Both cities were started thousands of years ago.

Cairo is the capital of Egypt. It is also the largest city in the Middle East and Africa. Cairo has almost six million people. West of Cairo, at Giza (*gee* zuh), one can view the world-famous pyramids. Tourists from all over the world are attracted to these sites.

The port city of Alexandria handles most of Egypt's international trade. Many beautiful beaches are located near Alexandria. Many Egyptians vacation on these beaches during the hot season.

Port Said (sah *eed*) is the Mediterranean entrance to the Suez Canal. Port Said is Egypt's third largest city. It is a fueling station for ocean-going vessels using the Suez Canal.

The Nile River contains many kinds of fish. However, the number of fish caught since the building of the Aswan High Dam has decreased. Why do you think this has happened?

Mining and Manufacturing in Egypt

Oil has been found near and under the Red Sea. Oil is also located in parts of the Sinai Peninsula in the Gulf of Suez. Deposits of oil have also been found in the Western and Eastern deserts. The deposits are not very large. Nonetheless, Egypt expects to be able to produce enough petroleum for its own use.

Some deposits of iron ore, phosphates, limestone, and salt are also located within Egypt. Some coal deposits exist in the Sinai Peninsula.

Oil refining and textile manufacturing are the largest industries in Egypt. Sugar refining, rice milling, and the production of soap, leather goods, fertilizers, and cement are also large industries. Some cigarettes, autos, trucks, and steel goods

President Sadat is greeted by the Egyptian people and navy as he sails along the Suez Canal.

are manufactured or assembled locally. The number of food processing plants has increased in recent years. Hydroelectric power from the Aswan High Dam has been useful to these industries. The cost of electricity has been lowered and more is produced.

The Suez Canal

The Suez Canal cuts through an isthmus that joins two continents. As you can see from the map on page 347, the Suez Canal connects the Mediterranean Sea and the Red Sea. Unlike the Panama Canal, the Suez is a sea-level canal. It is also about twice as long as the Panama Canal. The Suez Canal is about 60 meters (200 feet) wide and can take vessels requiring an 11-meter (37

foot) depth. The canal was opened in 1869.

The canal is a shortcut for ships sailing between European ports, the Persian Gulf, and the Far East. From England to Bombay (bahm *bay*), India, the trip through the Suez Canal is a distance of about 11 300 kilometers (7,000 miles). The other route around South Africa is quite a bit longer—about 19 500 kilometers (11,800 miles).

From 1967 to 1975, the canal was closed due to the Arab-Israeli crisis. Shipping firms were forced to use the route around Africa. Oil shipments are one of the main cargoes needing transport from the Persian Gulf to Europe and America. To overcome the disadvantages of having to use the longer ocean route, huge supertankers were built. These tankers can carry much more oil than the older, smaller tankers. Although the canal was reopened in 1975, many of the supertankers do not use the canal. Because of their giant size, many of them are too big to navigate the canal. Also, they can avoid the canal and still make a profit on their trip. Due to the gigantic volume that they are able to carry in one load, these supertankers can sail the longer distance and still make money.

Before the closing of the canal in 1967, Egypt earned about $250 million a year from canal revenues. The Arab League aided Egypt with funds to make up for the financial losses caused by the canal's closing. Now that the canal is reopened, some people are questioning the importance of the Suez Canal. Is it, some wonder, really needed?

Education in Egypt

According to Egyptian law, all children between seven and twelve years of age must attend school. The government has also encouraged the older, talented students to study subjects such as business, agriculture, industry, medicine, and engineering. Skilled Egyptians are needed to help the nation's future development.

The Future

Egypt is one of the major Arab nations involved in the dispute with Israel. Many peace plans have been discussed between representatives of Egypt, Israel, the United Nations, and several interested countries. However, it was not until 30 years after the conflict started that leaders from Egypt and Israel met to discuss their problems. The conflict in the Middle East is very complicated and may take many years to resolve.

The Middle East crisis has caused some of the nations involved to neglect social progress. A lot of the money earned from the sale of oil has gone to support the military. Until a lasting peace occurs money, time, and many lives may be spent on war and the weapons of war. If this happens progress may come very slowly to the people of Egypt.

REMEMBER, THINK, DO

1. Why do many Egyptian farmers plant crops of clover on their land?
2. What is the land of Egypt like beyond the Nile Valley?
3. Make an agricultural and manufacturing product map of Egypt. Compare it to the nations of the Middle East.
4. What do you believe were the main advantages and disadvantages of Egypt's closing the Suez Canal in 1967?
5. If peace really comes to the Middle East how would you suggest the governments spend their money?

Morocco

The Kingdom of Morocco is the most mountainous land in North Africa. Morocco has been an independent kingdom since 1956. It is located near the Strait of Gibraltar, south of the Iberian Peninsula in Europe. The western region of Morocco extends to the Atlantic Ocean. Find Morocco on the map on page 347.

The Land and Climate of Morocco

Three ranges of the Atlas Mountains are in Morocco. The highest range is known as the Great Atlas. It has some peaks more than 3900 meters (13,000 feet) high. This range extends from the Atlantic Ocean in southern Morocco northeast into Algeria. A number of the peaks in the Great Atlas are always snow covered. Forests cover much of the lower mountain sides. South of the Great Atlas Mountains are the Anti-Atlas Mountains. North of the main range are the Middle Atlas Mountains. The peaks of these mountains are lower in elevation. In northern Morocco along the Mediterranean Sea is another range of mountains, the Rif Mountains.

Between the mountain ranges of the Atlas, there are plains and high tablelands. Northwest of the mountains is a vast plain. The best agricultural land in Morocco is found in these regions. Deserts are located south of the mountains.

In winter the days are warm and sunny. Some snow falls in the mountains. Rain falls in the lower regions. Many tourists come to Morocco both summer and winter because of its varied land and comfortable climate. One can swim in the ocean one day and go skiing in the mountains the next.

The People and Their Work

Most of the people of Morocco are descendants of Arabs and *Berbers* (*buhr* buhrz). **Berbers** are those people who lived in the area of Morocco before the Arabs came. Many French and Spanish people also live in Morocco.

About three-fourths (¾) of the people in Morocco make their living by farming. Along the Mediterranean coast farmers grow olives, cork, and almonds. Inland, irrigation is necessary. The main crops include wheat, barley, citrus fruits, potatoes, and tomatoes. Much of the food is exported. The government of Morocco is trying to improve farming methods. Agricultural agents are teaching rural people how to use machines and newer farming methods. Tractors and combines are gradually replacing the hand tools and older farming practices.

Mining and Manufacturing in Morocco

Morocco is rapidly becoming an important industrial nation in North Africa. The land contains a variety of mineral resources. Morocco ranks third in the world in phosphate production. Large deposits of iron ore, coal, manganese, lead, sulfur, zinc, silver, salt, and cobalt are also mined. Some petroleum has been found, but it is limited in quantity.

In the past, Morocco exported most of its raw minerals to markets in Europe. However, factories and industries in Morocco now process the country's own resources. Steel and iron foundries make metal products. Mills also produce textiles, cement, sugar, and paints.

Food-processing plants are very important. Cheese products, chocolates, flour, and vinegar are now made in Morocco. The production of wine has grown rapidly in recent years. Much of the wine is exported.

Fishing industries have also been developed. Fisheries supply plenty of seafood for local use. Sardines, tuna, and anchovy are processed for export.

Handicraft items play an important part of the commercial world in Morocco. Leather products, silver, brass, copper, wrought iron, and pottery items are made by skilled artists.

Large dams have been constructed in the Atlas Mountains. Needed hydroelectric power is produced at these dams. Water is stored for irrigation. Most of the nation's industries are supplied with power from these dams.

In Morocco, as in Europe, open-air markets are the common place for people to buy fresh produce.

Major Cities of Morocco

Morocco's largest city is Casablanca (*kah* suh *blahn* kuh). It is a major port on the Atlantic Ocean. Rabat (rah *baht*) is located north of Casablanca. It is the capital of Morocco. At the Strait of Gibraltar is the port city of Tangier (tan *jeer*). Tourists from many nations of the world visit these interesting cities in Morocco.

Morocco is headed by a king who also serves as the nation's religious leader. He has a great deal of power. In 1963 the King of Morocco, King Hassan II, established a parliament to assist him in managing the affairs of the nation.

Conflicts with neighboring nations have been caused by disputes over national boundary lines. During the first half of the 20th century, most of the land in Africa was ruled by European countries. Boundary lines of colonial territories were often made in the interest of the colonial governments. After territories became independent, the new nations had to cooperate with one another over disputed boundaries.

The governments of Morocco and Algeria have not always agreed on their shared boundary lines. At times disputes over land areas have caused bad feelings between the governments of these two nations.

In 1969 the Spanish government decided to give its North African territory of Ifni (*ihf* nih) to the government of Morocco. Then, in 1976, Spain gave its territory of the Spanish Sahara to the governments of Morocco and Mauritania (*maw* rih *tay* nih uh). The government of Algeria did not agree with Morocco's claims to these new territories. The land of this region contains rich phosphate deposits. It is estimated that over one billion tons of phosphate rock is there. Maybe other valuable minerals will be found there too.

REMEMBER, THINK, DO

1. What are some of the major products manufactured in Morocco?
2. Where is it possible to ski in wintertime in North Africa?
3. Use an outline map of North Africa to complete the following:
   a. Locate and label the capital cities of Egypt, Libya, Tunisia, Algeria, and Morocco.
   b. Color the Mediterranean coastal plain and the desert lands of North Africa according to your map's landform key.
   c. Draw in (using a dashed line) and label the Tropic of Cancer.
   d. Locate and label the Atlas Mountains.
   e. Locate and label the Nile River. Also color the Nile Valley according to your map's landform key.
4. What mineral ore was discovered in the lands once called the "Spanish Sahara"?
5. In what ways are the governments of Morocco and Egypt similar? How are they different?

LEARN MORE ABOUT NORTH AFRICA

You may wish to extend your knowledge about the countries of North Africa. A few points of interest about each nation are included. Use the Study Guide on page 31 as you plan your own independent study of one or more of these other North African nations.

Libya

The country of Libya has undergone many changes since 1959. In that year, oil discoveries brought the nation sudden wealth. In 1969 the military took control of the government. A Revolutionary Command Council was organized. This group began to change the society of Libya by using the profits from oil to benefit the people of the nation.

A government school lunch program feeds more than 300,000 children. The government hopes to improve the health of the children by adding protein to their diet. Tuberculosis sanatoriums and hospitals have also been built. Doctors travel to remote desert oases bringing medical assistance to nomadic people. Housing, education, sanitation, and water resources have also been improved.

Libya has also used some of its oil money to aid Arab League members. The government has become very powerful and active in many international activities.

Not all elements of Arab society have been affected by modernization. Removing seeds from cotton plants may still be done by hand.

Tunisia

Northwest of Libya is the small Republic of Tunisia. It is about the size of the state of Louisiana. Though small in size, Tunisia has three times as many people as neighboring Libya. Most of the people are Muslim and speak Arabic. There are also some French and Italian people living in Tunisia. Locate Tunisia on the map on page 347. Notice how the land extends into the Mediterranean Sea towards Sicily. Tunisia extends farther north than any other nation of Africa. The majority of the people earn their living by farming. Wheat, barley, grapes, and olives are some of the main agricultural crops raised.

In 1957, Tunisia became a republic. A few years later, women were given the right to vote, and a national school system was established.

Algeria

Algeria is west of Tunisia and Libya. As the map on page 347 shows, Algeria is larger than Libya. It is larger than the Mountain States region of the United States.

Algeria became independent in 1962, following a bitter seven-year struggle for freedom with France. One reason France wanted to keep control of Algeria was because of its mineral and fuel resources. Major oil deposits were discovered in the Grand Erg Oriental southeast of Ouargla (*wahrg* lah). Pipelines carry the oil to the Mediterranean coast. It can then be easily shipped to the markets of Europe. Other minerals mined in Algeria include iron, high-grade phosphate, lead, zinc, and copper. A small amount of coal is mined.

The military took control of the government of Algeria in 1965. A Revolutionary Council manages the government. Many industries and oil companies have been nationalized by this government. This enables the country to receive most of the profits. Algeria has also been very active in its support of the Arab cause in the Middle East. Algeria still has much to improve at home. Vast problems of food supplies, health conditions, sanitation, education, and unemployment still plague this oil-rich nation.

Three of the world's major religions began in the Middle East. There are other important religions in the world. This map shows what religion the majority of people follow in a given area.

WORLDS MAJOR RELIGIONS

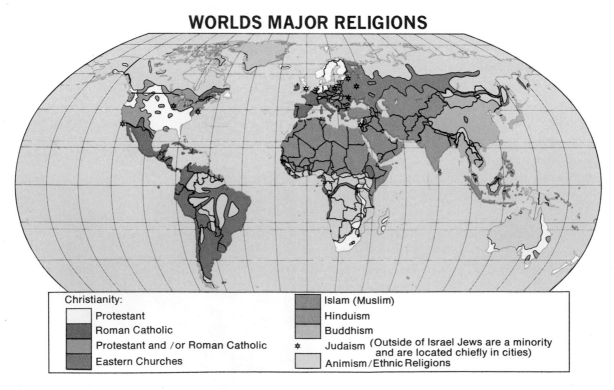

Christianity:
- Protestant
- Roman Catholic
- Protestant and /or Roman Catholic
- Eastern Churches

Islam (Muslim)
Hinduism
Buddhism
✿ Judaism (Outside of Israel Jews are a minority and are located chiefly in cities)
Animism/Ethnic Religions

A SUMMARY OF THE MIDDLE EAST
AND NORTH AFRICA

The lands of the Middle East and North Africa have been important in the history of the world. The ancient peoples of Mesopotamia and Egypt are still remembered for their contributions to the early development of civilization. Followers of Judaism, Christianity, and Islam trace their religions' beginnings back to the lands of the Middle East. Many of our laws and ideas about social conduct are based upon the teachings of the Bible, the Talmud, and the Koran. Some people living in the United States have ancestors that once lived in the Middle East and North Africa.

During the second half of the 20th century, many changes have occurred in the Middle East and North Africa. New nations have been created from older regional territories or colonies, such as Algeria, Yemen, and Libya. Some of the new countries are unfamiliar to many people in the United States. Kingdoms in some nations have been overthrown by military forces. Political leaders from the Middle East and North Africa have become well-known in the United Nations and other international organizations. Petroleum resources in this part of the world have created sudden wealth for many nations that once were poor. Political disagreements between some countries have caused destruction and loss of lives. At times the peace of the world has been threatened by conflicts in the Middle East.

Changes will probably continue to occur in this region of the world. Some changes may be very important to you and our country. You may want to continue to learn more about the lands of the Middle East and North Africa for many years to come.

WHAT HAVE YOU LEARNED?

1. How do the land and climate of the nations of the Middle East and North Africa differ from that of the Soviet Union?

2. What similarities and differences are there between the industries of Southern Europe and those of North Africa?

3. In what ways do governments differ among nations of the Middle East, North Africa, and Southern Europe?

4. How have nations of the Middle East, North Africa, and Europe encouraged industrial growth and trade between nations? Give some examples.

5. The Soviet Union, the Middle East, and North Africa all have regions of land that are not easy for people to settle upon. What are various nations doing to encourage settlements upon these "wastelands"?

BE A GEOGRAPHER

1. Which of the countries in the Middle East and North Africa has land closest to the equator?

2. What important parallel of latitude crosses Algeria, Libya, and Egypt?

3. Name the European capital cities which are almost straight north of the following cities:
   a. Derna, Libya
   b. Casablanca, Morocco
   c. Fez, Morocco
   d. Oran, Algeria
   e. Bone, Algeria
   f. Tunis, Tunisia

4. Name the cities that are located at the following places:
   a. 34°0′ N. and 35°30′ E.
   b. 32°0′ N. and 36°30′ E.
   c. 31°45′ N. and 35°15′ E.
   d. 36°0′ N. and 37°15′ E.
   e. 33°30′ N. and 36°30′ E.

5. Is the sun ever directly overhead at noon at:
   a. Alexandria?
   b. San'a?
   c. Aswan?
   d. Casablanca?
   e. Tripoli?
   f. Aden?
   g. Mecca?

6. If it is noon in Cairo, what time is it in:
   a. London, United Kingdom?
   b. New York, New York?
   c. Anchorage, Alaska?
   d. San Francisco, California?
   e. Tokyo, Japan?
   f. Igarka, U.S.S.R.?
   g. Aden, South Yemen?
   h. Fez, Morocco?

————QUESTIONS TO THINK ABOUT————

1. If you suddenly had to do without petroleum for one year, what changes might occur within your family's activities? What could you use instead of fuel-powered machines to carry out some of the same activities?

2. What do you think would happen to the OPEC nations if a new source of energy were developed, and the world no longer needed petroleum?

3. What are the real sources of conflict between the Arabs and the Jews over the land around the eastern end of the Mediterranean Sea? In your view, under what circumstances would Israel agree to return to the boundaries it had in 1967? What suggestions do you have for ending the conflict?

4. Why do deserts form some places on Earth and not at others? Is there much that people can do to change conditions so that a desert area will be productive?

5. Do you think waterways such as the Dardanelles, the Suez Canal, and the Panama Canal should be governed so that all nations can use them? If you do, how could this be done?

Unit Review

Students in Niger walk through a field on their way to school.

Unit

AFRICA FROM THE SAHARA TO THE CAPE OF GOOD HOPE

Unit Preview

- What do you know about these famous African people?

- What do these pictures tell you about African cities?

DO YOU KNOW?

- How many years people are believed to have inhabited the continent of Africa?

- Why it is possible for snow-capped mountains to exist along the equator in Africa?

- Why so many new nations have been created in the lands of Africa?

- Why the customs of people in Africa often differ greatly from one region to another?

- Why it has been difficult for some Whites and Blacks to live together peacefully in parts of Southern Africa?

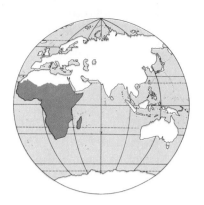

Unit Introduction

Africa from the Sahara to the Cape of Good Hope

The continent of Africa has had a very old and unusual history. Discoveries of human bones in Tanzania (*tan* zuh *nee* uh) and Ethiopia support the idea that people might have lived in Africa over two million years ago. In southern Africa the ruins of the ancient kingdom of Zimbabwe (zihm *bah* bway) have been located. It is believed that this kingdom existed more than 1,600 years ago. Empires in west Africa had once developed in Mali (*mah* lee) and Benin (*buh* neen). Many important trading centers and cultures existed in Africa long before Europeans knew about North and South America.

Years later, however, Europeans were an important element in two periods of history in Africa. One was the shameful period of slave trading. The second was the colonial period. It was during this second period that almost all of Africa came under the control of European nations. The colonial period lasted from about the mid-19th century to the mid-20th century.

A New Era in African History

After the Second World War, a third major period in African history began. New nations were formed from colonial territories. Within a 30-year period, almost all of Africa became independent of European rule. In June, 1977, the last colonial

Unit Introduction

government gave up its territory. The colonial period of Africa had ended. A period of independence had begun.

The Land and Climate of Africa

Africa is the second largest continent on Earth. The continent occupies about one-fifth (⅕) of the world's total land area. At its widest the continent is about 7500 kilometers (4,700 miles) from west to east. From Cairo in the north to Cape Town in the south, the length is about 8000 kilometers (5,000 miles).

Africa is surrounded by water. One finds the Mediterranean Sea to the north. Moving clockwise around the continent, one finds that the Suez Canal and the Gulf of Suez separate Africa from the Middle East. The Red Sea and Gulf of Aden meet at a peninsula known as the "horn" of Africa. The Indian Ocean is along the eastern coast of Africa south of 12° N. The southernmost point of the continent is known as the Cape of Good Hope. Waters off this cape are noted for their stormy seas and dangerous ocean currents. On the western side of Africa is the Atlantic Ocean. Use the map on page 369 to identify these important bodies of water.

Although Africa is large in land area, it is not heavily populated. Africa is three times the size of Europe, but it has fewer people. Most of the people live south of the huge desert named the Sahara (suh *hair* uh). The population map on page 408 shows that very few people live in the Sahara. This map also shows that many people live along the coast near the Gulf of Guinea (*gihn* ih). The largest population of any African nation is found in Nigeria (ny *jeer* ih uh). Heavily populated areas are also found near the east African coast and the lake regions of central Africa. The highlands in the east and the south are heavily populated as well. There are many large cities in Africa. Some are growing quite rapidly.

Many of you may be surprised to discover that most of Africa is *not* tropical jungle. The continent is crossed about midpoint by the equator. This leads some people to expect the land to be more tropical than it is. Look at the landform map on page 379. Notice that the majority of the continent is plateau, or tableland. The plateau is surrounded by a narrow band of low, coastal plains. Much of the plateau land is high above sea level. Grass grows on much of the plateau that lies south of the Sahara. In fact there is more **savanna**, or tropical grassland, in Africa than tropical rain forest. Much desert land is also found on the continent. The Sahara, Namib (*nah* mihb), and Kalahari (*kah* lah *hah* ree) deserts are the main ones. Between the deserts and savannas are semiarid steppes. Do you remember what steppes are?

The western and west-central regions of the continent are lower in elevation than eastern Africa. Most of the tropical rain forests are found at these lower elevations. High, rugged mountains are found in

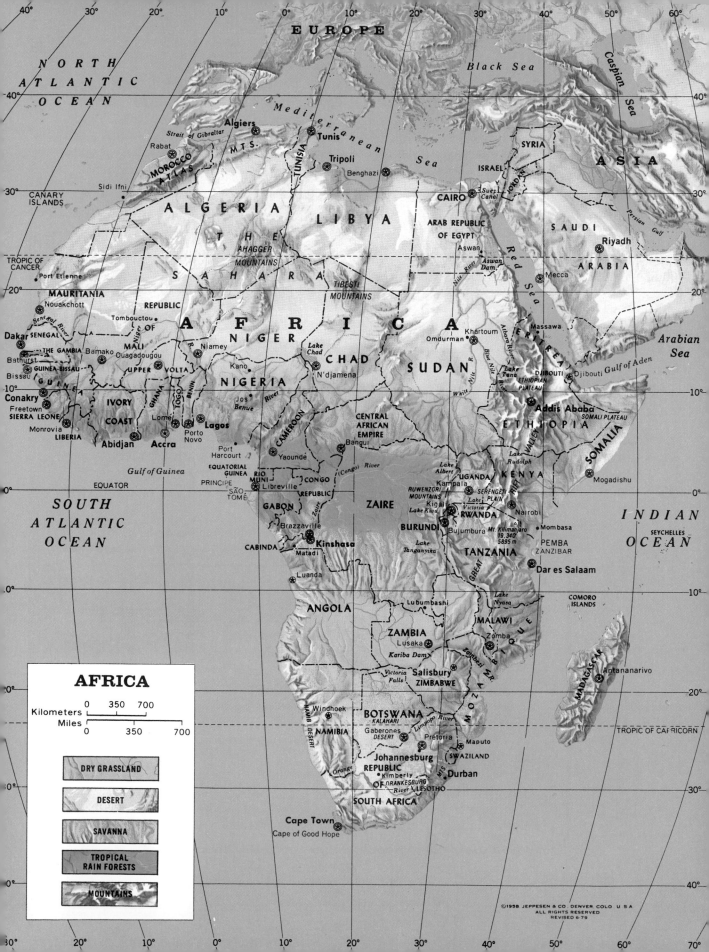

the eastern highlands. The snow-capped mountains near the equator are often a surprise to many people.

Three of the world's ten longest rivers are found in Africa. The longest river in Africa (and the world) is the Nile. The Zaire (zah *eer*), once commonly known as the Congo (*kong* goh) River, is the second longest in Africa. The Niger (*ny* juhr) River is the third in length. Many large lakes are also found in the highland area of eastern Africa.

The climate of Africa is as varied as the landscape. Near the equator, in western Africa, the direct rays of the sun cause air to rise during the daylight hours. As it rises, the air becomes cooler. Moisture in the air condenses into drops of water which fall as rain. As the Earth moves around the sun, the heavy rains move slightly north and south of the equator. From Lake Victoria westward to the coast, tropical rain forests grow on both sides of the equator. Such forests grow only at low altitudes and need a great deal of rain during the year. Rain forests are also found along the west coast of Africa near the Gulf of Guinea and along the eastern side of the island of Madagascar (*mad* uh *gas* kuhr).

In many regions beyond the rain forests, savanna lands have seasonal rains. Some places have two rainy seasons. The seasons are called the "long rains" and the "short rains." **Monsoons**, or seasonal winds, on the eastern coast of Africa also bring rainy seasons. Periods of dry weather occur between the rainy seasons. Sometimes a long drought results between rainy seasons. In the steppe and desert regions, not much rain falls at any time of the year.

REMEMBER, THINK, DO

1. How does Africa rank in size among the continents of the world?
2. Describe the major landforms of the African continent.
3. Use a desk-size outline map of Africa to do the following:
   a. Label the Tropic of Cancer, the equator, the Tropic of Capricorn, and the prime meridian.
   b. Locate and label the three longest rivers in Africa.
   c. Locate and label the following places: Cairo, Cape of Good Hope, Djibouti.
   d. Label the major bodies of water that surround the continent of Africa.
4. Identify on a map of Africa the area that has the greatest population.
5. Plan a debate on the topic: *Resolved:* That Africa has a population problem.

The Years Before Independence

In 1950 only four independent nations existed in Africa. They were Ethiopia, Liberia (ly *beer* ih uh), South Africa, and Egypt. The rest of Africa was under colonial rule. By the end of 1977 no colonial governments existed on the continent. Colonialism had ended. Yet colonialism has had a lasting effect upon many of the new nations of Africa. Before studying these new countries it might be useful to understand how colonialism began in the continent of Africa.

The Scramble for Africa

In the 19th century, the industries of Europe began to expand rapidly.

As business and trade grew, the industries needed more raw materials. Many European countries had some knowledge of African lands from their earlier years of slave trading. Some still had settlements in Africa. Europeans began to look to Africa as a source of raw materials, cheap labor, and new markets.

At first the Europeans were only interested in getting the raw materials that they needed to supply their factories. In the beginning they came as explorers who mapped and charted the lands. Later, such activities became the basis for the countries of Europe to claim the "explored land" as their own territory.

The competition for raw materials between European countries became greater in the second half of the 19th

The artistry of the African people is evident in much of what they create. These ladles, combs, and knives demonstrate this fact.

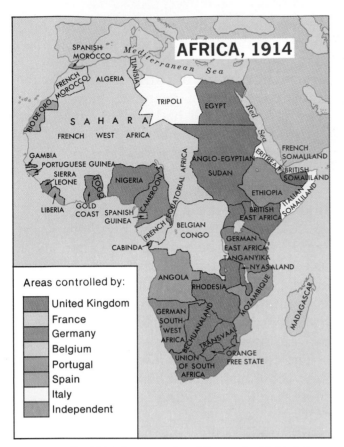

AFRICA, 1914

SPANISH MOROCCO
FRENCH MOROCCO
RIO DE ORO
ALGERIA
TUNISIA
TRIPOLI
EGYPT
SAHARA
FRENCH WEST AFRICA
GAMBIA
PORTUGUESE GUINEA
SIERRA LEONE
LIBERIA
GOLD COAST
SPANISH GUINEA
TOGO
NIGERIA
CAMEROON
FRENCH EQUATORIAL AFRICA
ANGLO-EGYPTIAN SUDAN
ERITREA
FRENCH SOMALILAND
BRITISH SOMALILAND
ETHIOPIA
ITALIAN SOMALILAND
BRITISH EAST AFRICA
BELGIAN CONGO
CABINDA
GERMAN EAST AFRICA
TANGANYIKA
NYASALAND
ANGOLA
RHODESIA
MOZAMBIQUE
MADAGASCAR
GERMAN SOUTH-WEST AFRICA
BECHUANALAND
TRANSVAAL
ORANGE FREE STATE
UNION OF SOUTH AFRICA
Mediterranean Sea
Red Sea

Areas controlled by:
- United Kingdom
- France
- Germany
- Belgium
- Portugal
- Spain
- Italy
- Independent

By 1914 the countries of Europe had laid claim to almost all of Africa.

century. In order to protect their supplies the countries of Europe began to claim areas of Africa as their own. Occasionally, the European nations disagreed about claims to lands in Africa. To prevent such disagreements, European leaders agreed to meet in Berlin, Germany, to talk about the problem. The Berlin Conference was held in 1884. Leaders of most European nations went to this conference.

The European leaders agreed on ways to divide the land of Africa. They insisted that a nation must build settlements on the land it claimed. If this was done, other nations would respect the colonial claims. This decision started what has been called "the scramble for Africa."

European nations quickly sent groups to Africa to settle upon their claimed territories. Within a few years after this meeting, the continent had begun its period of colonialism. France took most of northwestern Africa. The British had holdings both in west and east Africa. In fact, there were some attempts by the British to claim all the land in eastern Africa from Cairo to Cape Town. The King of Belgium claimed for himself the entire Congo area. Today this is the country of Zaire (zah *eer*). Germany claimed some land in east Africa and parts of west Africa, too. Italy claimed lands near Ethiopia. Spain and Portugal already had very old settlements in various parts of Africa. Only Liberia and Ethiopia remained independent.

During the first half of the 20th century, the African colonies provided much wealth to some European nations. Where agriculture was possible, vast *plantations* were started by the colonists. A **plantation** is a large farm which usually grows one or two main crops. Mines began to operate where minerals were found. Ships took the ores and agricultural products to factories in Europe. The materials were then processed and sold at large profits. African laborers were used to dig the ores out of the mines and to work the plantations. Little money, however, was paid the Africans.

Have you ever played a game similar to the one that these children from Ghana are playing?

The Fight for Freedom

During the Second World War, many African people helped the European nations. Some European nations promised independence to their colonies for their cooperation. However, after World War II, independence was slow in coming. African leaders began to protest. Freedom from colonialism became the cry heard over much of the continent. It took Algeria seven years of fighting to win its freedom from France in 1958. In other places independence came more peacefully. Ghana (*gah* nuh) became independent in 1957. Its leaders encouraged other Africans to demand their freedom, too. As the struggle continued, new nation after new nation appeared on Africa's map. Portugal and Spain had been the first to colonize African lands. They were among the last nations to end their colonial ties.

Colonialism is a thing of the past on the continent of Africa. However, some people are still not free. White people, who are a minority, still control the majority Black African population in some nations of southern Africa. In these lands the struggle for freedom continues.

In Rhodesia (roh *dee* zhuh), Black Africans were allowed to vote for the first time in 1979. A Black prime minister took office in May of that year. The name of the country was changed to Zimbabwe Rhodesia.

Whether majority rule will come about peacefully in Zimbabwe Rhodesia, Namibia (nuh *mihb* ee uh), and South Africa is a question that only time can answer. Most of the people of Africa, however, are at long last free. The years ahead will probably witness dramatic changes.

Unit Introduction

Wearing shirts that announced independence, these women were part of the official celebration that marked Djibouti's first day of freedom.

The Last Colony

On June 27, 1977, the French Territory of Afars (*ahf* ahrz) and Issas (ee *sahz)* became the independent Republic of Djibouti (jih *boo* tih). Representatives from many nations of the world came to Djibouti to witness the day's celebration. No people were happier than the hundreds of freedom fighters who lined the streets and parks of the capital city. For years, these men and women had fought a war against France for their freedom. Most of the people had wished to be free of French colonial rule. The people wanted to be self-governing.

Many people were dressed in traditional dress. Shops proudly displayed pictures of the nation's new leaders. The blue and green national colors hung across the major streets and crossings. Merry-making filled the day, while people waited for the official midnight ceremony.

Minutes before midnight, the leaders of the French colonial government and the new president of Djibouti stood at attention before a flagpole near the governor's residence. The tri-color flag of France was flying. The band began to play the national anthem of France. The flag was then gently lowered. A sudden stillness seemed to calm the warm night's breeze. The clock struck midnight as the new flag of the Republic of Djibouti rose to the top of the flagpole. The national anthem of Djibouti was played as fireworks exploded in the dark night. The green and blue flag with a red star on a white triangle could be seen as the fireworks exploded. Cheers, and some tears, greeted this historic moment. The Republic of Djibouti had been created. The last colony in Africa was free!

The People and Their Concerns for Rapid Change

Most of the people of Africa have ties to the land. Village elders can usually retell the history of their people. Numerous African languages are used to tell about the folklore, culture, and traditions of each group of people. For centuries, tribal loyalties united the people. South of the Sahara, the majority of the people are Black Africans. Only about two out of every 100 people living in Africa trace their ancestry to Europe. These White Africans are mainly in southern Africa.

Probably no place on Earth has changed as much in the past 30 years as the lands of Africa. Between 1950 and 1980, 40 new independent nations were created. Much improvement needs to be made in most of these countries. Many people need to be taught to read and write. Medical services are needed, especially in rural areas. Governments are trying to use their natural resources to help the people. Large sums of money are needed. At times the new governments have had dif-

ficulty meeting the wishes of all citizens. As you study Africa, you might wish to remember some critical issues about this continent.

● **Boundary Lines.** Most new nations in Africa had to accept the boundary lines set for them in the colonial era. The Europeans divided up the land without considering tribal settlements or the nomadic ways of local people. Nations today are sometimes in conflict with their neighbors because of disputes over boundary lines.

● **Language.** Most nations of Africa have a number of tribal groups, each speaking its own language. In order to create ties among the people of a nation, a national language has often been chosen. The language, it was hoped, would increase communication among people of the country. But the selection of one tribal language over another has sometimes caused problems. The use of a European colonial language, such as English or French, is often a problem as well. Many businesses and universities have used the colonial language for many years. Most books are printed in the colonial language. It is

Unit Introduction

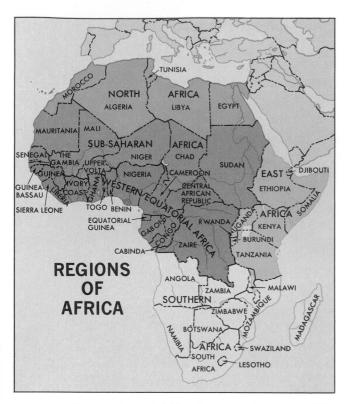

Four regions of Africa shown on this map will be studied in this unit. (North Africa is covered in Unit VII.)

not easy to give up past habits. However, the use of European languages hinders the development of the native languages.

● **Nationalism.** Most nations of Africa are trying to develop a spirit of nationalism among their citizens. Nationalism, as you have learned, is a feeling of pride in and loyalty to one's country. Before independence, people's loyalties were centered upon a tribal system of leadership. Many tribes exist in one country. It is difficult in some nations for people to change their loyalties from *tribalism* to nationalism. The strong feelings of pride and loyalty which people have toward other tribal members is called **tribalism**.

● **Single-Crop Economy.** The economy of many African countries is heavily dependent upon one or two agricultural cash crops. Under colonialism, many lands were developed into plantations that produced coffee, tea, cacao, or sisal. World markets fix the value of these cash crops. A nation's economy can change rapidly if world prices change. A sudden decrease in world prices can cause real hardships. Thus, many nations in Africa are trying to change their economies as rapidly as they are able.

● **Money and Labor Problems.** Under colonialism, most profits were not returned to African lands. Europeans rarely trained Africans for skilled positions. New nations, therefore, find themselves lacking not only cash but also skilled labor.

● **Pan-Africanism.** In a spirit of Pan-Africanism, some leaders of African nations have joined together to solve common problems. Many African leaders want to share experiences, both good and bad, with other leaders. Under this growing spirit of Pan-Africanism, the Organization of African Unity (O.A.U.) was created so leaders could share ideas and help one another.

● **A Third-World Power Bloc.** African nations have become aware of their power when they join together. Membership in the United Nations gives African countries a powerful voice in international politics. They often vote on issues as a group. The rest of the world cannot easily ignore the many nations of Africa.

Things You Might Like To Do As You Study About Africa

Africa may be one of the most interesting areas of study in the Eastern Hemisphere. There is much to learn about this continent. You may wish to begin some activities to help make your study of Africa more interesting and enjoyable. Here are some suggestions for you and the class.

1. Begin to collect news items about current events in the countries of Africa. Five groups could locate news items about the countries of: **a.** Southern Sahara, **b.** Western Equatorial Africa, **c.** East Africa, **d.** Island nations of the Indian Ocean, and **e.** Southern Africa.

2. You might be interested in learning about some leaders of the freedom movements in Africa. Groups might learn about Kwame Nkrumah, Jomo Kenyatta, Kenneth Kaunda, Julius Nyerere, Winnie Mandela. Groups could make oral presentations about their lives.

3. If you are interested in history, you might want to read about:

a. The discoveries at Olduvai Gorge in Tanzania, **b.** the ruins of Zimbabwe, **c.** Tombouctou, **d.** the art of ancient Benin. Oral reports could be shared with the class.

4. Select an interesting folk tale from the Southern Sahara region. Plan a puppet show to tell the tale. Tell the folk tale to primary classes in your school.

5. Find a cookbook in your school or local library or at home which has African recipes of interest to your class. Prepare a West African meal. Groups of students might each prepare a special food for the class.

6. See if it is possible for your teacher to show the film, *Search for the Nile*. Students might read about mid-19th century European explorers mentioned in this TV film about East Africa.

7. A student who is interested in science might conduct an experiment to explain why the Benguela (behn *gwehl* uh) Current causes fog along South Africa's western coast.

·┥ 20 ┝·
Chapter

THE SOUTHERN
──────SAHARA REGION──────

The nations of the Southern Sahara occupy one of Africa's unusual land areas. A similar landscape extends across the entire width of Africa. This region extends from Mauritania at the Atlantic to Sudan near the Red Sea. Within this vast region is much of the world's largest desert. Look at the map on page 369. Find Mauritania, Senegal (*sehn* uh *gawl*), and The Gambia (*gam* bih uh) at the Atlantic Coast. Inland, locate the four landlocked nations of Mali, Upper Volta (*vahl* tuh), Niger, and Chad. Completing this region is Sudan near the Red Sea. Notice how the vegetation/climate changes gradually from desert to steppe and savanna along the southern sections of the entire region. One of the worst droughts of the 20th century occurred in these steppes. The drought began about 1973 and lasted a number of years. Millions of people suffered from this disaster of nature. Many died.

Climate/Vegetation Regions

The Sahara. Three major regions are found in Southern Sahara. They are desert, steppe, and savanna. Most of the land is part of the world's largest desert, the Sahara. This desert is larger than the 48 mid-continental states of the United States. The Sahara stretches about 4800 kilometers (3,000 miles) across Africa, from the Atlantic Ocean to the Red Sea. From north to south the desert is about 1600 kilometers (1,000 miles) long.

Most of the land area of the Sahara is a fairly level plateau. Its elevation averages about 300 to 450 meters (1,000 to 1,500 feet) above sea level. The Ahaggar (ah *hag* uhr) Mountains in Algeria and the Tibesti (tih *behs* tih) Mountains in Chad, however, are two major mountain ranges located in this desert. Most of the surface of the desert is gravel or stone. Less than one-fifth (⅕) of the

378

land is sandy. In these areas, strong winds cause sand to drift and to form dunes.

At some locations in the Sahara *artesian springs* can be found. **Artesian springs** are places where water rises to the surface from underground without pumps. Such areas create an oasis.

The Sahara is one of the driest deserts in the world. Most of the barren land receives less than 2.5 centimeters (one inch) of rain a year. The northern and southern edges of the desert receive slightly more moisture. However, in no place is there more than 25 centimeters (10 inches) of rain a year.

Temperatures in the desert change rapidly from day to night. Daytime summer temperatures are often over 38° C (100° F). The evening temperatures are cooler. In the winter, the daytime temperatures may vary from 21° to 27° C (70° to 80° F). Nights can be chilly in this season. In the northern Sahara freezing night temperatures have been noted. But this never happens in the southern Sahara.

The Steppes. The steppes begin where the desert ends. These regions are quite dry. Usually short grasses and scrub brush grow on the land. Very little rain falls. Although some of the soil is fertile, crops need irrigation to grow. The steppe areas are much like the dry grasslands of the Great Plains in the United States. This sub-Sahara region is often called "*the sahel*" (sah *heel*). Severe drought has plagued this region since 1973. In some places it ended about 1977. However, much of the steppe lands have become desert land.

The Savannas. The savanna lands are areas where more rain usually falls and where tall grasses, shrubs, and clumps of trees grow easily. Between rainy seasons, there may be periods of drought.

Much of the savanna has been used as farmland. Years ago farmers burned the grasslands before planting crops. Many fine trees were killed in this process. Large herds of animals once roamed these savannas. Today the remaining animal herds live mostly in protected game parks in sections of the savannas.

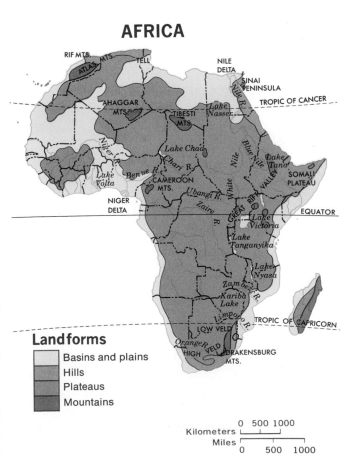

AFRICA

Landforms
- Basins and plains
- Hills
- Plateaus
- Mountains

Kilometers
0 500 1000

Miles
0 500 1000

The Sahel Club

On May 31, 1977, a club known as "Friends of the Sahel" began a three-day conference in Ottawa, Canada. Representatives from many nations attended this meeting. The club was studying a 25-year plan to fight drought in the steppes of sub-Sahara. The steppes in this part of Africa are called "the sahel." The word is Arabic, meaning "border."

The club is a very unusual one. It was started by people interested in helping the victims of a long-lasting drought that struck this region in the early 1970s.

Originally, only eight nations were members of this club. By 1977, 22 more nations had joined. Many national organizations also enrolled. The club was trying to collect over 850 million dollars during 1977 to aid drought victims in the sahel. More than ten billion dollars would be needed by the club to carry out the 25-year, anti-drought plan. The United States government promised some assistance to the club. You and your family may have already helped the needy people of the sub-Sahara in some way.

Lack of rainfall has caused this river to dry up. In order to get water the people of this region must dig into the riverbed.

Sudan

Sudan has the largest land area of all the countries in Africa. Its area is about as large as that of the United States east of the Mississippi River. Look for Sudan on the map of Africa on page 369. Notice that Sudan has part of its border on the Red Sea.

In the 15th century many people from the Middle East settled parts of Sudan. As a result most of the people in northern Sudan speak Arabic and believe in the Islamic religion today. Most of the southern Sudanese, however, are Black Africans with *animist* religious beliefs. **Animists** believe that there are many spiritual forces which rule over nature.

The Recent History of Sudan

From 1821 until 1955 the area that is now the Democratic Republic of Sudan was controlled by either Egypt or the United Kingdom. In the early 1950s the Sudanese people voted to become an independent nation. In 1956 Sudan became independent. A constitutional government, like the United Kingdom's, was established. In 1958 a group of military officers overthrew the government and took control of the country. The government was controlled by the military until 1965. In that year the military allowed elections to be held and the country was returned to civilian control.

The leaders of this new government wanted all citizens of Sudan to be able to read and to write Arabic. This angered many animist Sudanese. Some thought they would be forced to convert to the Islamic religion. Fighting, which had been going on since 1955, increased between people of the north and south. Again, the military took control of the government. In 1969 a Revolutionary Council was formed to run the government. The council selected a civilian premier and cabinet to start a socialist form of government. A peace treaty was signed with the people of the south in March, 1972. The fighting ended. It is hoped a lasting peace has been achieved.

The Land and Climate of Sudan

Sudan has three main regions. The northern third of the country is desert land. The middle third is dry steppe or grasslands. The southern third of the country is savanna land. Tall grass and patches of trees are scattered about the countryside. Much of the land along the White Nile is a vast marsh.

These three regions differ in climate. Two air masses tend to influence the northern and southern regions of Sudan. One air mass comes from the north. It blows in from the desert bringing hot, dry air. In northern Sudan summer temperatures run as high as 49° C (120° F). Rain hardly ever falls in this area. The second air mass is a humid one. It comes from the Indian Ocean. This air mass brings warm, humid air as well as rain to southern Sudan. Along the Uganda (yoo *gan* duh) border, about 125 centimeters (50 inches) of rain fall annually.

Most of Sudan is a fairly level plateau. Some high mountains are found along the western and southern borders. Lower peaks and hills extend along the Red Sea coast.

Agriculture in Sudan

Many of the people who live in northern Sudan are herders. These people graze their animals as do Arab herders in North Africa and in the Middle East. Many herds have suffered from the great drought of the sahel.

Most of the farmland in the Sudan must be irrigated. Cotton of high quality has been the nation's main cash crop. Huge quantities of Sudanese cotton are exported each year. However, Sudan's economy has suffered at times due to its dependency upon this one main export. The production of synthetic fabrics has often lowered the world market price for cotton. The government has asked farmers to grow a variety of crops on new irrigated farmland. New roads have been built from farms to city markets. Reservoirs and wells are being constructed to provide more water for irrigation. Cash crops such as sugar cane, coffee, tobacco, jute, and sisal are sometimes grown on new farmlands.

Most farmers grow enough food to feed themselves. Millet, peanuts, wheat, beans, peas, and dates are grown. Fruits such as mangoes and bananas add variety to the diet.

Another important export from Sudan is **gum arabic**. It is made from the resin of the acacia (uh *kay* shuh) trees that grow in the savanna lands. Gum arabic is used to make the glue used on the backs of postage stamps and envelopes.

Sudan and the Nile River

The waters of the White Nile and Blue Nile join at the city of Khartoum (kahr *toom*) to form the great Nile. From there, the river flows north into Egypt. Water rights along this river have been a concern to both the people of Sudan and the people of Egypt. During the colonial era, the British agreed that Egypt should control the waters of the Nile. As a result, the Sudanese were permitted to store water behind their dams only during flood seasons. You can understand why the government of Sudan might not like this rule.

In flood season—late August—the Nile peaks north of Khartoum. It expands to 16 times its size during flood season. Over half the total volume of the year's supply of water from the Nile passes through Sudan between mid-July and September. In the flood season, the Blue Nile and its tributaries provide more than three-fourths (¾) of the lower Nile's total water.

After independence, the government of Sudan agreed with Egypt on a system for sharing the Nile's water. Three main irrigated farming areas have been developed in Sudan. Two of these areas are between the main branches of the Nile River. The other area is between the Atbara (*aht* bah rah) River and Khartoum.

Khartoum and Omdurman

The capital city of Sudan is Khartoum. It was built by the British as a colonial headquarters. It is now a modern city with tree-lined, paved streets and avenues.

A major industry in Khartoum is textile manufacturing which makes quality cotton fabric. A number of smaller industries provide goods such as shoes, matches, soft drinks, and some food products. Most of these items are sold locally.

Across the White Nile from Khartoum is Sudan's largest city, Omdurman (*ahm* duhr *man*). As in many Egyptian cities, most people have built their homes from sun-dried bricks. Houses in Omdurman tend to be smaller than the ones of more modern design in Khartoum. Homes are often surrounded by a high brick wall or courtyard. The wall shields the home from some of the sun's heat. The thick walls act as cooling insulation in the hot, dry climate.

The government of Sudan has been an active member of the Arab League and of the OAU. The political leaders in Sudan sometimes seem to be more interested in the activities of the Middle East than in those of the landlocked nations to its west. The Sudan, therefore, is often included in studies of the Middle East.

Modern refrigeration is not available to all the people of Africa. In Sudan many people depend on ice delivered to their homes to keep food fresh.

1. What are the three main regions in the Southern Sahara?
2. What causes some of the differences in climate between northern and southern Sudan?
3. Prepare a display of fabrics, comparing Sudanese cotton with a variety of synthetic fabrics. Note the strengths and weaknesses of each fabric in your display.
4. Use a desk outline map of Africa to complete the following activities:
   a. Locate and label the following countries:
      Sudan, Upper Volta, The Gambia, Mauritania, Niger, Mali, Senegal.
   b. For each nation of the Southern Sahara region;
      (1) Place a blue mark on each nation that is east of the prime meridian.
      (2) Place a red mark on each nation that is west of the prime meridian.
      (3) Place a green mark on each nation that has land both east and west of the prime meridian.
5. Why do you think many people of the sahel want to return to their drought-stricken land?

LEARN MORE ABOUT THE SOUTHERN SAHARA REGION

Due west of Sudan are four land-locked, sub-saharan nations. West of them are three nations with borders that touch the Atlantic Ocean. You might wish to extend your knowledge of Africa by studying more about these countries. Use the Study Guide on page 31 to plan your additional work.

Chad, Niger, Upper Volta, and Mali

These countries of the Southern Sahara share many similar characteristics. By looking at the map on page 369 you can see that they have no direct access to the sea. These countries must depend upon neighbor nations for an outlet to the sea.

All four nations were colonized by France in the 19th century. They were all granted independence by France in the summer of 1960. France's influence is still an important factor. French is the official language in these nations. Also, the four countries trade mainly with France.

The people of these countries depend upon livestock as their main source of cash income. As a result of the long drought in the sahel these four nations suffered huge financial losses. Thousands of cattle died.

Chad, Niger, and Mali all have about the same land area. Upper Volta, however, is much smaller than its three landlocked neighbors. Chad and Niger have fewer people (a little more than 4 million) than Mali and Upper Volta (more than 6 million).

Mauritania, Senegal, and The Gambia

Along the Atlantic Ocean are the countries of Mauritania, Senegal, and The Gambia. These lands are also quite interesting for further study.

The Islamic Republic of Mauritania. This nation is known as "the land of the Moors." Four out of five people are Moors, a mixture of Arab and Berber people. These people live mostly in the northern part of the country. Only about one person in five is Black African.

In 1960 the economy began to change from its traditional pattern. In the north nomadic Moorish herders still tend their goats, sheep, and camels. However, in the south rich iron ore deposits were discovered in 1960. Up to that time the southern part of this nation had only been used for farming. Many changes began because of the new resources. The profits from iron ore helped the nation to build its first heavy industry, mining. The profits from mining helped the country to pay for its

Based on per-capita income, Africa contains some of the poorest and some of the richest countries in the world. What other parts of the world have rich countries? Poor countries?

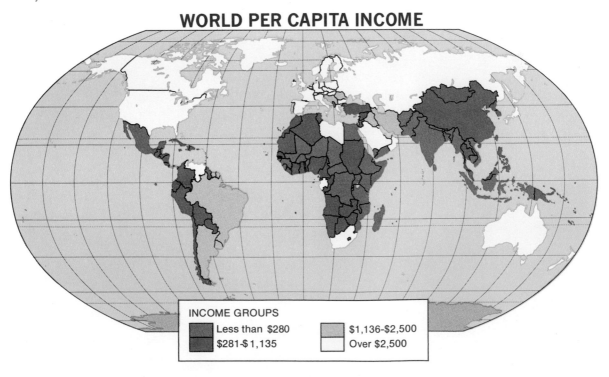

WORLD PER CAPITA INCOME

INCOME GROUPS
Less than $280
$281-$1,135
$1,136-$2,500
Over $2,500

Southern Sahara

first railroad. Mauritania was able to improve the ocean harbor at Nouadhibou (noo *ah* di boo).

Senegal. This country was the "hub" of French colonial rule in western Africa. Under colonial rule, major transportation systems were built from Senegal to other French territories. As a result a major harbor is located in Dakar (duh *kahr*). A major rail terminal is there as well. International air service has been in operation at Dakar for many years. Part of Senegal's economy has been based on services and commercial needs of its neighbor nations.

The Gambia. This nation has an unusual location. It is a thin sliver of a nation, surrounded on three sides by Senegal. This nation's land area consists mainly of the Gambia River Valley. The Gambia's economy depends largely upon its peanut crops for export revenues. The government began to grow rice as an extra crop. By 1975, The Gambia was growing enough rice for its own use. The Gambia is hoping to export rice in addition to peanuts. The Gambia's coastal area contains excellent beaches. Many tourists from Scandinavia vacation at the ocean resorts.

The nomadic Tuareg (twah rehg) people of Mali and Niger depend on the scattered wells and oases of the Sahel.

WHAT HAVE YOU LEARNED?

1. Why are places like Tanzania and Zimbabwe important to the history of Africa?

2. What happened at the Berlin Conference of 1884? Why did the nations of Europe want territory in Africa?

3. What are some of the problems faced by new nations of Africa?

4. What is the difference between steppes and savanna land?

5. Which is the largest nation in all of Africa, in terms of land area?

BE A GEOGRAPHER

1. Use a desk outline map of Africa to do the following tasks:
   a. Identify three nations of the Southern Sahara that have Atlantic Ocean coastlines.
   b. Identify four landlocked nations of the Southern Sahara.
   c. Identify the Southern Sahara nation with a Red Sea coastline.

2. On a desk outline map of Africa locate and label the following:
   a. The White Nile, the Blue Nile, and Khartoum.
   b. The Niger River. Indicate the direction of the river's flow.
   c. The Tibesti Mountains, the Senegal River, Lake Chad, and the Gambia River.

QUESTIONS TO THINK ABOUT

1. Why do you think some people might have shed tears when Djibouti became independent?

2. Why do you think some people in the new countries do not like a national system of government?

3. What do you think might happen if Sudan were to decide to use all the water of the Nile River to irrigate its desert land?

4. Which of the major problems faced by many African nations do you think are the easiest to solve in nations of the Southern Sahara?

WESTERN
———— EQUATORIAL AFRICA ————

Western Equatorial Africa includes much of the land between the Sahara and the Gulf of Guinea, as well as the land from the Cape Verde (vuhrd) Islands eastward to Zaire. Look at the map on page 376. Notice the many nations included in Western Equatorial Africa.

Locate the Cape Verde Islands in an atlas. These islands are near the countries of Guinea-Bissau (bihs *ow*), Guinea, Sierra Leone (sih *air* uh lee *ohn*), and Liberia. This section of land is sometimes called the "western bulge" of Africa. Continuing eastward along the Gulf of Guinea are a number of interesting countries. The coastal strip was called the Gold Coast because much gold was once mined there. Find the Ivory Coast, Ghana, Togo (*toh* goh), Benin, and Nigeria. At about 10° E. the coastline turns almost directly to the south. On a map it looks like a right angle. Some people call this the "hinge" of Africa. Can you locate the nations of Cameroon (*kam* uh *roon*), Equatorial Guinea, Gabon (ga *bohn*), Central African Republic, and Zaire?

You have probably noted that most of this land is tropical rain forest. Some savanna lands are located in the northern section of the region. Find the equator. Most of the West African countries are slightly north of the equator. However, the equator does cross through Gabon, Congo, and Zaire.

Western Equatorial Africa has had a long and varied history. The art, music, and drama of West Africa are a reminder of this continent's rich cultural heritage. The achievements of the people of West Africa today continue to make one aware of the importance of this region and continent.

Angie Brooks—Liberia's Contribution to the United Nations

In 1969 the delegates to the United Nations were ready to elect the year's President of the General Assembly. Regions of the world take turns presenting candidates for the elected

Angie Brooks opened the 24th annual session of the United Nations in 1969.

office. It was Africa's turn for a candidate. When the votes were counted, Angie Brooks from Liberia was elected President of the General Assembly. The delegates rose to their feet to applaud Ms. Brooks. She rushed to the podium to take her place as president. Her warm smile seemed to express her happiness for receiving this international honor. Angie Brooks was the third African to hold this office. She was the second woman to be elected President of the General Assembly.

Angie Brooks was born and raised in Liberia. She grew up in a foster home in Monrovia (muhn *roh* vih uh). As a teenager she paid for her schooling by doing part-time work. Having taught herself to type, she earned money as a typist. Angie Brooks wanted to become a lawyer. However, there were no law schools in Liberia at that time. She was able to come to the United States to study law. After years of hard work, Ms. Brooks returned to Liberia as a lawyer.

In 1954 she was appointed as a delegate from Liberia to the United Nations. Ms. Brooks worked on many UN committees. She was very dedicated to the goals of international cooperation and peace. Her election to this high post in 1969 shows how other UN members felt about the contributions of Ms. Brooks.

Western Equatorial Africa is one of the most densely populated regions of the continent. Seasonal farm workers also flock to these lands during harvest time. Laborers from the Southern Sahara have added to this population since the drought of the sahel in the early 1970s. Look at the population map on page 408. Notice the densely populated lands of this region.

Climate in Rain Forest Regions. The climate of Western Equatorial Africa is tropical. The maps showing land use, on page 398, and rainfall on these two pages will help you discover some features of this region.

Most of the land receives heavy rainfall. Yearly averages vary between 100 and 200 centimeters (40 and 80 inches). Warm currents in the Gulf of Guinea also affect the climate of this land. Notice on the map on page 369 that much of the land is covered by tropical rain forests. Plantation agriculture and old ways of farming are found throughout this region. The rains and temperatures of this region enable certain kinds of agriculture to exist.

Temperatures in coastal forest regions tend to be warm and humid all year long. Temperatures average about 25° to 30° C (75° to 85° F). Humidity is often about 80 percent. There is little change in temperatures between day and night.

Two rainy seasons occur with dry seasons between. This provides some changes in the yearly weather.

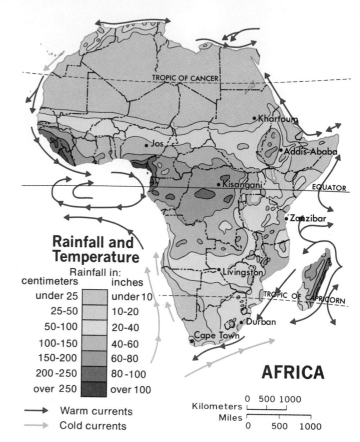

Rainfall and Temperature

Rainfall in:

centimeters	inches
under 25	under 10
25-50	10-20
50-100	20-40
100-150	40-60
150-200	60-80
200-250	80-100
over 250	over 100

→ Warm currents
→ Cold currents

AFRICA

Kilometers 0 500 1000
Miles 0 500 1000

The "long rains" begin about March. These rains might occur a month or two later in regions north of the equator. The "long rains" bring a heavy volume of water and last about three months. The "dry season" lasts from June to September. However, light rains may occur in this season. The "short rains" follow for about two months. The longer "dry season" then lasts from late November to March.

Climate in the Savanna Lands. On the savanna lands of the interior, temperatures vary more. There is a greater difference between day and nighttime temperatures. There also is less humidity. Some areas have experienced daytime highs of about 50° C (120° F). Seasonal lows of 4° C (40° F) have been reported. Savanna

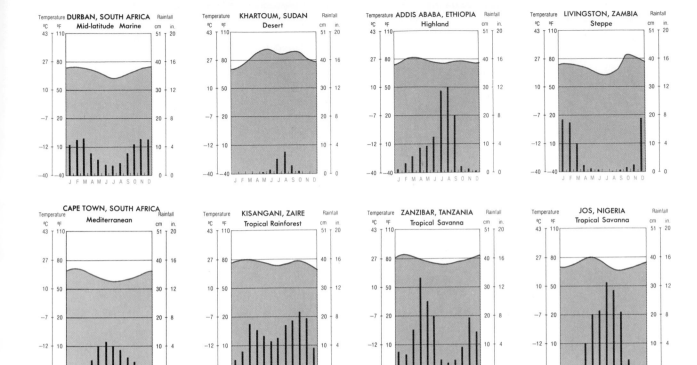

lands usually have two seasons: rainy and dry. The rainy season is normally from April to September. Savanna lands receive less rainfall than the coastal forested areas.

The savanna lands have occasional "**harmattan winds**" (*hahr* muh *tan*). These are hot, dry, and dusty winds that sweep off the Sahara. These winds often raise the temperatures of the area rapidly. Harmattans create a red haze in the sky due to the vast amount of sand which they collect as they blow along. The hot dry winds absorb moisture from plants, animals, and humans. They can be very uncomfortable and damaging.

The Rivers. You may have noticed on the map that many rivers drain the lands of Western Equatori-al Africa. Two of the three largest rivers on the continent are located in this region. They are the Zaire (Congo) and the Niger. Trace these rivers from source to mouth.

The People. In the past, most people lived off the land as farmers or herders in West Africa. Since independence, new governments are working hard to improve the way of living for their citizens. Schooling for children and adults usually receives a large part of the government's money. Improvements in health care have been made. Many new industries and businesses are providing employment to workers. As you study more about this region, you may discover many other changes occurring in Western Equatorial Africa.

Western Africa

Nigeria

Nigeria is located along the eastern end of the Gulf of Guinea. It is the largest country on the Gulf. Nigeria is almost as large as the states of Texas, Oklahoma, and Arkansas combined. However, Nigeria contains more than four times the number of people that live in these three south central states. More than 68 million people were estimated to live in Nigeria in 1978. Nigeria has more people than any other nation in Africa.

The Land and the Rivers

Nigeria can be divided geographically into three main regions: northern, eastern, and western Nigeria. Notice from the map on page 369

how the Niger and Benue (bay *nway*) rivers divide the land. North of the rivers is a rolling plateau. The plateau region varies in elevation from about 300 to 1500 meters (1,000 to 5,000 feet) above sea level. The rivers join together in south-central Nigeria. The river then divides the land into the eastern and western regions.

Near the coast much of the land is swampy. It is covered by dense mangrove forests. Mangrove trees grow in shallow ocean salt water. A great deal of rain falls along the coast. In the southeast as much as 380 centimeters (150 inches) of rain fall in a year!

Farther inland, the land is slightly higher in elevation. Tropical rain forests grow on much of this land. To the north are savannas. The rains

The dense mangrove forests of Nigeria are used by the people of that area as a source of timber.

there decrease to as little as 65 centimeters (25 inches) per year. Mountains are found only along the Nigerian-Cameroon border.

The two main rivers of Nigeria, the Niger and Benue, form one of the main river systems of the continent. The Niger-Benue River has created a huge delta at its mouth in the Gulf of Guinea. Unfortunately these rivers are not navigable all year long. During the dry season, the water is usually too shallow for deep-bottomed boats. Shifting sand bars also make the waterways difficult for navigation.

Independence and Civil War

Nigeria became independent from Britain in 1960. People thought that Nigeria would advance rapidly. The country had many agricultural and mineral resources. These resources could provide a rich economy for the people. However, there were many problems that faced the new leaders of Nigeria.

The boundary lines for independent Nigeria included a number of tribal territories. Some of these tribal groups were very different from one another. People distrusted and feared one another. In the north were the **Hausa** (*how* suh) and the **Fulani** (foo *lah* nih) people. Many of these people were devout Muslims. In the west lived the **Yoruba** (*yohr* uh buh). The **Ibo** (*ee* boh) people were a majority of the population in the east. The people of the east and west were either Christian or animists. Religious differences tended to separate many people. More than 250 different languages are spoken throughout Nigeria. These language differences became another block to national unity.

During the early years of independence, government jobs were held by people from a variety of tribes. The prime minister was a Hausa from the north. Most of the important government positions were held by Ibos from the east. In January 1966 a group of Ibo army officers overthrew the government. But the powerful northerners, who outnumbered the Ibos, objected to being governed by the Ibo minority. Great violence and tribal persecutions began to occur. Thousands of Ibos lost their lives in the conflict. About 250,000 Ibos fled the northern land of the Hausa to the safety of their tribal land in the east. Six months later, in July, 1966, the northerners overthrew the Ibo-led government.

However, the conflict did not stop. The new government tried to solve the problems by writing a new constitution and dividing the regions into 12 states. The leaders of the Ibos, however, thought that their people would lose control of the eastern region. A major ocean port in the east was very important to the Ibos. Of even greater value was the wealth of oil reserves discovered near the Bay of Biafra (bee *af* ruh). The Ibos decided to take control over all the land of the eastern region. They created a new nation called *Biafra*. This started Nigeria's civil war in 1967.

The government of Nigeria did

Damaged during the civil war in Nigeria, this schoolhouse is again put to use.

not wish to lose this rich eastern territory. Bitter fighting began between national troops and the Ibos. Many homes, businesses, farms, and roadways were destroyed. Hundreds of thousands of people became victims of the civil war. After three years the Nigerian national troops regained all the land. The civil war ended. Since the end of the war much has been done to settle the problems and fix the damage from the fighting. However, the memories of the bitter civil war will remain in the minds of Nigerians.

Agriculture in Nigeria

Many people depend on the land for their livelihood. In the southeast rainfall is abundant. Many root crops are raised. Cassava and yams are the chief ones. The roots are usually ground into flour and cooked as a main part of each day's meal. Corn is grown in most of this area. The main cash crop in this area is **palm kernels**. These are the fruit of the oil palm tree. Palm oil is used in making soap and oleomargarine. Nigeria is the world's leading exporter of palm oil and kernels.

In the southwest where the land is slightly drier, cacao is grown. Only Ghana exports more cacao than Nigeria. Peanuts are also a major export.

Northern Nigerian farmers raise many grains, cotton, and peanuts. Herds of cattle and goats are also kept. In the drier steppes of the extreme north many nomadic herders graze their animals. However, many of them have lost their animals due to the sahel drought.

The Mineral Wealth of Nigeria

Nigeria is a land of great mineral resources. In northern Nigeria is a high plateau about 2000 meters (6,000 feet) above sea level. It is near the city of Jos (jahs). Locate this area on the map on page 369. Many valuable mineral deposits of Nigeria are mined in this location. Tin is one of the major ores mined here. Tin smelters process most of the ore before it is exported. **Columbite** is a by-product of the tin mining. It is a rare mineral. From this mineral, some metals are obtained which are used in jet engines and nuclear reactors. Huge deposits of iron ore, coal, and limestone are also found in Nigeria.

Nigeria's most important mineral is petroleum. Oil was discovered in 1959. Several oil companies have continued the development and search for oil in the eastern shore area. Nigeria is rapidly becoming one of the major oil-producing nations of the world. Profits from the oil revenues are helping the government to finance new industries, improve transportation systems, and provide more help to local people. Plans have been made to start an auto assembly plant and expand into iron and steel manufacturing. Power from the waterways is being used to meet the electrical needs of the nation. Hopefully, future wealth and lasting peace will make Nigeria a major nation of the world.

Zaire

Zaire occupies a large portion of the land in west-central Africa. Zaire is almost a landlocked nation. It has a very narrow outlet to the Atlantic Ocean between the two parts of Angola (an *goh* luh). Zaire has more nations surrounding it than any other country in Africa. Locate Zaire on the map on page 369.

A Rich Nation Makes a Slow Start

Zaire has many natural resources which might make the country one of the most prosperous nations in all of sub-Saharan Africa. However, in the early days of independence it was not easy to start a new nation. A bitter civil war occurred shortly after independence. Many nations of the world became involved in the crisis between 1960 and 1965. Why was it so difficult for the people of Zaire to manage peacefully their new nation? Knowing something about the colonial history of this territory will provide you with some of the answers.

The "Belgian Congo." For over 60 years the area was under the rule of Belgium. The region was commonly called the "Congo." The name "Belgian Congo" was also used to identify it from the neighboring French territory which was also called "Congo." The Belgian colonists used African laborers to work their plantations and mine the rich minerals. However, the colonial rulers did not allow Africans to improve themselves through education. Africans were not allowed to work in the highly skilled jobs of industry. Europeans were brought to the Congo to fill these important jobs. Africans were not allowed to take part in the governmental affairs of the colony. The Africans had few, if any, social, economic, or political rights.

Tribalism vs. nationalism. All through the colonial period Africans in the Congo kept their tribal traditions. By doing this the Africans were able to keep their social and political unity. However, there were many different tribal groups through the land. Up to 200 have been counted. Tribal customs differed greatly. There were no reasons for tribes to work together until independence day in 1960. At that time nationalism was thrust upon the people. This meant that all of the different tribal groups now had to unite as one group of citizens. It was not an easy idea for the millions of people to understand.

Distrust quickly occurred among some of the leaders of the new nation. The fear and dislike among different tribes sometimes made it very hard for the government to control the country. Some leaders tried to take control of the government. Other leaders declared their tribal lands independent countries. A civil war was underway.

The United Nations sent troops to aid the government in ending the civil war. Four years after independence, the UN troops were still there. No one thought it would take so long to form a united government in Zaire. Peace-keeping effort became so costly it almost caused the UN to go bankrupt!

A strong national government was finally formed to establish peace throughout the nation. By the end of the 1960s the nation was at peace.

Keeping the machinery running at open-pit copper mines is important to the economy of Zaire.

The president changed the name of the country from the Congo to Zaire in 1971. He hoped to create a new feeling of nationalism among the people. Many people also replaced their Christian-European names with African names. Many cities of Zaire also had their names changed. Map makers have had a difficult time keeping up with changes in Zaire.

In 1977 there was fighting again in Zaire. Rebels tried to take away the mineral rich region of Shaba (*shah buh*) Province near Angola. With

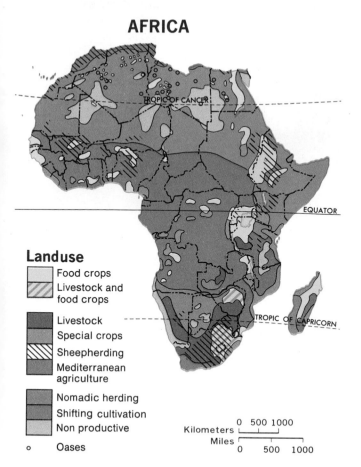

AFRICA

Landuse
- Food crops
- Livestock and food crops
- Livestock
- Special crops
- Sheepherding
- Mediterranean agriculture
- Nomadic herding
- Shifting cultivation
- Non productive
- ○ Oases

TROPIC OF CANCER

EQUATOR

TROPIC OF CAPRICORN

Kilometers 0 500 1000
Miles 0 500 1000

Only the Amazon River has a greater yearly flow of water.

Mountains along the extreme eastern border near the equator rise to more than 5000 meters (17,000 feet). These peaks are snow-capped all year long. The lakes of the Great Rift Valley form part of the eastern border as well. High plateaus and low mountains extend along the southern border. Notice these features on the map on page 379.

Most of the rest of the land along the equator is covered by tropical rain forests. An undergrowth of ferns and bushes creates a low ground cover. Rising through them to heights of 45 to 60 meters (150 to 200 feet) are gigantic hardwood trees. Many vines and plants also grow on these trees. Some of the land in the north and south is savanna.

Rainfall averages more than 100 centimeters (40 inches) a year. Along the narrow coastal strip on the Atlantic, a heavier volume of rain falls. Along the equator rains fall most of the months of the year. Some dry seasons occur for one to five months in the savanna lands.

international help, the Zaire government was able to keep its national territory. Lasting peace and nationalism sometimes develops slowly.

The Land and Climate of Zaire

Zaire contains about one-fourth (¼) as much land as the 50 states of the United States. Most of this area is a large basin drained by the Zaire (Congo) River and its many tributaries. The Zaire (Congo) River is one of the world's largest rivers. It flows for about 4800 kilometers (3,000 miles) through Zaire or along its border.

The Wealth of the Land

Vast mineral wealth exists in Zaire. About nine-tenths (⁹⁄₁₀) of all exports from the country are minerals. Most of them are found near Lubumbashi (*loo* boom *bahsh* ee) in the southern province of Shaba. Some of the world's richest copper mines are located in this region. Zaire is among the leading nations of the world in the production of cobalt and industrial diamonds. Natural gas fields and off-shore oil may add revenues once they are developed. Other valuable ores found in Zaire are gold, coal, manganese, zinc, and tungsten.

Agriculture

Some fertile land has been developed for agricultural use in Zaire. Farmers have been taught to rotate crops and use fertilizers. Experimental stations are developing improved seeds for the farmers' use. Main export crops are vegetable oils, coffee, cotton, and rubber. Crops used locally include sisal, sesame, tobacco, wheat, rice, cacao, corn, yams, cassava, and sugar cane. Some fishing is done in the many lakes and streams. Lumber products include valuable hardwoods of ebony, teak, and mahogany.

Some herding is done on the savannah lands that are free of the *tsetse* (*tseht* see) *fly*. The **tsetse fly** spreads a deadly disease among cattle. This insect is found in parts of central Africa, and is difficult to control. The tsetse fly spreads sleeping sickness to humans. Few people could live in tsetse-fly regions until medicines were developed to help control the disease. Insecticides are used to kill the flies that carry the germs.

Manufacturing

Industries in Zaire process either mineral or agricultural products.

Most ores are refined or smelted before exporting. Chemical plants have been built. Metal shops produce wire and steel products. Items made from plastic are manufactured locally: shoes, dishes, curtains, spoons, forks, and other household items. Some bicycle and auto assembling is done in Zaire. Textile factories produce cotton textiles and blankets. Vegetable products, cooking oils, soaps, and cigarettes are made for local use. New factories are developing each year.

Zaire is very fortunate to have an abundant source of hydroelectric power. Within the boundaries of the country is about one-sixth (⅙) of the world's possible hydroelectric power. Electricity has already been produced from many of the rivers in the country. One of the largest hydroelectric plants has been developed on the lower Zaire (Congo) River. A part of the Inga (in *gah*) Hydro Scheme began operating in 1972. When completed, it will produce about twice as much power as the Grand Coulee Dam in the state of Washington. Other dams on the Lualaba (*loo* uh *lah* buh) River supply power to smelters near the copper fields.

Transportation

It is necessary to transport huge quantities of goods to and from the mineral regions of the interior of Zaire. Recent construction has improved both river and railway systems of the nation. Many railroad spurs help by-pass rapids on the

Most of the people in Zaire live along or near the Zaire River, either in modern cities or in traditional villages.

huge Zaire (Congo) River. A railroad by-passes the dangerous rapids of the river between Kinshasa (kihn *shahs* uh) and the seaport of Matadi (muh *tahd* ee). All-weather roads are also being built into the interior of the country. River steamers continue to provide much transport for goods and people along this lengthy river. Some minerals were shipped by rail through neighboring Angola. However, disputes between the governments have occasionally closed this rail line. This has forced Zaire to improve the rail system within its boundaries. Some transport to eastern Africa is possible on Lake Tanganyika (*tan* gan *yee* kuh). However, this is not as heavily used as is the Atlantic seaport on the west coast. Airports link all major regions of the country. An international airport is located near Kinshasa.

Cities and Places of Interest

The capital of Zaire is the beautiful city of Kinshasa. It is located east of the rapids that block ocean traffic from entering the Zaire (Congo) River. At this point the Zaire River is very wide. It looks like a lake. This water area is called Malebo (muh *lee* boh) Pool (formerly Stanley Pool). Many offices of the capital have been built on the southern side of this pool. Modern buildings and housing developments, shops, theaters, and restaurants are to be found in sections of the main city. Radio, television, and movies are quite popular with many people in Kinshasa. On the opposite shore is Brazzaville (*brahz* uh *veel*), capital of neighboring Congo.

Albert National Park is located near Lake Kivu (*kee* voo). Many wild animals live in eastern Zaire. Park wildlife includes lions, elephants, giraffes, gorillas, antelopes, hyenas, hippos, and the rare okapi. Several active volcanoes are found in the east. The Ruwenzori (*roo* wuhn *zohr* ee) Mountains are sometimes called the "Mountains of the Moon." The snow-capped peaks on the equator are an unusual sight in Western Equatorial Africa.

LEARN MORE ABOUT WESTERN EQUATORIAL AFRICA

Western Equatorial Africa contains more nations than any other region of Africa. A wealth of information about western Africa could be gathered by doing extra study of nations in this region. You might wish to do a report on one other nation of Western Equatorial Africa. Groups might be formed to study the nations of a locality within the region. Use the Study Guide on page 31 to help you plan your extra study of Western Equatorial Africa.

Countries of the "Western Bulge"

The countries along the "western bulge" might be one locality for a group to study. The Cape Verde Islands are the only island-nation near the western coast of Africa. Guinea and neighboring Guinea-Bissau, though similar in name, have very different histories. Guinea was made independent from France in 1958. Guinea-Bissau fought a war to free itself from Portuguese rule. It became independent in 1974. Sierra Leone and Liberia were both once used as settlements for freed slaves returned to Africa. Liberia, the oldest republic in Africa, has been independent since 1847. Many Liberians are descendants of people who were once slaves in the United States.

Countries of the "Gold Coast"

The countries of the Gold Coast could be a second section for group

The sisal plant is grown in many countries in the low latitudes. Its tough, stringy leaves are used to make ropes, bags, and insulation.

study. The coastal regions of the Ivory Coast have become a very popular vacation center in western Africa. Ghana's early independence in 1957 was a model for other nations to follow during their struggle for freedom. Togo and Benin, though small in size, have had to use the military to form a peaceful government.

Countries of the "Hinge of Africa"

The nations near the "hinge of Africa" could be a third section for group study. The plantation farming system of Cameroon may be of interest to some students. Discoveries of uranium ore in landlocked Central African Republic had added a bright note to the poor economy of this nation. Equatorial Guinea is located on both island and mainland territories. These lands were under Spanish rule for more than 180 years. Gabon has been using its variety of oil-mineral resources to improve the ways of living for many citizens. The People's Republic of Congo is much smaller than neighboring Zaire. The land has not been blessed with the variety of mineral resources found in neighboring nations. Once you begin to study about these nations, you will have much information to share with students and friends about Western Equatorial Africa.

CHAPTER 21 REVIEW

WHAT HAVE YOU LEARNED?

1. What is the climate like where tropical rain forests grow?

2. What are harmattan winds?

3. What are the three main geographical regions of Nigeria?

4. Why is the area around Lumbumbashi so important to Zaire?

5. Why have short railroads been built at certain places along rivers in Zaire?

BE A GEOGRAPHER

1. Which has more of its land within middle latitudes:
   a. Africa or South America?
   b. Africa or North America?
   c. Africa or Europe?

2. Generally, does the average rainfall increase or decrease as you journey either north or south from the equator in Africa?

3. What nations in Western Equatorial Africa are crossed by the prime meridian?

4. What country would you be in if you were at these locations?
   a. 10° N., 10° E.
   b. 10° N., 0° Longitude
   c. 10° N., 10° W.

5. Name the direction asked for in each of the following:
   a. Is Ghana east or west of Togo?
   b. Is Benin east or west of Ivory Coast?
   c. Is Liberia east or west of Nigeria?

QUESTIONS TO THINK ABOUT

1. Why do you think most early people of Western Equatorial Africa were either farmers or herders?

2. In what ways do you think the seasons in Western Africa might be helpful or harmful to the herders and the farmers?

3. What do you think are some ways other than civil war to settle disagreements between citizens of a nation?

4. If Zaire has so many valuable natural resources, why do you think it called for foreign assistance to help fight rebel attacks in its territory?

Chapter Review

·✦[22]✦·
Chapter

──── EASTERN AFRICA ────

Eastern Africa is the region south of Sudan and east of Zaire. Locate this region on the map of Africa, page 369. Notice the three land-locked nations east of Zaire. Can you find the nations Burundi (buh *roon* dee), Rwanda (ruh *ahn* duh), and Uganda? East of Sudan is Ethiopia, Djibouti, and Somalia (soh *mahl* ee uh). This section of land is called the "Horn of Africa." Notice how the horn-shaped land mass extends into the Gulf of Aden and the Indian Ocean. South of the horn is Kenya (*kehn* yuh) and Tanzania, completing the nations of Eastern Africa. The equator crosses Eastern Africa at about its center.

The Land of Eastern Africa

The lands of Eastern Africa are very different from the regions of Africa already presented. Look carefully at the landform map on page 379. High mountains and long lakes separate eastern Africa from the deserts and tropical rain forest lands to the west. Most of the land to the east is a high plateau. Land elevation varies from 900 to 3000 meters (3,000 to 10,000 feet) above sea level. Trees and tall grasses cover much of the savanna lands of the plateau.

The Great Rift Valley. Cutting through Eastern Africa's plateau is one of the unusual landforms in Africa, the Great Rift Valley. The Rift Valley is believed to have been caused by a volcanic *fault*. The **fault**, or break in the earth's crust, caused part of the land to drop, and another section to rise. This "crack" seemed to occur all the way from Jordan in the Middle East, along the Red Sea southward through Eastern Africa. It ends at about Malawi (muh *lah* wee). The Valley is most easily noticed in Kenya. A sudden drop of 600 to 900 meters (2,000 to 3,000 feet) occurs, creating a deep valley about 80 kilometers (50 miles) wide.

The plateau descends to sea level quite rapidly near the Indian Ocean. A narrow tropical lowland extends along the coastline of the Indian Ocean. Beautiful white sandy beaches are found along most of the coast.

A coral reef parallels the coastline a brief distance away.

The Climate of Eastern Africa

The climate of Eastern Africa tends to be moderate in temperature. Although Eastern Africa is along and near the equator, the high elevation of much of the land has a cooling effect upon the temperatures. In the plateau areas 1200 meters (4,000 feet) above sea level average yearly temperatures are about 21° C (70° F). In the higher plateau lands in Ethiopia and parts of Rwanda and Burundi, yearly averages are as low as 15° C (60° F).

High mountain regions experience some cold temperatures. Many snow-capped mountains are seen in Eastern Africa. Africa's highest mountain, Mount Kilimanjaro (*kilh uh mun jahr* oh), is located in Tanzania. The peak stands 5895 meters (19,340 feet) above sea level.

Rainy and dry seasons are common in most lands of Eastern Africa. Two rainy seasons are separated by two dry seasons. The "long rains" occur between March and May. These rains are brought by monsoons, or seasonal winds. The "short rains" arrive between September and November. Rainfall is much less severe then.

Some of the most beautiful natural landscapes and wildlife herds in the world are found in the region of Eastern Africa.

Local farmers bring their produce to the market in Harrar, Ethiopia. Markets similar to this one are common in many parts of Africa.

Have you ever lived near the equator? One of the authors of this text once lived in Mombasa (mahm *bahs* uh), Kenya. Many people along the coast spoke the Swahili (swah *hee* lee) language. The Swahili system for telling time there was based upon the area's location near the equator. There are an equal number of hours of daylight and darkness along the equator. The sun rises about 6 A.M. It sets about 6 P.M. This remains true all year long. There is just about always 12 hours of daylight and 12 hours of darkness. According to the Swahili custom, people told time by counting the number of hours that had passed since the rising or setting of the sun.

For example, Dr. Hughes would eat his breakfast at 7 A.M. In Swahili one would say it was "*sa moja*," or one o'clock in the morning. To a Swahili speaker, they understood this meant one hour after the sun had risen. At 2 A.M. Dr. Hughes went to work (8 A.M.). In the evening the sun set about 6 P.M. If Dr. Hughes ate supper at 7 P.M., one might say in Swahili time that it was one o'clock after the sun had set. At first this time system seemed confusing to Dr. Hughes. But after he had lived along the equator for several months, the Swahili system of telling time seemed very sensible.

Ethiopia

Ethiopia is the oldest independent nation in Africa. It was, until 1974, the world's oldest kingdom. Ethiopia was never colonized. However, in 1936, Italian military forces bombed and occupied parts of Ethiopia. Emperor Haile Selassie (*hy* lee suh *las* ee) went to the League of Nations in Geneva. He asked the Court of Justice for help. He also warned the leaders of western nations, "It is us today, it will be you tomorrow." He received no immediate help. Years later during World War II many people remembered his warnings.

In 1974 a military regime overthrew the Emperor and ended his rule. He was imprisoned in his home at the capital, Addis Ababa (*ad* is *ab* uh buh). The following year, the 82-year-old emperor died.

The Land of Ethiopia

Ethiopia is almost as large as the states of Texas, Oklahoma, Arkansas, Louisiana, and Mississippi combined. About 24 million people live in Ethiopia.

More than half of Ethiopia is mountainous plateaus. Winding their way between many plateaus are deep river valleys. The largest of these plateaus is known as the *Ethiopian Plateau*. It extends from the southern border with Kenya northward to the Red Sea. The Great Rift Valley separates this plateau from the *Somali Plateau* in southeastern Ethiopia. Most of the land on these plateaus is from 1200 to 3300 meters (4,000 to 11,000 feet) above sea level. Mountain peaks rise to elevations of 4500 meters (15,000 feet) on the Ethiopian Plateau. Some peaks measure 4200 meters (14,000 feet) on the Somali Plateau. Volcanoes were once found in parts of Ethiopia.

Not all of Ethiopia is highlands. In northeastern Ethiopia, for instance, there is a low basin in the Rift Valley which is 116 meters (381 feet) below sea level. Southeastern Ethiopia is also lowland.

In the north of Ethiopia is the province of Eritrea (*ehr* uh *tree* uh). This land was given to Ethiopia by the UN after World War II. Its lands provide Ethiopia with an access to ports on the Red Sea. However, many people in Eritrea wish to be independent. There has been fighting going on there for many years. In addition to these ports, Ethiopia often depends heavily upon the port at Djibouti.

The People of Ethiopia

Many different tribal groups live in Ethiopia. A majority of Ethiopians have the same ancestors as the Berbers in northwest Africa and Egypt. A group of people in Ethiopia claim to be descendants of the "lost tribe of Israel." Many Ethiopians are Cop-

These priests of the Coptic Christian church are preparing for the Christmas feast. Although there are many Coptic Christians in Ethiopia, the country also has large numbers of Muslims and animists.

tic (*kahp* tihk) Christians. Most of these belong to the Amhara (ahm *hahr* uh) group. Under the rule of the emperor, the leaders of the Coptic Church had great power in the country. Since the overthrow of the emperor, this has changed. About half the members of the Galla (*gahl* luh) group are Muslim. Many Somali nomads have also lived from time to time in Ethiopia.

The Climate of Ethiopia

Even though Ethiopia is in low latitudes, the climate in most of the country is pleasantly cool. In what part of Ethiopia do you suppose the climate is hot? If you thought "the lowlands" you were right. In the

AFRICA

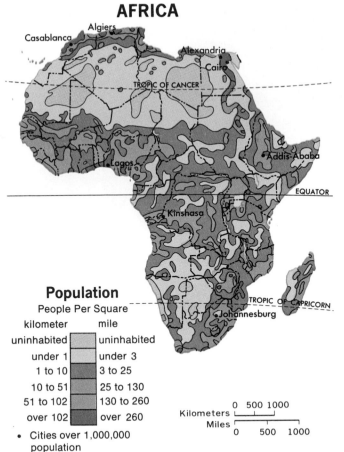

Population

People Per Square

kilometer	mile
uninhabited	uninhabited
under 1	under 3
1 to 10	3 to 25
10 to 51	25 to 130
51 to 102	130 to 260
over 102	over 260

• Cities over 1,000,000 population

```
         0  500 1000
Kilometers └──┴──┘
Miles   ┌──┴──┐
        0   500  1000
```

Nine out of ten people in Ethiopia make their living by farming. Most farmers still use very old farming methods. Low crop yields often result. Some efforts have been made by agricultural experts to teach farmers new methods. On most of the plateaus the soils are fertile and irrigation is not needed.

The two main food crops raised by the farmers are the grains called *teff* and *durra*. A white flour is made from **teff**, while **durra** is used mainly for feeding livestock. These plants are hardy and can be grown at high altitudes.

Other crops include wheat, barley, and oats. Also sugar cane, tea, cotton, and spices are raised in some regions. Many oil-bearing seeds such as castor beans, sunflowers, linseed, peanuts, and sesame are also raised.

Ethiopia's main agricultural export for many years has been coffee. Most of it is gathered from trees which grow wild on the high plateaus in the southwest.

Ethiopian herders raise many cattle, sheep, goats, horses, donkeys, and chickens. Camels and bees are also kept by some people. However, the drought in the early 1970s killed most of the herds. Many nomadic herders left their lands in the north to seek a livelihood farther south.

Not many mineral resources have been discovered. Small industries make some leather and hide products, cotton textiles, sugar, flour, cement, lumber, and soap.

highlands, temperature averages about 18° C (65° F) yearly. By comparison, lowland areas like Massawa (muh *sah* wuh) on the Red Sea become as hot as 49° C (120° F).

The map on page 369 shows several large rivers flowing from the Ethiopian and Somali plateaus. Those which flow northward join the Nile River. Those which flow southward empty into the Indian Ocean through Somalia. Lake Tana on the Ethiopian Plateau is the source of the Blue Nile River. The heavy rains that fall on this plateau also feed the Atbara and Sobat (*soh* bat) rivers. These rivers flow into Sudan and empty into the Nile.

Government and the Future

In 1974 a military take-over ended the reign of the emperor. The purpose of the military take-over was to end unemployment, rising inflation, and famine from the sahel drought. Increased resistance from Eritrean rebels caused the government major problems. Widespread unemployment, high prices, and famine are not problems that are easy to solve. The government in Ethiopia will be facing them for a long time.

REMEMBER, THINK, DO

1. Use a desk outline map of Africa to locate and label the following countries:
   a. the landlocked nations of Burundi, Rwanda, and Uganda;
   b. the nations in the "horn of Africa": Ethiopia, Somalia, and Djibouti;
   c. the East African nations of Kenya and Tanzania.
2. What are the main seasons in Eastern Africa?
3. Who was the last Emperor of Ethiopia?
4. Why did the military take over the government of Ethiopia?
5. What makes the Great Rift Valley an unusual landform? In what country is it most easily noticed?

In parts of Ethiopia water is a valuable resource. In many communities houses do not have their own water supply. Water is often delivered in barrels.

Tanzania

Tanzania is located just south of the equator on the eastern coast of Africa. It is south of Kenya and east of Zaire. Find Tanzania on the map on page 369. Lake Tanganyika and Lake Victoria form natural boundaries along the west. Lake Malawi forms part of the nation's southern border. The east boundary is formed by the Indian Ocean.

Tanzania contains two separate regions of land. One is the mainland, called Tanganyika. The second region is an island area opposite the mainland in the Indian Ocean. The islands are Zanzibar (*zan* zih bahr) and Pemba (*pehm* buh). They are located about 6° S. and 39° E. Find them on the map on page 369.

Find Tanzania on the map on page 369.

A Brief History

People have lived in eastern Africa for over two million years. Discoveries of bones of human beings at Olduvai (*ohl* duh *vy*) Gorge in Tanzania lead some people to believe that the human race may have originated in Eastern Africa. Arab traders are known to have visited the coastal lands about 1,000 years ago. Many African settlements and villages were already in Tanzania at that time. In the 16th century Portuguese traders had to fight the Arabs for control of ports along the Tanzanian coastline. The Sultan of Muscat (*muhs* kat) took control of Zanzibar in the 16th century. Sultans ruled the island of Zanzibar until the mid-20th century.

Following the Berlin Conference in 1884, Germany took possession of mainland Tanganyika. They created huge plantations to raise sisal and tea. During World War I the Germans lost this territory to the British. Britain united Tanganyika with its neighbors Uganda and Kenya to form British East Africa. Tanganyika became independent in 1961. Three years later it joined with the offshore island of Zanzibar to become Tanzania.

The Land and Climate of Tanzania

Most of the land of mainland Tanzania is a plateau. In some ways, Tanzania's plateau is shaped like a saucer. The edges of the plateau tend to be higher in elevation than the center. Outer areas average about 1200 meters (4,000 feet) above sea level. The central plateau is only about 900 meters (3,000 feet). The central plateau is drier and warmer than the surrounding higher lands. Only about 75 centimeters (30 inches) of rain fall annually in the central plateau. Much of this land is savanna grasslands. The southwest, northeast, and coastal lowlands receive between 100 to 250 centimeters (40 to 100 inches) of rain. The coastal lowlands have a hot, humid, tropical climate. The islands of Zanzibar and Pemba have climates similar to the coastal lowlands.

North of Lake Malawi the Great Rift Valley divides the mainland into two parts. The western valley includes the Lakes Tanganyika, Kivu, Edward, and Albert. The eastern part of the valley extends north into Kenya. Lake Rudolph is part of this Rift Valley. Lake Victoria is located to the northwest. This gigantic inland lake is the second largest fresh water lake on earth.

During the mid-19th century European explorers were seeking the source of the Nile. Tanzanian Africans showed the explorers this lake. The explorers found what they were seeking. Lake Victoria's waters form the source of the White Nile River.

Mountains are found along the southern border of Malawi and Zambia (*zam* bih uh). High peaks, including snow-capped Mount Kilimanjaro, are found along the Kenya border. Some African legends say that Mt. Kilimanjaro was built as a "heavenly throne" for God. According to legend, once the Earth was created God rested on the heavenly throne to view the world.

"Ujamaa"— The People and Their Work

After independence the government wanted to improve the ways of living for all citizens. Under the colonial system only a few African people were given employment. Thousands of others were living at a very poor level. In 1967, President Julius Nyerere (ny *rer* ay) introduced a new system of work. The plans were written in his Arusha (ah *roo* shuh) Declaration. Some called this system African socialism. Others called it Ujamaa (oo *jah* mah), which means familyhood.

At a height of 5895 meters (19,340 feet) Mount Kilimanjaro in Tanzania is Africa's tallest mountain.

Under the system of **ujamaa**, all members of the family are expected to respect one another. All members of the family are expected to work. The possessions of the family are shared. No person receives more benefits than another person until everyone has received some share. Any surplus profits may be given to those people who have contributed more work.

Agriculture in Tanzania

Agriculture has been one of the main sources of livelihood for people in Tanzania. Many cooperative

Many small boats can be seen along the wharves in the port of Dar es Salaam. These boats transport goods to and from the interior of Tanzania.

farms have been started. By working together, farmers are able to increase the quality and quantity of their crop yields. Modern equipment and tools have also been given to the cooperative farms. Main exports have been sisal, cotton, coffee, tea, tobacco, and meat and hides.

Sisal was once a valuable crop. However, the production of cheap, strong synthetic fibers has decreased the value of sisal fibers. This has almost ruined many of the huge sisal plantations of Tanzania.

Zanzibar's major crops are two spices, cloves and cinnamon. The islands grow more cloves than any other nation in the world.

The government is also encouraging farmers to grow crops high in food value. Nutritious food has been grown for the general health of children and adults. Food—such as wheat, barley, rice, sorghum, millet, cassava, and sweet potatoes—is grown in greater quantities. Green vegetables, meats, and fruits are also being added to the diet of many people. The government also helped people to build irrigation projects and control insects and diseases that harm plants, animals, and people.

Mining and Manufacturing in Tanzania

Tanzania has limited mineral resources. Some industrial diamonds are mined in the northwest. Gold and mica are also found. A few precious stones like rubies and sapphires are found occasionally. A new gem stone called *tanzanite* was discovered near Mount Kilimanjaro in 1968. A substance called *meerschaum* (*mihr* shum) is also mined. It is used in making tobacco smoking pipes. The meerschaum pipes are sold all over the world.

Food processing and light industries are main manufacturing activities. Textile factories produce many usable fabrics for local use. Other factories make shoes, soaps, razor blades, tires, and light appliances. Some bicycles, autos, and trucks are assembled in Tanzania, also. Powdered coffee is one of Tanzania's major food exports. Tea, cinnamon, cloves, and cashew nuts are also major food exports. Oil refining has become more important since independence.

Tourists from all parts of the world visit Tanzania. The Serengeti (sehr uhn *geht* ee) Plains are a home to some of the biggest herds of wildlife in all of Africa. Many beautiful lodges are found near the game parks. These have been built by the govern-ment. Dar es Salaam (*dahr* ehs suh *lahm*) is the capital and main port of the country. This city's name means "Haven of Peace." Its beautiful natural harbor and beach resorts are very popular.

Education and Government

The educational system of Tanzania has changed since the Arusha Declaration. School subjects include lessons about the government's work. The lessons include work projects for the school children. Students are taught skills that might be useful in improving local village life. All students are required to learn the national language, Swahili.

The government of Tanzania has been a very active member of the Organization of African Unity and the United Nations. The government sponsors many "Pan-African" projects. Tanzania also has given much assistance to Southern African freedom fighters.

REMEMBER, THINK, DO

1. What caused the formation of the Great Rift Valley?
2. Which peak in Eastern Africa is the highest mountain on the continent?
3. Use an outline map of Africa to trace the approximate location of the Great Rift fault from Jordan to Malawi.
4. Using the same outline map, locate and label the following interior lakes and rivers in Eastern Africa:
   a. Lake Tanganyika
   b. Lake Victoria
   c. Lake Tana
   d. Lake Malawi
   e. White Nile River
   f. Blue Nile River
5. Why are the discoveries at Olduvai Gorge important to the history of Africa?

Eastern Africa

LEARN MORE ABOUT EASTERN AFRICA

Additional information about Eastern Africa might be of interest to you. The history, art, crafts, music, and ways of living in Eastern Africa are very different from that of other regions you have studied. Use the Study Guide on page 31 to help you plan further studies. You might wish to include other items of interest about these countries that you find yourself.

Other Nations of the Mainland

Burundi and **Rwanda** are two of the smallest nations of Eastern Africa. They are sometimes called "the Switzerland of Africa" because of their location near the "Mountains of the Moon."

Uganda is a landlocked nation near Lake Victoria. Its government caused great concern to many people in the world during the 1970s.

Traffic congestion in the urban areas of Africa is often a problem.

That government was overthrown in 1979. Perhaps there are better days ahead for the people of Uganda.

Djibouti was the last colony to become independent, in 1977.

Somalia is shaped like the figure "7" outlining part of the "horn" of Eastern Africa. Border disputes with Ethiopia in the late 1970s became a small war.

Neighboring **Kenya** has developed its capital city, Nairobi (ny *roh* bee) into one of the major commercial centers of Eastern Africa. The tourist business provides one of the major sources of income for this beautiful country.

Island Nations of East Africa

The island nations in the Indian Ocean might be of interest to you. The population of the islands includes people with ancestors from Asia, Africa, and Europe. The **Seychelles** (say *shehlz*) **Islands** are about 1600 kilometers (1,000 miles) east of Zanzibar. Some of the islands of this country are still uninhabited. The **Comoro** (*kahm* oh *roh*) **Islands** became independent on July 6, 1975. The main industry of this nation is making perfume.

Madagascar is the world's fourth largest island. **Mauritius** (maw *rish* us) is about 800 kilometers (500 miles) east of Madagascar. Growing sugar cane and processing it have been the main business on Mauritius for years. Try to find all these islands on a globe or map of the world.

WHAT HAVE YOU LEARNED?

1. Why is the land of Eastern Africa often called "the horn of Africa"?

2. About how much of the land of Ethiopia is mountainous?

3. Why is the climate of Ethiopia cool, even though it is located within the low latitudes?

4. Describe the main features of the "ujamaa" system of Tanzania.

BE A GEOGRAPHER

1. Use a map to match the following capitals with their countries.

Capitals	Countries
a. Dar es Salaam	1. Kenya
b. Addis Ababa	2. Somalia
c. Nairobi	3. Tanzania
d. Mogadiscio	4. Ethiopia
e. Kampala	5. Uganda

2. What Eastern African nation is completely north of the equator?

3. Which Eastern African countries are totally south of the equator?

4. Use a globe to locate the island nations in the Indian Ocean.

QUESTIONS TO THINK ABOUT

1. Do you think the Swahili method of telling time would be useful for the area in which you live? Why do you think it might be impractical?

2. Why do you think the government of Ethiopia has tried to keep the province of Eritrea as a part of the nation?

3. Do you think that physical features and cities in Africa should have names given to them by European explorers, or should they have African names?

4. Discuss the following statement:

   "Africans usually helped Europeans locate areas of land which were unknown to the explorers. The European explorers then made the regions known to westerners. Thus, Europeans made many discoveries."

   Were these true discoveries? If so, why? If not, why not?

5. Why do you think the League of Nations refused to help Ethiopia in 1936?

23

Chapter

——— SOUTHERN AFRICA ———

Southern Africa is a region torn with racial conflict. In both South Africa and Zimbabwe Rhodesia much less than half the people are White African. However, White Africans have controlled the governments for many years. Until recently they have refused to allow Black Africans to take part in the government and society equally with White Africans. In fact, South African laws separate the Whites and the Blacks. Black Africans are not allowed the right to vote.

THE LANDS OF SOUTHERN AFRICA

Southern Africa includes all the land of the continent south of Zaire and Tanzania. Look at the map on page 369. Notice that much of this land is savanna. Along the west coast are desert lands. Near the southeast coast in South Africa high mountains rise more than 3000 meters (10,000 feet). Inland from the mountains is a plateau which varies from 900 meters (3,000 feet) to 1800 meters (6,000 feet) above sea level.

Changes of seasons are more noticeable in the lands here, for they are more distant from the equator. The lands south of the equator are in the Southern Hemisphere. Winter occurs in the months opposite those of the Northern Hemisphere.

South of Zaire is Angola. Locate this nation on the map on page 369. Notice the territory of Cabinda (kuh *bihn* duh) just north of the Zaire (Congo) River's outlet at the Atlantic Ocean. Angola has controlled the lands of Cabinda for almost three centuries. East of Angola find three landlocked nations: Zambia, Zimbabwe (Rhodesia), and Malawi. Mozambique (moh zam *beek*) contains a lengthy coastline along the Indian Ocean. South of Angola is Namibia (South-West Africa). Botswana (baht *swahn* uh) is the landlocked nation due east of Zambia. Within the boundary of South Africa are two landlocked countries, Swaziland (*swahz* ee *land*) and Lesotho (luh *soh* toh). Two other countries have been declared independent near Lesotho. South Africa made Bophuthatswana (ba put *hat* swana) and Transkei (*trans* ky) independent homelands for Blacks. Only South Africa has recognized Transkei and Bophuthatswana as independent countries. The Republic of South Africa is located at the southern end of the continent. The countries of this region complete our survey of Africa from the Southern Sahara to the Cape of Good Hope.

Zambia

Zambia is a landlocked country south of Zaire and Tanzania. The shape of the nation somewhat resembles a "bat." Locate Zambia on the map on page 369. Notice the number of nations that surround this country. Zambia's area is slightly larger than the state of Texas. Over five million people live in Zambia.

The Land and Climate of Zambia

Most of Zambia is part of a plateau 900 to 1200 meters (3,000 to 4,000 feet) above sea level. The plateau is cut by several rivers, one of which is the Zambesi (zam *bee* zee) River. The Zambesi River is one of Africa's larger rivers. In southwestern Zambia this river forms most of the boundary line with Zimbabwe Rho-

Victoria Falls, located along the Zambesi River, on the Zambia-Zimbabwe border is one of the world's largest waterfalls. Once every second 10,000 cubic meters (38,500 cubic feet) of water go over the edge.

desia. At a point along this part of the river is one of the natural wonders of the world, beautiful Victoria Falls. The waters of the Zambesi River drop about 100 meters (350 feet) into a canyon. The falls are more than two kilometers (1¼ miles) long. So much mist and spray rise from the waterfall that a permanent rain forest grows nearby. The original inhabitants were very impressed with the roar and mist from the falls. They named the falls Mosiotunya (mohs *ee* oh *tuhn* yah)—"the smoke that thunders."

Zambia is not a mountainous land. Hills on the plateau are about 1800 meters (6,000 feet) above sea level. The southern tip of Lake Tanganyika is in northeastern Zambia. The lake formed behind Kariba (kah *ree* bah) Dam adds another interesting water feature to Zambia. The huge dam was built on the lower Zambesi River between Zambia and Zimbabwe Rhodesia. It produces a great deal of hydroelectric power.

The climate in Zambia is pleasant most of the year because of altitude. Rain falls in the warm season which lasts from about October to April. Most of the rest of the year it is fairly dry.

The Recent History of Zambia

In colonial times Zambia was part of the British Federation of Rhodesia and Nyasaland (ny *as* uh land). The area was called "Northern Rhodesia." A majority-rule government was created at independence. In 1964 the Black majority established the independent nation of Zambia.

The White Africans in neighboring "Southern Rhodesia" did not wish to allow majority rule in their land. The colony declared itself independent in 1965. A government controlled by the White minority was established. As a result, Great Britain asked nations of the world to **boycott**—that is, not to trade with Rhodesia. Then the United Nations asked countries not to buy or sell goods to the rebel government. The government of Zambia believed that people living in Rhodesia should have a government based on majority rule. Zambia joined the boycott of Rhodesia. At that time most roads and railways connecting Zambia to the sea went through Rhodesia. How could Zambia continue to export materials and receive needed imports?

Neighboring Tanzania supported the action of the government of Zambia. Almost immediately work crews began building a road from Zambia to Tanzania's port at Dar es Salaam. The United States assisted in the road building project. Hundreds of oil and supply trucks traveled this rough road day and night to keep supplies moving to and from Zambia. By 1968 an oil pipeline had been laid. This ended the problem of transporting oil hundreds of miles by truck. The People's Republic of China assisted, also, by building a railway from Tanzania into Zambia. It was completed in 1975. The Tan-Zam Railway provides what some people call the "All-Black Outlet to the Sea".

Agriculture

Farming is the main occupation for most people in Zambia. Many people work small plots of land, raising only enough food for their own needs. Food crops grown are corn, cassava, wheat, citrus fruits, and a variety of green vegetables. However, larger farms produce cash crops for export. Tobacco and peanuts are the most valuable cash crops grown. Some sugar cane and cotton are also sold.

Mining and Manufacturing

Zambia is fortunate to have extensive mineral deposits. The main industry of the country is mining. Most of the minerals are located in north central Zambia. The most valuable mineral by far is copper. About one-fourth (¼) of the world's known copper reserves are found in Zambia. Almost all of Zambia's income from world trade comes from copper. Cobalt, zinc, lead, manganese, and silver are also mined.

In 1969 the government of Zambia nationalized the mines. It did so by taking over half ownership of the mining companies. Some banks and insurance companies were nationalized as well.

Many industries produce metal products from the nation's ores. Factories make copper wire and cable, metal window frames, metal furniture, nuts and bolts, wire, and explosives. Food processing is also done. Some textiles, cement, glass, and beverages are also manufactured.

The Kariba Dam provides electricity for the industrial needs of Zimbabwe Rhodesia and Zambia. However, when the dam was built, the generators for the power plant were not located on the Zambian side of the river. Zambia began to worry because Zimbabwe Rhodesia might decide to shut off the power supplied to Zambia. So Zambia built generators on its own side of the dam.

Industrial development has made steady progress since independence. However, the boycott against Rhodesia slowed some development in Zambia.

REMEMBER, THINK, DO

1. Why did Zambia build an outlet to the sea through Tanzania?
2. Find out about David Livingston's first visit to Mosiotunya and his renaming of the falls.
3. See a film about the lands of Southern Africa. Try to get one that shows Kruger National Park and Victoria Falls.
4. Using an outline map of Africa:
   a. Locate and label the nations of Southern Africa
   b. Name the nations surrounding Zambia

Ian Smith congratulates Bishop Muzorewa after Mr. Muzorewa became Zimbabwe Rhodesia's first Black African prime minister.

Rhodesia's "Udi"

Zimbabwe Rhodesia was a colony of the United Kingdom until 1965. Whites who had come to the area as colonists were only about one-twentieth (1/20) of the population. Yet the colonial government was operated mainly by the White African population. Many White settlers had originally come to Zimbabwe Rhodesia from Great Britain and South Africa. They developed large tobacco and wheat farms, as well as cattle and sheep ranches. Copper mines were also opened and managed by White Africans. Black Africans were allowed to work on the farms and in the mines. But their pay was very low. Black Africans could not advance to higher positions or salaries.

In the early 1960s the United Kingdom was trying to arrange for the independence of the colony. Britain wanted the colonial leaders to make a plan for an independent government that would include all citizens of the country. Many White Africans in the colony were afraid they might lose their lands and businesses if Black Africans ran the government. Since most of the people were Black, they felt that Black Africans probably would be elected to run the government. The White African leaders were not willing to agree to the United Kingdom's rules for independence. Therefore, they declared the country independent from Britain.

On November 11, 1965, Rhodesian leaders issued a "Unilateral Declaration of Independence". This is commonly called "UDI." The new nation continued to be governed by the White leaders. They refused to allow Black Africans much of a voice in the government. A group of Black people decided to fight to force majority rule in their country. They became guerilla fighters. They trained and lived near the borders in Mozambique and Zambia. They made raids into Zimbabwe Rhodesia, trying to force a change in the government.

In 1979, for the first time, Zimbabwe Rhodesia elected a government with a majority of Blacks in Parliament. And it had its first Black prime minister, Abel Muzorewa (moo zohr ay uh). Before his election he had been a Methodist bishop. However, not everyone was happy with these events. Check to see what has happened in Zimbabwe Rhodesia since 1979.

The Republic of South Africa

The Republic of South Africa is one of the world's richer nations. Its agricultural resources alone are enough to provide a fairly high standard of living for all its citizens. In addition, South Africa has very rich mineral deposits. The country has developed many kinds of industries, using raw materials from both agriculture and mining. A variety of finished products is manufactured. Many items are sold within southern Africa. Others are exported throughout the world.

The riches of South Africa, however, have only been made available to the White African minority population. *Apartheid* (uh *pahr* tyt) is the official racial policy of the government in South Africa. Under **apartheid**, which means separateness, there is almost a complete separation of the races in South Africa.

Black Africans are referred to as **Bantu** (*ban* too) in South Africa. The White minority considers all Bantu as immigrants. Black Africans, or Bantu, are treated as if they were not citizens of the country. They must carry a passbook for travel and identification. The apartheid laws in South Africa classify people by color. There are four classifications: White, Asian, Colored, and Bantu. "Colored" are people of mixed racial parentage. The Asians are mostly people of Indian descent. The laws strictly segregate and regulate the lives of all. The White minority make up only about one-fifth (⅕) of the total population. However, the non-whites—who make up four-fifths (⅘) of the people—are the main labor force. These people support the rich economy of the nation. The White minority controls and keeps for itself the major share of the nation's wealth. The apartheid system is of great concern to many people in the world.

A Brief History of South Africa

The first Europeans to settle in South Africa were the Dutch. They came originally in the mid-17th century. They found Bushmen and Hottentots (*haht* en tahts) living on the coastal land near the Cape of Good Hope. Both the Bushmen and Hottentots were Black Africans. The Dutch fought them for the land. The story is much like what happened to the Indians in North America.

The Dutch people continued to control South Africa for the next hundred years. Then British settlers began to arrive in this area. After they took possession of part of this land, British settlers began to make changes in the laws of the colony. Some of these changes included better treatment of the native Black Africans.

The *Afrikaners* did not agree with the changes. **Afrikaners** is the name given to the people descended from the Dutch settlers. Rather than live by the changes, more than five thousand Afrikaners moved northward. They resettled in the provinces now known as Natal (nuh *tahl*), Orange Free State, and Transvaal (trans *vahl*) in South Africa. Find these areas on the map on page 369. The journey northward by the Afrikaners is known as "The Great Trek." Once again the Afrikaners had to fight for the land. The Zulu (*zoo* loo) tribes almost forced the Afrikaners back to the south.

In 1867 diamonds were discovered at Kimberly (*kihm* buhr lee). Twenty years later gold was discovered near Johannesburg (joh *han* uhs buhrg). The Afrikaners and the British both wanted control over this mineral rich land. Many new White settlers came to prospect in these lands. Fighting broke out in 1899 between the British and the Afrikaners. After winning the Boer War in 1902, the British united the provinces into the Union of South Africa. Years later the country became a republic. The British and Afrikaner settlers together developed a government dominated by the White minority. In 1948 the apartheid system became official government policy in the country.

The Land of South Africa

The Republic of South Africa is almost completely in middle latitudes. It has about as much land as the four states of California, Nevada, Utah, and Arizona. Although some of the resources and land are similar to the southwestern United States, the winter climate in South Africa is generally warmer.

Except for a narrow coastal lowland, almost all of South Africa is a high plateau. Inland from Cape Town is a series of step-like terraces leading up to the plateau. Mountain ranges form the edge of these terraces. The plateau in the west is about

900 meters (3,000 feet) above sea level. The central plateau of the country is about 1350 to 1800 meters (4,500 to 6,000 feet) high. The Drakensberg (*drahk* uhnz buhrg) Mountains in the east have some peaks as high as 3000 meters (10,000 feet).

The Orange and Vaal (vahl) Rivers flow from the Drakensberg Mountains westward to the Atlantic Ocean. The Limpopo (lihm *poh* poh) River flows around the north end of the mountains and empties into the Indian Ocean. These rivers drain a high plateau called the *high veld* (vehlt). **Veld** is a Dutch word which means field. The cool plateau is covered with scattered trees and tall grass. The high veld is in the provinces of the Orange Free State and Transvaal. The northern Transvaal, along the Limpopo River, is much lower in elevation. It is called the low veld.

The Climate of South Africa

The climate in South Africa is generally cooler than in North African countries. You should notice, however, that these regions are about the same distance from the equator. There are several reasons why South Africa is cooler than North Africa. First, more of South Africa's land is nearer the ocean. No part of it extends inland as far as in the Sahara. Second, most of the land is higher in elevation and is therefore cooler. Third, a cool ocean current flows northward along the western coast. The Benguela Current often causes dense fog along

Molten gold is being poured into molds in order to form gold bricks.

the western coastal region of South Africa.

Southeastern *trade winds* bring rain to the eastern half of South Africa during summer months. **Trade winds** are winds which blow steadily in the same direction towards the equator. Most of the rain falls along the coast and along the high ridges of the Drakensberg Mountains. The high veld, beyond the mountains, usually gets from 35 to 75 centimeters (15 to 30 inches) of rain a year. This amount is normally enough to raise crops without irrigation. The rains must fall during the planting season, however.

Unfortunately, droughts sometimes occur in the veld. The western half of the country usually gets from 10 to 40 centimeters (5 to 15 inches) of rain yearly. Irrigation is needed for farming in this region. Cloud-bursts often occur in the west. Such storms often cause erosion due to the heavy runoff of water on the hillsides. The southern tip of South Africa near Cape Town has more rain than inland areas. Near the Cape, rains come during the winter rather than summer. Summers are often dry and hot.

Agricultural Products

Around the Cape Town area, the crops are similar to those raised near the Mediterranean Sea in North Af-rica. Citrus fruits, grapes, and some winter wheat are grown. Along the more humid east coast, sugar cane and pineapples are major crops. Inland, on the cooler hills, a shrub called **wattle** is grown for its valua-ble bark. The bark produces **tannin**, which is used in making hides into leather. Some cattle and sheep are grazed on the hilly slopes.

On the low veld in the Transvaal, the soil and climate are suitable for raising crops. Tobacco, wheat, alfal-fa, fruits, and a variety of vegetables are grown. On the cooler high veld, corn, peanuts, tobacco, and wheat are also raised. However, the main crop grown is corn. It is used for fattening beef cattle and other live-stock. Modern methods of farming and irrigation help to produce excel-lent crop yields on most South Afri-can farms.

Livestock

A great deal of land in South Africa is used to graze livestock. Only about one-sixth ($\frac{1}{6}$) of the total land area of the country is suitable for growing crops. Much of the rest of the land is excellent for grazing. Along the coast where much rain falls, dairy cattle are kept. Inland, beef cattle and sheep are raised. South Africa exports great quantities of wool, beef, mutton, and dairy products.

Fishing

Fishing has also proved to be an important industry in South Africa.

AFRICA

Mineral and Fuel Resources

Bx	Bauxite	P	Potash
Cr	Chromite	Ag	Silver
Cu	Copper	Sn	Tin
D	Diamond	W	Tungsten
Au	Gold	Zn	Zinc
Fe	Iron	⬭	Coal
Pb	Lead	▫	Natural Gas
Mn	Manganese	△	Uranium
Hg	Mercury	○	Petroleum
Ni	Nickel	⬭	Major
Pt	Platinum		Industrial Regions

Kilometers 0 500 1000

Miles 0 500 1000

The Indian and Atlantic oceans provide vast varieties and quantities of sea foods. South Africa has the tenth largest fishing fleet in the world. The government has been concerned about the number of foreign fishing fleets that enter South African waters to fish.

Mineral Wealth

Mineral wealth has created a 20th-century industrial revolution in South Africa. South Africa ranks as the greatest industrial nation on the continent. Pretoria (prih *tohr* ee uh) is one of the major diamond centers of the country. Diamonds were once found close to the surface of the earth. Now, rock must be blasted underground and the ore brought to the surface. The ore is lifted by an endless conveyor belt. The ore is then crushed and carefully sorted to find diamonds.

Gold is the most valuable mineral mined in South Africa. More gold is mined in South Africa than anywhere else in the world. About three-fifths (⅗) of all gold produced in the world comes from South Africa. Most of the mines are in the Transvaal near Johannesburg. Some new deposits have been found in the province of the Orange Free State.

Around the gold mines there were great piles of rock which had been brought up from below the ground. Gold had already been removed from these rocks. Mineralogists discovered that this waste material from the gold mines contained uranium. South Africa now ranks third in the world in production of uranium.

The Republic of South Africa has other important minerals for its industries. Coal, iron, manganese, nickel, copper, platinum, limestone, asbestos, chrome, and phosphates are among some of the most important. Coal deposits are vast and easy to mine. Iron ore reserves are so extensive that at the current rate of use the ore could last a thousand years!

Manufacturing

As you might expect, steel is a major industry in the country. Pretoria and Johannesberg are centers for the steel mills. Major products made from steel in the mills are steel beams, galvanized iron for roofs, drills, boilers, railway cars, ships, automobiles, trucks, and assorted electrical appliances.

Many industries also process food products. Meats and fish are canned or frozen and shipped all over the world. Factories make textiles, clothing, furniture, shoes, and other leather goods. Jewelry from the fine minerals of South Africa is also produced. Durban (*duhr* buhn) on the east coast is the nation's main ocean port. Railways crisscross the nation, allowing goods to move in and out of the interior. The transportation system in South Africa has also been linked to neighboring nations to the north. Air service is readily available within the country, as well as to major international airports of the world.

Southern Africa

Before the arrival of the Europeans, the Zulu ancestors of this family controlled an empire that covered a large portion of South Africa.

Kruger National Park

Kruger National Park is located near the eastern border of the Transvaal Province. It is a game reserve about as large as the state of Massachusetts. The park is famous for its variety of wild animals that roam protected in their own natural environment. Many international tourists and vacationing White South Africans enjoy the beauties of this national park.

The Government of South Africa

The Republic of South Africa has a parliamentary type of government. It includes a Senate and an Assembly. Some members of the Senate are appointed, and some are elected. All members of the Assembly are elected. The leader of the largest political party in the Assembly becomes the prime minister. All members in this government are White Africans.

About 27 million people live in South Africa. Almost seven-tenths ($\frac{7}{10}$) of the people are classified as Bantu, or Black. Less than two-tenths ($\frac{2}{10}$) of the population are White African. About one-tenth ($\frac{1}{10}$) of the people are classified as Colored and as Asian. Among the White minority there are two groups of Europeans. The largest group is of Dutch descent, commonly known as Afrikaners. They speak their own Afrikan language and have their own political party. The other group is mainly of British descent. Not all White Africans approve of the government's apartheid policies. However, the opposition party in South Africa's government has never had enough voting power to change the policies of the government.

There is a separate advisory **Indian Council**. Some of its members are elected and some are appointed by the South African government. The Council tries to solve problems of Asians in South Africa. There is also a **Colored People's Representative Council.** Its members are elected. Such councils have been created in each of the four provinces of the country. Since the government does not consider Black Africans to be citizens, there is no provision for them to have a voice in the government. Black Africans are considered to be "temporary migrants."

Bantus and Separate Development

The South African government believes in separate development of the races. It has set aside one-tenth (1/10) of the land in South Africa for "homelands" or "reservations" for the Bantu population. The homelands for the Black Africans are called **Bantustans** (ban too *stans*). The South African government will allow the Black Africans to manage the Bantustans as their own independent nations. However, the governing groups in the Bantustans must have South African government approval.

Transkei and Bophuthatswana are Bantustans. These Bantustans are mostly wasteland, containing few of the mineral resources of the nation. The land is not good agricultural land, either. These areas are mainly "bedroom communities" surrounding the major industrial cities of South Africa. Africans living on the Bantustans have to work as laborers in nearby mines, factories, and farms owned by White Africans in order to earn a living. The government of South Africa has said that the Bantustans are the answer to the race problem in the nation.

Peaceful protests and riots have increased in number in South Africa during the past few years. As you might expect, White Africans view the Bantustans much differently than do the Black Africans. They are demanding more freedom and more opportunities.

REMEMBER, THINK, DO

1. What was "The Great Trek"?
2. Why do cloudbursts cause erosion in western lands of South Africa?
3. Give three reasons why South Africa is cooler than countries in North Africa that are about as far from the equator.
4. What are Bantustans?
5. How do you think the fishing resources of the ocean should be controlled?
6. Use an outline map to locate and label the following places in Southern Africa:

a. Lusaka	e. Lake Malawi	i. Kalahari Desert
b. Cape Town	f. Namib Desert	j. Johannesburg
c. Victoria Falls	g. Pretoria	k. Cape of Good
d. Kariba Dam	h. Durban	Hope

LEARN MORE ABOUT SOUTHERN AFRICA

Southern African countries will no doubt undergo vast changes in racial policies and governments in the final quarter of the 20th century. You will need to know more about this region of Africa so you can better understand the problems of all people. Use the Study Guide on page 31 as you plan additional study about Southern Africa.

Angola became independent from Portugal in 1975. Rival Black African leaders fought to control the new government of this large country. Offshore oil was discovered in Cabinda province. Since that happened, some residents of Cabinda wished to become independent.

Zimbabwe Rhodesia became the name of this country in 1979. Zimbabwe was the name used for this land by many Black Africans. The name Zimbabwe was the name of an ancient African kingdom. Rhodesia was the name of the country when it was ruled by White Africans. Now the country uses both names.

Malawi is a landlocked nation between Tanzania and Mozambique. Lake Malawi forms a natural border along the entire eastern side of the country. Most people of Malawi earn their living as farmers.

Mozambique has a coastline on the Indian Ocean that extends for about 1900 km (1,200 mi.). The ocean port at Maputo (mah *puh* toh) is the capital city. On many maps it is still shown as Lourenco Marques. It was used to import and export goods for landlocked Malawi, Rhodesia, and Zambia before the fighting began in Rhodesia.

Namibia was a United Nations Trust Territory managed by South Africa. The UN changed the territory's name from South-West Africa to Namibia in 1968. Plans were made for Namibia's independence. However, South African leaders delayed Namibia's promised independence.

Botswana is the landlocked nation east of Namibia. The Kalahari Desert is located in southern Botswana not far from Namibia's border.

The **Kingdom of Swaziland** is surrounded by South Africa and Mozambique. It is smaller in size than the state of New Jersey.

The country of **Lesotho** is a kingdom. Its land is entirely within South Africa. The country is located near the Drakensberg Mountains. Heavy snow is common in the higher elevations of these mountains in wintertime.

The nations of Malawi, Botswana, Namibia, Swaziland, and Lesotho share a common characteristic in their way of living. These countries supply laborers to the mines, farms, and industries of South Africa. Laborers "contract out" for a number of months each year. Some nations send about half their male workers to South Africa each year. Although the workers earn a cash salary, they are forced to live away from their home and family. Until the smaller nations are able to develop their own economy they are quite dependent upon South Africa.

SUMMARY OF AFRICA FROM THE SAHARA TO THE CAPE OF GOOD HOPE

At the beginning of this unit, seven items were presented for you to keep in mind. They were to help you understand why the countries in Africa have developed as they have. Do you remember what the seven items were? (See pages 375-376.) They serve well as a review.

You may be interested in learning more about the Organization of African Unity. This international group tries to solve some of the boundary disputes that sometimes occur between African countries. The OAU also tries to get the countries working together to improve the standard of living. Leaders of African nations work together in the United Nations, too. Imagine how members of the OAU must feel when they think about the problems that the nations of the second largest continent have.

Now in ruins, the ancient city of Zimbabwe was once a major center of the Bantu people in southern Africa. This picture shows the remains of the Temple or Great Enclosure.

1. What are some of the reasons given in this text for the racial conflict in Southern Africa?

2. In what ways are the goals of the Organization of African Unity and the Arab League alike? How are they different?

3. What is meant by "apartheid" in South Africa?

4. Prepare a chart showing the major mineral deposits on the continent of Africa. Compare the chart to ones for the Middle East and the Soviet Union. In what ways are the mineral resources of these regions similar?

5. Why have Bantustans been created in South Africa?

BE A GEOGRAPHER

1. Prepare a climate chart for the five major regions of Africa. Include North Africa, Southern Sahara, Western Equatorial Africa, Eastern Africa, and Southern Africa. Note the seasonal and climatic facts of each region. Think about similarities and differences among regions after completing the chart.

2. Where on the equator in Africa does the least rain fall?

3. Judging from its position on the Earth, where would you expect a milder climate; at the Cape of Good Hope or at Cape Horn?

4. Which of the capital cities of Africa south of the Sahara are *not* river, lake, or ocean ports?

5. Which of the following lakes are not in the Great Rift Valley?
   a. Lake Nyasa
   b. Lake Rudolf
   c. Lake Tana
   d. Malebo Pool
   e. Lake Tanganyika
   f. Lake Chad
   g. Lake Victoria
   h. Lake Albert
   i. Lake Malawi

6. Is the sun ever seen directly overhead at noon at:
   a. Kinshasa
   b. Durban
   c. Accra
   d. Addis Ababa
   e. Salisbury
   f. Tananarive
      (tun *nan* uh *reev*)

7. Which countries on the continent of Africa have land in:
   a. low latitudes
   b. middle latitudes
   c. high latitudes

8. On a desk outline map of Africa locate and label the following:

   a. Mount Kilimanjaro
   b. Ruwenzori Mountains
   c. Eritrea
   d. Addis Ababa
   e. Nairobi
   f. Victoria Falls
   g. Zambezi River
   h. Cape of Good Hope

--------- QUESTIONS TO THINK ABOUT ---------

1. What advantages do countries have which are located on major rivers; such as the Niger, the Zaire (Congo), or the Zambezi? Are there any disadvantages to having a major river(s) in your country?

2. If dams are built on the Zaire (Congo) River as they have been on the Nile River, what differences would there be in the purposes for which the dams are built?

3. Think about the vast area in Africa south of the Sahara. What advantages exist that may help the people raise their living standards and their productivity? What disadvantages are there which may keep the people from higher standards of living?

4. If people continue to kill wild animals in the next hundred years as they have in the past century, what is likely to be true in Africa then?

5. Should the UN do anything about the treatment of Blacks in South Africa? What action could the UN take without seeming to interfere in the government of this country?

6. How are the problems of the Palestinian refugees of the Middle East and the Bantus of South Africa alike and/or different?

7. Do you think an All-African Common Market should be started like COMECON and the European Community? What advantages and difficulties might there be in operating a common market in Africa?

8. Why do you think so many Europeans settled and stayed in parts of Southern Africa?

9. Who should have control over the power produced at the Kariba Dam, Zambia, or Zimbabwe? Why?

Shop signs along a street in the city of Kowloon in the British Crown Colony of Hong Kong.

Unit

ASIA

Unit Preview

Can you • identify these famous mountains in Asia?

• identify these places of worship:
Buddhist Stupa, Islamic Mosque,
Hindu Temple, Christian Church?

DO YOU KNOW?

- How much of the Earth's land area is taken up by the continent of Asia?

- What some of the major religions of Asia are?

- Why monsoon rains are so important to many people in Asia?

- Why, for many years, there were two "Chinas"?

- Why some people think that Asia and Australia were at one time connected by land?

Unit Introduction

The Continent of Asia

Look at the map of Eurasia on pages 8-9. Notice how large Asia is. Indeed, Asia is the largest continent on Earth. Including the many islands off the mainland of Asia, this continent occupies 44 747 194 km² (17,276,909 sq. mi.) of land. That is about one-third (⅓) of all the land on Earth. Note also that Asia is separated from Europe only by an imaginary line. On a map the line is drawn from the Arctic Ocean in the north through the Ural Mountains in the U.S.S.R.. It follows the Ural River to the Caspian Sea. It then turns westward along the Caucasus Mountains to the Black Sea. From there it continues through the Bosporus Straits, and the Sea of Marmara to the Aegean Sea and the Mediterranean. The Suez Canal and Red Sea separate Asia from Africa.

Notice on the map the major oceans that surround this large continent. In the north is the Arctic Ocean. The Indian Ocean is south of Asia. East of Asia is the Pacific Ocean. These oceans have much influence on the lives of many Asian people.

More than half the people on Earth live in Asia. It is estimated that more than 2⅓ billion people live on this continent. At present rates of population growth, Asia will have twice as many people in 2013 as it did in 1978!

Asia also has a great variety of cultures. Some of the world's richest nations are found in Asia. So, too, are some of the world's poorest nations. The people of Asia follow Hindu, Buddhist, Muslim, Hebrew, and Christian religions. Some of the governments are run as monarchies. Others are run as democracies. There are also communist and socialist nations. A vast number of languages and cultural traditions are found in Asia.

Why Study About Asia?

The continent of Asia is one of the most interesting of all the continents. It may also be one of the most important. True, most of Asia is thousands of miles away from North America. However, changes that occur in parts of Asia sometimes change our way of life in the United States.

For example, the economic conditions of one country can affect the economy of another nation. Japanese color television sets are sold in the United States. During the late 1960s and early 1970s these sets cost less than television sets made in American factories. Many Americans, therefore, bought the lower priced Japanese sets. What effect do you think this had upon the American companies and their employees? The United States government then reduced the number of Japanese television sets that were imported to help American companies. What do you think happened to television manufacturing in Japan? Do you know where most television sets are made today?

The political conditions within one nation can, at times, affect another nation. When Communist forces started to take over South Vietnam (vee eht *nahm*), thousands of Americans were sent to fight in Southeast Asia. However, many Americans did not believe that the United States should fight a war in Vietnam. Many protests were made to the government demanding that the military withdraw from Southeast Asia. Finally, the United States' government ordered its troops home.

In recent years the Communist government of the People's Republic of China has become more powerful. Both the United States and the Soviet Union, therefore, have had to change their policies toward China. For many years the United States voted against admitting the People's Republic of China to the United Nations. Finally the United States decided not to vote against China's entry any longer, and it became a member.

Developments in other parts of Asia may also have an effect upon our lives in the future. Knowing more about Asia may help us to solve some international problems in the future.

Problems of Third World Asian Nations

Asia includes a large number of Third World countries. Within the Third World, many problems exist which prevent the nations from developing more rapidly. Among these problems are the following:
● Many nations have difficulty controlling their population growth. At times there is not enough food to feed the people already born. If increases in population continue, a major food crisis could occur in parts of Asia.
● Many dangerous diseases still occur in parts of Asia. Dysentery, cholera, typhoid fever, and tuberculosis take countless lives each year in

many Third World nations. **Malnutrition**, which comes from lack of a balanced diet, is also a problem. Some people are starving in Asia.

● Some countries lack educational programs for their adults and children. Efforts to improve reading, writing, health care, and vocational skills are needed. Better educated people should be able to help solve some of the problems in Asia.

● In some places, natural resources have been found, but the country may lack the money to develop these resources. Some countries, therefore, need better income-producing industries.

It has become necessary for many Third World nations to ask for assistance from the developed nations. The Third World nations need help in solving some of these problems. Some Asian nations formed the **Colombo** (kuh *luhm* boh) **Plan** as one way to try to help themselves. Each of the 27 member nations has its own development plan. Ideas and help are provided by other nations. The United States, as well as many other nations of the world, also contribute to help these Asian Third World nations. The United Nations is also hard at work in Asia. Hopefully, such assistance will one day solve many of these problems. It is hard, however, to reduce the economic and social differences between developing Third World nations and the more developed nations.

This map shows the three regions of Asia that you will study in this unit.

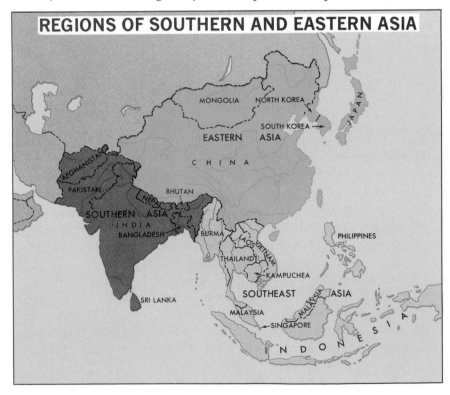

REGIONS OF SOUTHERN AND EASTERN ASIA

MONGOLIA   NORTH KOREA

JAPAN

SOUTH KOREA →

EASTERN   ASIA

C H I N A

AFGHANISTAN

PAKISTAN

NEPAL   BHUTAN

SOUTHERN   ASIA

I N D I A

BANGLADESH   BURMA

PHILIPPINES

LAOS VIETNAM

THAILAND

KAMPUCHEA

SOUTHEAST   ASIA

SRI LANKA

MALAYSIA

MALAYSIA

SINGAPORE

I N D O N E S I A

Unit Introduction

There are many different landforms on the Asian continent. Asia has the highest point on earth. Mount Everest—also called Mount Sagarmatha (*sag* ahr *mah* tah)—is 8846 meters (29,028 feet) above sea level. This famous mountain peak is but one of many high peaks in the rugged Himalaya (huh *mahl* uh yuh) Mountains of Asia. Other high mountains are found in many places in Asia.

Asia also contains the lowest point on Earth not covered by water. The Dead Sea in Jordan is about 400 meters (1,300 feet) *below* sea level. The Dead Sea is near a vast desert region in Southwest Asia. You already learned about it in the chapter on the Middle East.

Asia not only has high mountains and deserts but also has fertile river plains, grasslands, rain forests, and high plateau regions. Not all of Asia is suitable for human use. Some areas have no settlements at all! As a result, some areas are very crowded. People tend to group together in areas that can best provide a way of living. As you study Asia, think about where the people live and how many live in an area. Too many people now live in countries such as China, India, Bangladesh (*bang* luh dehsh), Japan, and Singapore (*sing* guh pohr). Some nations have been able to adapt to the environment to meet their needs. Other nations have at times misused their resources and overused their lands.

Remember that Northern Asia was already presented in this book in the unit on the U.S.S.R. Most of western Asia was included in the section on the Middle East. This unit will focus on the rest of Asia.

REMEMBER, THINK, DO

1. Give at least two reasons why it is important to know more about Asia.
2. What are some of the different religious groups found on the continent of Asia?
3. Using the map on page 27, find out how many time zones there are on the continent of Asia.
4. Approximately how many degrees of latitude are there between the northernmost point of Asia and its southernmost point? (Do not include islands.)
5. Using a globe figure out about how far (in kilometers or miles) Asia is from the United States at their closest points.

Things You Might Like To Do As You Study About Asia

As you begin your study of Asia, you might like to begin some activities to make the unit more interesting. Here are some suggestions.

1. Use an opaque projector, if one is available, to make enlarged outline maps of each country in Asia for use as each country is studied. Place on the outline maps the major cities and major rivers.

2. Use the card catalog in the library to find books about the people and the countries of Asia. Make a list of good books about each country. Post it on the bulletin board so that everyone in the class will know what books are available.

3. Read books about climbing the high mountains in Southern Asia. Especially recommended are: Sir John Hunt's *The Conquest of Everest* and Maurice Herzog's *Annapurna*.

4. Look in the *Reader's Guide to Periodical Literature* for recent articles on the people and the countries of Asia. Read some of the articles that sound most interesting and report on them.

5. Make a chart or graph showing the height of the world's 10 highest peaks. On the chart or graph also show the country in which each peak is located. How many of them are in Asia?

6. Plan and present to the class, using the opaque projector, a travelogue to some spot in Asia, such as:
  a. Mt. Everest;
  b. a jute mill;
  c. Kashmir;
  d. a village in India;
  e. Taj Majal;
  f. a tea plantation;
  g. Bombay and Calcutta;
  h. the island of Sri Lanka.
  Use maps and pictures from the textbook and from other books and magazines. Write a script to read aloud as the pictures are shown.

7. Search in the library and in your music textbooks for music, games, and folk dances from Asia. If your school has a good record collection, search for examples of music from this area. Plan to play some of the music you are able to find, and show pictures of the instruments on which the music is played. Learn several of the folk dances or games.

8. Find out all you can about the Community Development programs in India. Find out how other countries, including the United States, are helping to improve living conditions for the villagers.

9. Make a chart of the major rivers in the Eastern Hemisphere. Show their length, how much of each river is navigable by oceangoing ships, and the major cities built near the rivers.

10. On an outline map of Southeast Asia, draw in what you think the shape of Asia might have been at one time. Assume that Australia and Asia were connected.

11. Make a special report on José Rizal, the George Washington of the Philippines.

Chapter

—————— SOUTHERN ASIA ——————

India occupies most of the land in Southern Asia. To the east and west of India are Bangladesh and Pakistan. Both of these countries were once part of India. In the Indian Ocean southeast of India is the island-nation, Sri Lanka. West and north of India are the landlocked mountain nations of Afghanistan, Nepal (nuh *pawl*), and Bhutan (boo *tahn*). Use the map on page 445 to locate each of these nations.

India

A Brief History of India

The history of India can be traced back about 5,000 years. Archaeologists have discovered records of human activity in India as early as 3000 B.C. Most Indian people are very proud of their history and culture.

The early history of India has had a great influence upon other nations in Asia. The Hindu and Buddhist religions began in ancient India. Hinduism is believed to have started in India about the 14th century B.C. Buddhism also had its origin on the Indian subcontinent about 400 B.C. By the 20th century about half the people in all of Asia were either Hindus or Buddhists.

The Islamic Era. During the 11th and 12th centuries a great change occurred in India. Invaders from the Islamic lands of Turkey and Afghanistan took control of India. They divided up much of the land among their own people. Small areas of land were controlled by Islamic leaders. They tried to force the Indians, who were Hindus, to accept the Islamic religion. Bitter feelings began to develop between Hindus and Muslims living in India. Nevertheless, the Muslim rulers kept control of India for almost 500 years.

The Coming of the Europeans. Another great change occurred with the discovery by Europeans of a water route around Africa to India. After this the Portuguese, Dutch,

French, and English became rivals for trade with India. No single ruler held all of India. Therefore, the European traders had to deal with many people in carrying on trade.

In 1600 a group of English merchants organized the **British East India Company**. During a period of 200 years this company built trading settlements along the coast of India. To defend their settlement against other Europeans and Indians the company hired troops. From these settlements the company began to extend its control inland. By 1857 the British East India Company ruled over half of India. In 1858 the British government took control of this region from the company.

The section of India ruled directly by the British was known as **British India**. The rest of the land was divided into more than 500 states, each with a native ruler. Two of these states, Hyderabad (*hyd uh ruh bad*) and Kashmir, were as large as the states of Kansas and Idaho. Others were no bigger than a large farm. How does the history of India compare with Africa's history?

Independence. After World War II, the British government began to make plans to give India self-rule. However, the Indian people disagreed with one another about their independence. Muslims living in India insisted on having their own country. They felt that the Hindus, who outnumbered them about three to one, would have too much power. After much discussion and argument, British India was finally **partitioned**. That is, it was divided into two countries, India and Pakistan. Pakistan received the land where most of the Muslims lived. Areas

When the Europeans arrived in India, they brought Christianity with them. Akbar, the Emperor of India (1542-1605), is shown here discussing Christianity with two priests from Portugal.

Southern Asia

where mostly Hindus lived became part of India. As a result, two Muslim territories, a thousand miles apart, formed the nation of Pakistan. The two territories were known as East and West Pakistan. Independence for these two countries came on August 15, 1947.

There were bitter feelings over the partitioning of India. Many people were killed in fighting which broke out during the division of the land. Millions of Muslims in India left their homes and property to move to Pakistan. Many Hindus moved from East Pakistan and West Pakistan into India.

In 1971 East Pakistan declared itself independent. It is now the nation of Bangladesh.

King Asoka

In ancient days the lands of India were ruled by kings. At times invaders threatened the land. Alexander the Great extended his empire from Greece to the borders of India. However, Alexander's death in 323 B.C. ended this threat to India.

In 272 B.C. an era of peace began. This peaceful period was due to the efforts of King Asoka (ah *so* kah). Asoka had been king for ten years when he took part in a war to expand his empire into southern India. He was horrified at the suffering caused by his war. He saw no value in the damage and destruction of war. At that point he decided he would not fight any more wars.

The king was a follower of the teachings of Buddha. The Buddhist religion taught that people should live in peace. The king tried to teach people how to live peacefully. For 30 years there were no wars in India. Asoka found peaceful ways to solve disagreements. He traveled throughout the land helping other people learn peaceful ways. His lessons of peace were carved in stone pillars and placed around the countryside for all to read. Many of India's current leaders still use the ideas of King Asoka in their speeches.

The Land of India

India is quite often called a **subcontinent**—a large area smaller than a continent. The Indian subcontinent is a peninsula separated from Asia by very high mountains in the north. The land of India occupies about 3 275 000 square kilometers (1,264,555 square miles). This includes the lands of Kashmir and Sikkim (*sihk* uhm). More than 637

King Asoka and his ideas of peace have greatly influenced the people of India. These carved lions sat atop one of the stone pillars that King Asoka had erected in Sarnath, India. They have become the national symbol of India.

million people live in India. Only one country has more people. Use the map on page 445 to locate the nations that have borders with India. Also locate the Arabian Sea and Bay of Bengal (behn *gawl*). These bodies of water along the coasts of India are important to the climate and way of life in India.

The land of India can be divided into four main parts. One is the high Himalayas. Another is the northern plains along the Ganges (*gan* jeez) River. The Deccan (*dehk* uhn) Plateau of south-central and southern India is a third part. Assam (uh *sam*), in the extreme east, makes up the fourth part. Find these areas on the map on page 445.

The Himalayas separate India from the rest of Asia. They are the highest mountains on the Earth. The Himalayas are almost impassable, except at the eastern and western ends of the range. They keep out of India the cold winter winds which blow southward from the high interior plateaus of Asia. In the summer, they provide water from the melting snows for the rivers of India and its neighbors. In the past the Himalayas made it difficult for armies to invade India. Today these mountains are a major concern for security. India and China face each other across the Himalayas.

The Northern Plains reach across northern India just south and west of the Himalayas. Most of this lowland is drained by the Ganges River and its tributaries. This level lowland is the most densely populated and developed part of India. Most of this land is used for farming. India's larger cities are also located in this area. Calcutta is located at the mouth of the Ganges River. It is India's largest seaport. New Delhi (*dehl* ee), the capital of India, is also in the northern section of the lowland plains.

The Deccan Plateau covers most of the southern half of India. As the map on page 476 shows, this is a hilly area. Near the west coast is a low range of mountains called the Western Ghats (gawts). Near the eastern coast is another range called the Eastern Ghats. The Deccan Pla-teau lies between these two ranges. The plateau is cut by many rivers. Many of them flow from the Western Ghats across the peninsula and through passes in the Eastern Ghats to the sea.

Assam appears to be cut off from India by the nation of Bangladesh. The northern part of Assam is low-land along the Brahmaputra (*brahm* uh *pyoo* truh) River. This river, as the map shows, flows from the northern slopes of the Himalayas and around the range to the east. It then flows southwesterly into India and Bangladesh.

REMEMBER, THINK, DO

1. Why is King Asoka an important person in the history of India?
2. How are the Himalaya Mountains helpful to India?
3. Learn more about the Muslim and Hindu religions. Invite speakers from both religious groups to your class to explain the difficulties faced by Muslim-Hindu Indians living together in India in 1946 and today.
4. See a film on the land of India.
5. Be sure you are familiar with the location of important regions and cities of India. Use an outline map of India to mark the following places:
   a. Arabian Sea
   b. Indian Ocean
   c. Bay of Bengal
   d. Himalaya Mountains
   e. Ganges River
   f. Western and Eastern Ghats
   g. Calcutta
   h. New Delhi
6. What parts of the United States are located along the same latitudes as parts of India?

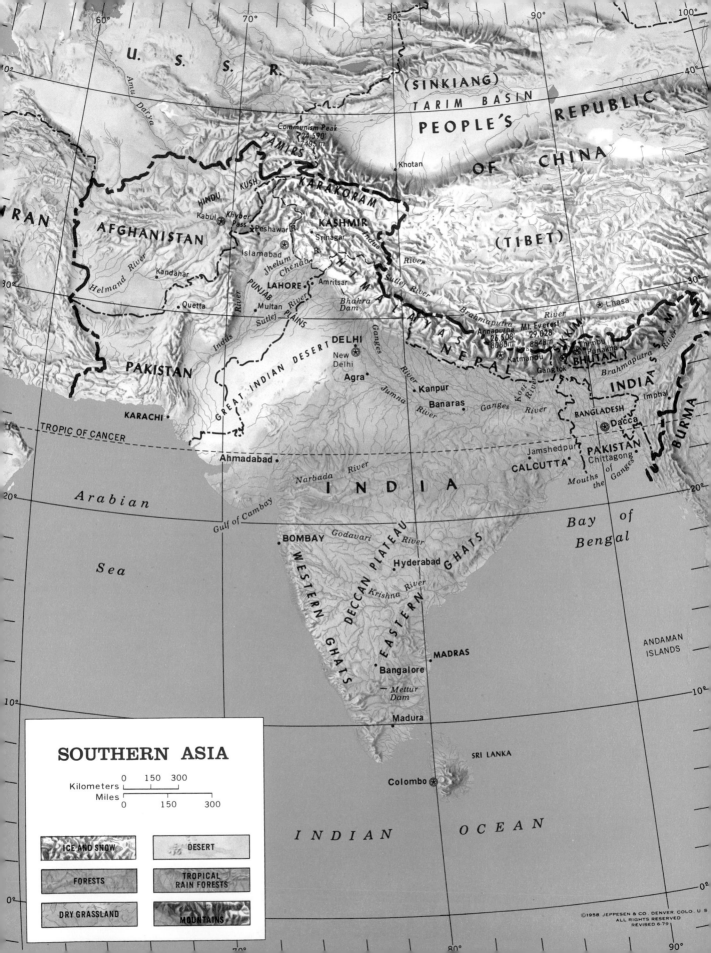

U. S. S. R.

60° 70° 80° 90° 100°

Amu Darya

(SINKIANG)
TARIM BASIN
PEOPLE'S REPUBLIC

Communism Peak
24,590'
7495 m

PAMIRS

OF CHINA

Khotan

KARAKORAM

HINDU KUSH

Kabul

Khyber Pass

Peshawar R.

KASHMIR

Srinagar

Islamabad

(TIBET)

AFGHANISTAN

Lhasa

IRAN

Helmand River

Kandahar

Jhelum

Chenab

R.

Indus River

LAHORE

Amritsar

Sutlej River

Brahmaputra River

Annapurna
26,508
8080m

Mt. Everest
29,028
8848m

Thimbu

Quetta

PUNJAB

Multan

Bhakra Dam

PLAINS

SIKKIM

Punakha

Sutlej River

HIMALAYAS

NEPAL

Gangtok

BHUTAN

PAKISTAN

Indus

River

GREAT INDIAN DESERT

DELHI

New Delhi

Ganges

Katmandu

Brahmaputra River

ASSAM

INDIA

Imphal

Agra

River

Kanpur

Kosi River

BURMA

KARACHI

Jumna River

Banaras

Ganges

River

BANGLADESH

Dacca

TROPIC OF CANCER

Jamshedpur

PAKISTAN

Chittagong

Ahmadabad

Narbada River

I N D I A

CALCUTTA

Mouths of the Ganges

Arabian

Gulf of Cambay

BOMBAY

Godavari

DECCAN PLATEAU

River

Bay of Bengal

20°

Sea

Hyderabad

WESTERN GHATS

Krishna River

EASTERN GHATS

ANDAMAN ISLANDS

MADRAS

Bangalore

Mettur Dam

10°

Madura

SRI LANKA

Colombo

INDIAN   OCEAN

SOUTHERN ASIA

Kilometers
0   150   300

Miles
0   150   300

ICE AND SNOW   DESERT

FORESTS   TROPICAL RAIN FORESTS

DRY GRASSLAND
                      MOUNTAINS

SUMMER MONSOON

Prevailing winds

Evaporation

Dry

Water for crops

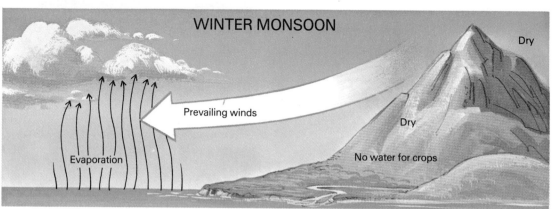

WINTER MONSOON

Prevailing winds

Evaporation

Dry

Dry

No water for crops

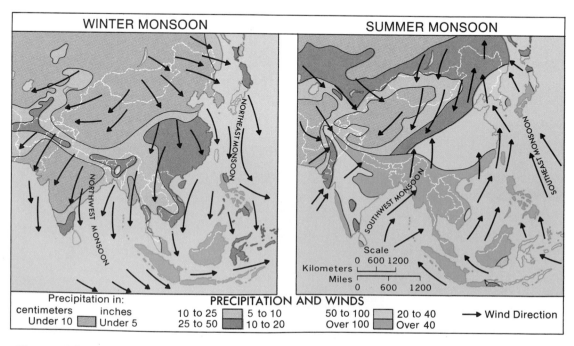

WINTER MONSOON

SUMMER MONSOON

NORTHEAST MONSOON

NORTHWEST MONSOON

SOUTHWEST MONSOON

SOUTHEAST MONSOON

Scale
0 600 1200
Kilometers
Miles
0   600   1200

Precipitation in:

centimeters	inches		
Under 10	Under 5		

PRECIPITATION AND WINDS

10 to 25	5 to 10
25 to 50	10 to 20
50 to 100	20 to 40
Over 100	Over 40

→ Wind Direction

The Climate of India

There are three major seasons in India. The hot season extends from about early March to mid-June. This is followed by the wet season, or **summer monsoon**. The rains last from late June to late August or early September. The cool season then extends about six months, completing the cycle of seasons. Temperatures and rainfall vary considerably during these seasons. Note the seasonal temperature and rainfall on the chart, page 498.

The Monsoons. As you have learned, monsoons are seasonal winds that blow in different directions at different times of the year. Monsoons occur in Asia mainly because land heats and cools more rapidly than water. The very large Asian continent, with its high mountains, cools very quickly in the fall. Since the oceans do not cool as quickly, the air over the water is kept warmer than the air over the land. As this warm, lighter air rises, the cool, heavier air over the land flows toward the ocean. **Winter monsoons** in Southern Asia blow from the land toward the seas. They bring months of cool, dry weather to most areas.

In the spring, the land warms up more rapidly than the water. There are many places in Southern Asia where the land gets very hot before the summer monsoon starts. As the air over the land becomes hotter, it rises higher because it becomes lighter. In time, cooler, heavier, and moisture-bearing air flows from the ocean toward and across the land.

In India, the summer monsoon blows from the Indian Ocean toward the western and southern coasts and on across the land. Because the summer monsoon gathers moisture as it blows over the ocean toward the land, it brings heavy rains to India. The winter monsoon, however, is a drying wind because it blows over the land toward the ocean. People refer to the winter monsoon as the cool season. The summer monsoon, of course, is the wet season.

The "Wind That Means Life to India." The summer monsoon has also been called the "wind that means life to India." It brings most of the rains needed for plant life and food crops. Prior to the monsoon, much of India is hot, dry, and dusty. Trees are sometimes bare, and the grasses are withered from the long cool and hot seasons. It may have been many months since any rain has fallen. As the time for the monsoon nears, people look anxiously to the southwest for signs of rain. If the rains are late, it can be very bad for the farmers. Only irrigated lands can escape the damages of a severe drought.

When the rains begin, the people are happy. Much work is then begun. Crops need planting. It is necessary for the rains to continue throughout the season. Too little rain can damage crops. Too much rain can also cause flooding. Usually, the summer monsoon brings enough rain for the farmers to successfully work their land. After crops are harvested, many happy festivals take place.

With the arrival of the summer monsoon farmers in Nepal transplant rice seedlings into paddies.

Water: Too Much and Too Little. As the population of India has increased, so too has the demand for food. At times, a late monsoon or a drought will result in poor crops. Then, India is forced to import huge amounts of grain. Sometimes too much rain causes flooding in the lowland plains. Crops then are also ruined. So, too, are many of the earthen canals and dikes used for irrigating the rice paddies. Many homes are often destroyed and people killed in floods.

Most of the rivers of northern India have their sources in the Hi-malaya Mountains. During the monsoon season, many of these rivers become raging torrents as they flow from the steep highlands to the Ganges plains. It has become important to slow down the flow of the rivers before they flood the plains. However, India does not have control over all of the highlands. Afghanistan, Pakistan, and Nepal own parts of the highlands. Thus, water management for flood control, as well as for irrigation, has become a major international problem for India.

To control the flood waters during the rainy season the government has built many large dams. The stored water enables millions of hectares (acres) of land to be irrigated during the dry season. The dams also store water from melting snow during the hot season.

The Main Food of India—Rice

Throughout India most people live in villages and farm the nearby land. Rice is the main food crop of most people. Of every 100 sacks of rice harvested on Earth each year, about 90 of them are harvested in Southern and Eastern Asia. Not much of this rice is exported. A few countries, such as Burma (*bur* muh) and Thailand (*ty* land) do raise enough rice to sell to neighboring countries. Most people in the United States eat about 4 kilograms (8 pounds) of rice a year. People living in Southern and Eastern Asia, however, eat about 97 kilograms (215 pounds) of rice a year!

A majority of the people in Asia eat rice for several reasons. People tend to depend most upon the food crop best suited to the soil and climate of the region in which they live. Since almost half of the people on Earth live in Southern and Eastern Asia, much food must be grown to feed them. Much of the land is hilly and mountainous. Many farmers have only a small plot or terrace on which to grow crops. Rice can be grown in greater quantities on one hectare of land than any other grain, if there is plenty of water. The summer monsoons usually bring rain to this region about the time that water is needed for the rice crop. Rice grows best in a warm climate, and in fields which are kept flooded for about two and one-half months of the year. Thus, rice became the main food crop of Southern and Eastern Asia.

Rice Farming. Do you know how rice is grown? In most areas of Southern and Eastern Asia rice is grown in *paddies*. A **paddy** is a field with a low bank of earth around it to hold water in the field. Usually an outlet in the lowest part of the field enables the water to flow gradually from the field.

When the land is almost level, as on a river plain, paddies can be made quite large. On hillsides where considerable rice is grown, terraces have to be built. On steep hills, the paddies may be only a few feet wide, but very long. The farmers build the terraced paddies so that water will flow slowly from the highest paddy to the lowest paddies.

The new rice plants are grown in a seedbed before the monsoon rains come. The seedbed is kept irrigated by water from a nearby well or stream. After the rains come, the paddies begin to fill up with water. The farmer, sometimes with the help of a water buffalo, pulls a plow through the mud. Then the farmer's buffalo pulls something that looks like a large comb through the field. The "comb" pulls all weeds and grass from the paddy. In terraces high on hillsides most work is done by hand. Once the field is prepared, the seedlings are then transplanted

In Japan, as in all rice-growing areas of Asia, the rice plants are cut, tied, and left in the fields to dry.

EASTERN HEMISPHERE: FOOD PRODUCTION

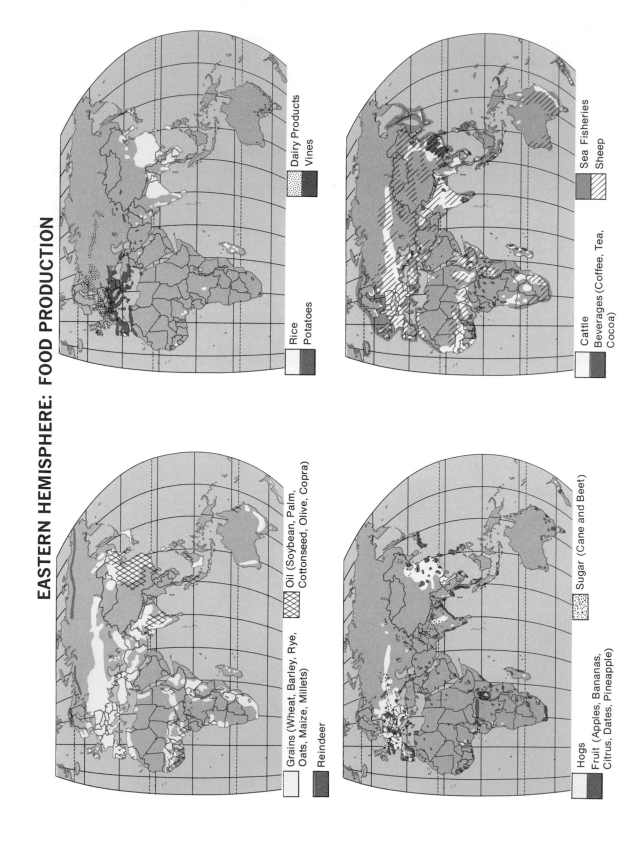

Grains (Wheat, Barley, Rye, Oats, Maize, Millets)

Reindeer

Oil (Soybean, Palm, Cottonseed, Olive, Copra)

Rice

Potatoes

Dairy Products

Vines

Hogs

Fruit (Apples, Bananas, Citrus, Dates, Pineapple)

Sugar (Cane and Beet)

Cattle

Beverages (Coffee, Tea, Cocoa)

Sea Fisheries

Sheep

into the paddies. Each plant is placed in the mud by hand. The plants are placed about a foot apart in long rows.

After the rice has grown for about two and one-half months, the paddies are allowed to dry out. The grain then gradually ripens. As it dries, the plants turn from bright green to a light tan. It looks very much like ripe wheat. Then it is ready to harvest.

Most rice plants are cut by hand. Often they are tied in bunches, and left in stacks to dry. The grains are threshed by knocking the heads of grain from the stalk. Sometimes threshing is done by simple machines. It is also done by hand. Each grain of rice is covered by a husk. The husk protects the grain so it can be stored for a long while. Usually, the grain is put into large burlap bags. People store their own grain in large containers, and sell what they don't need in markets.

Other Crops. Inland on the Deccan Plateau and in northern India, where less rain falls, most farmers raise wheat, millet, corn, or barley. In many places wheat can also be grown during the cool season in dried-out paddies.

Experiments in raising cereal grains have produced hardier strains of wheat and rice plants. Some of these new seeds produce plants that are more resistant to disease than previous plants. Some require less water for growth. Others produce greater quantities of grain per plant. A farmer can harvest more grain per hectare (acre) using some of these newly developed plants. Hopefully, such scientific advancements will help the small farmers of India increase their crops. The development and distribution of cheap fertilizer will also aid the farmers.

The main cash crops of India differ from place to place. Near Calcutta, in the lower Ganges Valley, jute is the main money crop. Sugar cane is a major crop along the coasts, where irrigation is possible. India is one of the world's main producers of cane sugar. Spices, tobacco, coffee, peanuts, and oil seeds are also valuable crops.

Many fruits and vegetables are also raised in India. Mangoes, bananas, oranges, pineapples, lemons, and tangerines are commonly found in the markets during their season. Green leafy vegetables such as spinach and mustard greens are also popular. Cucumbers, eggplant, and onions are raised on many farms. Most of these vegetables are used locally. Some are used in the food processing industry. Some canned and packaged goods are shipped to neighboring nations. Cotton is also raised in the upper Ganges Valley and in parts of the northern Deccan Plateau. Tea is a major product from the hills of Assam. Darjeeling (dahr *jee* lihng) tea is world famous. Have you ever tasted tea from India?

Animals in India. There are more cattle in India than in any other country, and for a good reason. Almost 9 out of 10 people in India are Hindus. According to the teachings of their religion, the cow is sacred.

Why Are Cows Sacred?

Many visitors to Hindu nations such as India and Nepal are surprised at the respect and freedom given cows. One of the authors of this text lived in Kathmandu (kaht mahn *doo*), Nepal for two years. Cows were allowed to roam freely in the streets and sidewalks of the city. Motorists had to be very careful to avoid the slow-moving animals as they grazed and sunned themselves along the roadways. It was not uncommon to find a cow or bull blocking downtown sidewalks. Why are cows sacred? Here are a few of the reasons given.

According to one legend, the Hindu Goddess Luxmi (*lux* mee) once disguised herself in the form of a cow. Luxmi is worshipped by Hindus as the Goddess of Wealth. The cow provides milk for food. The cow's manure is used as fertilizer for rice crops. These are important products to the local people. Therefore, the cow too, becomes a symbol of wealth, and is given a place of respect in Hindu customs.

In Hindu belief once a person dies he or she must cross a raging river to reach heaven. According to legend, one Hindu god was trying to cross the river but the current was too swift. A cow appeared and let the god hold its tail. Then the cow crossed the river, pulling the god safely to heaven. Some people say the cow is respected because it may help people to reach Heaven.

Cattle are not killed by the Hindus. Even when the cattle are old and useless, they are still left to die a natural death. Hindus do not eat beef. Many Hindus and Buddhists are strict **vegetarians**. They will eat no meat of any kind. Non-vegetarians, however, will eat chicken, mutton, pork, buffalo meat, and goat —but not beef.

Many other domestic animals can be found on farms in India. Chickens and ducks are popular. Goats, pigs, sheep, donkeys, and horses may also be found. Male cattle called **bullocks** are often used as work animals. It is very common to see pairs of bullocks hitched to small high-wheeled carts pulling products to and from the local villages. Water buffalos are also common work animals in parts of India. The elephant is found in India doing its share of labor as well.

REMEMBER, THINK, DO

1. Why do monsoon rains occur in India?
2. Make a chart, comparing the annual rainfall and temperatures of your area to that of Delhi, India. (Use the chart on page 498 for data on Delhi.)
3. Make a diorama showing the stages of growth of a rice plant.
4. Make a product map, note the major crops grown in the four regions of India.
5. Read some folk tales of India. You might wish to tell some of the interesting ones to students from other classes.

The Mineral Wealth of India

India is fortunate to have minerals needed for industries. In colonial times the minerals were either poorly developed or shipped out of the country. Since independence, India has developed many of its own mining companies.

India has huge deposits of good iron ore. It is possible that India leads the world in its reserves of iron ore. Eight billion tons of high-grade ore are known to exist. Lower grades of iron are also abundant. Much of the ore is located near Jamshedpur (*jahm shehd poor*), west of Calcutta.

The country also has major deposits of manganese. This metal helps make steel hard. India also has considerable coal. However, much of it is not of the quality needed for smelting iron ore.

Other minerals which have been found in important amounts are uranium, gold, aluminum, chromite, and gypsum. Building materials such as granite, slate, marble, and sandstone also are available. India's supply of mica represents nine-tenths (9/10) of the world's supply. Limestone, used in both fertilizers and cement, is also quite abundant.

Manufacturing in India

A number of steel mills have been built in India. The Tata Iron & Steel Works in Jamshedpur is one of the largest steel mills in Asia.

India's major manufacturing industry is textiles. This is not a new industry to India. Long before the 19th century, India was a leading exporter of handmade textiles to Europe. Today, textile plants produce machine-made cotton, woolen, and silk products. Jute manufacturing is also an important industry. Chemical plants have been built to produce drugs and fertilizers. Other factories manufacture paper, cement, sugar, matches, typewriters, bicycles, locomotives, sewing machines, telephones, machine tools, leather goods, and gasoline.

For many centuries the Indian people have been known for their skill in handicrafts. **Cottage industries**, in which cloth, leather, and metal goods are made with the use of simple hand tools are still important to India. Cottage industries are sometimes called home industries. They furnish employment to millions of Indian workers.

Energy for industry has been a major worry in India. Water resources for hydroelectric power are not always sufficient for year-round use. Petroleum is also limited. However, India has basic minerals needed for atomic fuels. Efforts have been made to develop nuclear energy. A nuclear power plant was built near Bombay in 1970. The United States helped India in this effort.

Transportation

India has a good railway system. Most of it was built during the British colonial period. Over 57 000 kilometers (36,000 miles) of railways cross the nation.

International air service is available from New Delhi, Bombay, and Calcutta. Domestic flights serve most major cities within India.

Roads are being improved quite rapidly in India. About 965 500 kilometers (600,000 miles) of roads have been developed. Of these, almost 290 000 kilometers (178,000 miles) are all-weather, surfaced roadways. But many small villages lack roadways leading to the nearest city. Villagers from these locations rarely go far away from their birthplace. Few own automobiles. Most villagers either walk, ride animal-pulled carts, or use bicycles whenever they wish to go from one place to another. On the main roads, bus service is often available to transport people to and from major places.

Major seaports are located at Calcutta and Bombay. Many passengers and much freight move to and from these ports.

Living in India

Cities and Villages. India has many large cities with high-rise buildings, wide streets, electricity, good transportation facilities, and many activities for tourists. Traffic jams at rush hour look quite similar to those you might see anywhere in the United States. Bombay and Calcutta are the two largest cities of India. Both contain important manufacturing centers, in addition to seaport facilities. New Delhi was built to serve as the nation's capital. It is a beautiful city, having much of the charm of Indian art and culture. Television is quite popular in the New Delhi area.

As in most cities, there are areas where many poor people live. Calcutta is especially known for its large population of homeless people. In the evenings, after shops have closed, thousands of homeless people place their bedding upon the sidewalks. They sleep in doorways and under the awnings of many shops. The government is trying to improve conditions for the poor. But millions of people have flocked to urban areas like Calcutta in hopes of finding employment. There have not been enough jobs, however, to take care of the thousands of homeless people.

Most of the people of India live in small villages. These villagers earn their living by farming. Most of these farmers only own a small piece of land. More than half of all farms are less than 2 hectares (5 acres) in size. Much work has been done to improve village life. Water taps have been built at central locations. This helps people in their daily search for water. Adult education programs have helped village people learn more about health and sanitation. Many new schools have been built. Medical clinics have been set up in some of the larger villages. Electricity has been brought to some villages

as well. Roadways have been built between the large villages. Although village life has improved for some people, there is still much that needs to be done for the millions of villages in India.

Health and Population. India's population has grown, in part, due

Supplying water to the people of India is a major problem. In large cities there is often not enough water for all the people. These workers are installing part of a new water supply system for Calcutta.

to improvements in health care. In recent years, inoculations have ended smallpox in India. Polio vaccines have also been given to people all over the nation. Improvements in child care and diet have helped save many children from numerous diseases and malnutrition. Medical care has been extended to adults as well. The number of elderly people and infants has increased since India became independent.

If the population continues to grow, some people fear that the nation will not be able to care for its people in the future. However, many Indian people do not want family planning programs. It is estimated that India's population may grow by 100 million in 20 years. This may create drastic problems for India in the future.

The Caste System. Although now against the law, the *caste system* once had a major effect upon the way of life for people in India. Under this system, people were divided into major groups, or **castes**. Each caste performed a particular kind of work. A person was born into the caste. One did not eat with people of a different caste. Marriage was only permitted between members of the same caste.

The highest caste were the priests and scholars. Below this caste were the rulers and warriors. Then came the merchants and farmers. Below them were the unskilled laborers. The lowest of all castes were the untouchables. The untouchables were often considered outcastes. These people were assigned the dirtiest jobs, such as cleaning the streets.

The caste system came into India in ancient times. It is believed to have been introduced as a part of the religious system. According to their religion, a person who lived a good life would be reborn into a higher caste in one's next life.

Although outlawed in India, many people still consider their caste to be important. Social feelings about caste have been slow to disappear. In many ways, caste still affects the way of life for many people in India.

Education. When India became independent few people could read or write. About 180 languages with 700 dialects were spoken by people throughout the nation. Almost at once the government started to teach basic reading and writing to adults. **Hindi** (*hin* dee), spoken by more than half the people, became the official national language. English is another official language used in some places. Thousands of new schools were opened. Many new teachers were trained to teach in these schools. Education is free for people up to 14 years of age.

In remote areas many classes are held out of doors. There are few books, and often the teacher's copy is the only one available. In spite of such difficulties progress in education is being made. People in India want their children to be educated.

There are more than 2,500 colleges in India. Most of these offer business and vocational subjects. Eighty universities throughout India offer a variety of professional programs. More than half of these universities were built in the 10-year period between 1958 and 1968.

India—Its Government and Border Problems

India is a democratic republic with a parliamentary form of government. Its prime minister is usually the leader of the elected majority party. The government of India has been active in Asian affairs, the Commonwealth of Nations, and the United Nations.

Kashmir is located in northwestern India. The region is claimed by both Pakistan and India. In this high mountainous region that borders China are the headwaters for the Indus (*ihn* duhs) River. In the past, India and Pakistan have fought each other for the rights to the land and water of Kashmir. The United Nations helped these two nations settle the dispute over this land. Under an agreement Pakistan took control of the Indus River and one-third (⅓) of the Kashmir territory. India took the rest.

China also claimed part of the eastern border of Kashmir. At one time, military forces of India and China faced each other at the border of this remote mountainous land. Agreements were finally reached between these two nations over the boundary lines. However, India still has great worries over its northern mountain border lands.

Sikkim was an independent nation, under India's protection, until 1975. The Himalayan nation, near Nepal and Bhutan, was made the 22nd state of India on May 16, 1975. This land does not contain known valuable resources. However, its location on the northern Himalayan border is very important to India. Some neighboring countries protested to the United Nations when India took control of Sikkim. In late 1979 China sent troops into Bhutan. Check to see what has happened. Small nations near India and China have reason to fear these two large countries.

REMEMBER, THINK, DO

1. If possible, visit a crafts shop where Indian handmade goods are displayed. A film about India's arts and crafts might also be available for your class to see.
2. What are some of the important industries of India?
3. What do you think might be done to help the homeless people living in Calcutta? How does your state or community try to solve the problems facing poor, homeless people?
4. Why do you think India might want to enlarge its Himalaya border lands?

Dr. Chris Chandler, a member of the American Bicentennial Everest expedition, stands atop Mount Everest on October 8, 1976.

Climbing Mount Everest

Have you ever climbed a high hill? How did you feel as you neared the top? Imagine climbing the world's highest mountain. Mount Everest, on the border between Nepal and Tibet, China, is 8848 meters (29,028 feet) high. It is always covered by snow and ice. During parts of the year storms are so bad that climbing it is impossible. Even during good weather, strong winds and constant cold make climbing difficult. Special clothing has to be worn, including face masks. Above 7900 meters (26,000 feet) almost no one can climb without oxygen. Only people who have lived in high altitudes all their lives can climb that high without oxygen. (One of the authors once climbed Long's Peak in Colorado. Although it is only 4345 meters [14,256 feet] high there was not enough oxygen to provide good breathing near the top. Mount Everest is twice as high!)

Many people have tried to climb Mount Everest. Some people ask "Why?" Mountain climbers usually say, "Because it is there!" Not until 1953 did anyone successfully climb this highest of all mountains. Then, Edmund P. Hillary of New Zealand and Tensing Norgay of Nepal made it to the top. They carried oxygen bottles as have almost all successful climbers since. Their story with many beautiful photographs is told in *The Conquest of Everest.* Your library probably has a copy.

Two members of an American climbing group climbed Everest for the first time up the west face in 1963. They returned the more common way through the South Col (a pass through the mountains). The story of their struggle up and down, including one night above 7900 meters (26,000 feet) without oxygen or a tent is told in *Americans on Everest.*

LEARN MORE ABOUT SOUTHERN ASIA

You may wish to learn more about the nations of Pakistan and Bangladesh since they were both once part of India. The landlocked Himalaya nations of Afghanistan, Nepal, and Bhutan also offer interesting topics to study. Sri Lanka is the only island-nation in this region of Asia. Use the Study Guide on page 31 to help you plan your studies.

Pakistan and Bangladesh

The creation of the independent nations of Pakistan and Bangladesh is a bit unusual. As you know, in 1947 the two Muslim territories of British India became the independent nation of Pakistan. This two-state division of West and East Pakistan, however, created many problems for the new nation. The two states were about 1600 kilometers (1,000 miles) apart. The government had a difficult time managing the development of these distant and quite different states.

After 20 years people in East Pakistan began to lose faith in their government. Many people believed that West Pakistan was being favored over East Pakistan. It was noted that only one-third (⅓) of the national budget was spent in East Pakistan. Most of the foreign aid received in Pakistan was spent in West Pakistan. Most of the military and government jobs were held by people from West Pakistan. Bitter disagreements arose between the West and East over the choice of a national language. Finally, people in the East decided it would be better to become an independent nation.

In 1971 fighting began between the people of West and East Pakistan. For about six months much killing and destruction occurred in East Pakistan. India then entered the dispute, forcing the end of the war in East Pakistan. At that time, East Pakistan became the country of Bangladesh.

Sri Lanka

Sri Lanka (Ceylon) is an island-nation located in the Indian Ocean south and east of the southern tip of India. This country was originally called Ceylon. However, the people preferred to rename their country with a word from their local language. The name Sri Lanka means "shaped like a pearl." Look at the shape of this island on the map on page 445. You may see why the name was selected.

North of the Indian subcontinent are the majestic mountains of the Hindu Kush (*hihn* doo koosh) and the Himalaya Mountains. Within these rugged mountains and high plateaus are the countries of Afghanistan, Nepal, and Bhutan. In part, the mountains have helped to isolate these areas from the rest of the world. At times, some governments have preferred their isolation and closed their borders to all foreigners.

Afghanistan has been the least isolated of these three nations. Situated in the Hindu Kush Mountains,

Because much of the land in Nepal is steep, the farmers have had to make terraces in the hillsides.

more than half of this nation's land is more than 1800 meters (6,000 feet) high. Some of the land is 6000 meters (20,000 feet) high. Between high peaks in the mountains are deep valleys. The Khyber (*ky* ber) Pass is a low area in the rugged mountains between Afghanistan and Pakistan. It has been a major trade route for hundreds of years.

Nepal had been one of the most isolated of all Himalaya nations. Foreigners were not free to enter Nepal until after 1951. Before this, many Westerners only knew of Nepal and surrounding nations as "the mysterious, forbidden kingdom of Shangri La (*shang* gree *lah*)."

The location of Nepal is considered to be very important. It separates two of the largest nations of Asia: India and China. Perched in the highest part of the Himalayas, Nepal has one of the most rugged landscapes of any nation in the world. Elevation varies from 60 meters (200 feet) on the lowland India border to the 8848 meters (29,028 feet) of Mount Sagarmatha—more commonly known as Mount Everest.

Bhutan is east of Nepal. It is a semi-independent kingdom. The king of Bhutan handles the internal affairs of his country. The government of India controls all international relations for this nation. Bhutan remains quite isolated from most of the outside world. A road has been constructed linking eastern and western Bhutan to India. However, the government of India does not permit foreign visitors to use the road.

WHAT HAVE YOU LEARNED?

1. What effect did Muslim rule have over India during the Islamic era?

2. What are the four main land areas of India?

3. Why is water sometimes a problem in India?

4. How have agricultural experiments aided India's crop production?

5. What is life like for many Indian villagers? What changes have taken place in the villages?

BE A GEOGRAPHER

1. Between what parallels of latitude is India located?

2. Between what meridians of longitude is India located?

3. What does the dashed line show that crosses central India?

4. Which is farther north:
   a. Delhi or Washington, D.C.?
   b. New Delhi or Miami, Florida?
   c. Calcutta or Mexico City?
   d. Bombay or Mexico City?

5. How many kilometers (or miles) is it from:
   a. New Delhi to Agra?
   b. Bombay to Calcutta?
   c. Srinagar, Kashmir, to Madras?
   d. Amritsar to Calcutta?

6. What is the main tributary of the Ganges River?

7. What river in addition to the Ganges flows into the Bay of Bengal east of Calcutta? What are these rivers building?

QUESTIONS TO THINK ABOUT

1. Why do you think the people of India wanted their freedom from British colonial rule?

2. How do the seasons of India compare to the seasons where you live?

3. Many people in India have to haul water from a pump to their home. If you had to do this, how much water would be needed for yourself in one 24-hour period? How many trips to the pump might be required for your water needs?

4. In your opinion, is there any type of caste system in the United States? Explain your reasons.

25

Chapter

EASTERN ASIA

More than one-fourth (¼) of all people on the Earth live in Eastern Asia. Until the 20th century the majority of Eastern Asians lived off the land, growing just enough food for the family. The 19th century industrial revolution that changed Europe so much did little to cause major changes in Eastern Asia. In this century, however, much change has occurred. Japan became an important industrial nation before World War II. It is now one of the world's most productive nations. It was not until after 1945 that important changes began to occur rapidly in the rest of Eastern Asia. The People's Republic of China has challenged the Soviet Union's leadership among Communist nations. It has become a major power in international affairs. Many changes have also occurred in the countries of Taiwan, North and South Korea (kuh *ree* uh), Mongolia, and the two colonies of Macau (muh *cow*) and Hong Kong. Use the map on page 464 to locate these nations of Eastern Asia.

Japan

Japan is today one of the major industrialized nations of the world. This is a remarkable achievement for two reasons. First, Japan has rather limited land and natural resources. Second, at the end of World War II Japan was a war-torn, defeated nation. Equally as startling is the change in the way of life in Japan. Until about 1940 most Japanese worked on farms or in small industries. Within 25 years Japan changed that way of life. Now it equals or surpasses the production of most European nations. City life today barely resembles that of Japan 50 years ago. Many people consider the developments of modern-day Japan as a true wonder in the world. Find the islands of Japan on the map on page 464. The islands are located off the Pacific Coast of Asia near 40° N.

The History of Japan

Not much is known about the beginning of life in ancient Japan.

Japan was the first nation in East Asia to establish an industrialized, modern society. Urban areas, such as the port of Nagasaki, reflect this development.

There is little doubt, though, that one of the first peoples to live on these islands were the **Ainu** (*eye noo*). Only about 15,000 Ainu still live in Japan. Most of them live on the island of Hokkaido (hah *kyd* oh). By the first century, A.D., several of the Japanese clans or tribes had been united under one ruler. By the 12th century a number of powerful warriors, called **shoguns** (*shoh* guhns) gained actual control of Japan.

Recent History. During the period of European exploration in the 16th century, ships from Portugal, Spain, The Netherlands, and England visited the harbors of Japan. Some trading occurred. Christian missionaries also came to teach their religion to the Japanese people. The shogun ruling Japan at that time began to fear that Europeans planned to take control of Japan. He made all the missionaries leave the island of Japan. Except for one Dutch ship each year, all outside trading was stopped. Thus, little contact with the rest of the world occurred. Japan remained fairly isolated for about 200 years.

In 1845 the United States sent Commodore Perry with a fleet of ships to Japan. Some American sailors had been shipwrecked on the Japanese coast. The United States government felt that these sailors had not been treated properly. Commodore Perry was directed to demand protection for any Americans who might be shipwrecked there in

EASTERN ASIA

Kilometers

0 200 400
|———|———|———|
0 200 400
Miles

©1958 JEPPI EN & CO. DENVER COLO U.S.A
J.L RIGHTS RESERVED.
REVISED 6-79

DESERT

DRY GRASSLAND

GRASSLAND

SAVANNA

MOUNTAINS

ICE AND SNOW

FOREST

the future. He was also to demand permission for American ships to take on supplies and to trade in Japanese ports. Commodore Perry's ships were loaded with goods ready to trade. They also had guns ready to enforce their demands if they met with resistance. Within a short period of time and without war, Perry signed a treaty with Japan. The treaty opened Japanese ports to trade with the United States. Soon other countries worked out treaties, too.

World War II. In 1931 armies from Japan attacked and conquered Manchuria (man *choor* ee uh). Manchuria is a province in northeastern China with both coal and iron ore. In 1937 Japanese armies invaded the rest of China. By 1941 the Japanese had gained control of most of the major cities and main farming areas of China.

On December 7, 1941, the Japanese began a war to seize all of Southeast Asia. They attacked Pearl Harbor in Hawaii to destroy the U.S. Pacific fleet. At that time Pearl Harbor was the major headquarters for the United States Navy in the Pacific. Japan's surprise attack brought the United States into the war. The attack was followed by a Japanese invasion of the Philippine Islands. Later other areas of Southeast Asia were taken by Japan. For four long years Australian, British, Chinese, Dutch, Filipino, and American forces fought the army, navy, and air forces of Japan. The Japanese were finally defeated in 1945 after atomic bombs were dropped on the cities of Hiroshima (here uh *she* muh) and Nagasaki (*nahg* uh *sahk* ee). Find these cities on the map on page 464.

American forces occupied Japan after the war. A democratic government, much like that of Great Britain's, was started in Japan. Laws were passed so that farmers could buy the land they were farming from the rich people who owned it. Power was taken from the Emperor of Japan, and he was no longer considered a god. He now represents the government of Japan much as the Queen of England does today in Britain.

After the war, Japan had to rebuild many of its damaged cities. The United States gave the Japanese government millions of dollars to help them do so. All children were soon attending schools to receive a better education. Farmers began to learn better ways of farming. Industries and businesses began to rebuild. New and better factories were designed to make what the world needed and wanted.

The Land and the Climate of Japan

About 115 million people live on the islands of Japan. Most of the people live on four main islands—Honshu (*hahn* shoo), Hokkaido, Kyushu (kee *yoo* shoo), and Shikoku (shih *koh* koo). Find these islands on the map on page 464. These islands together have about as much land as the state of Montana. Much of the land on the islands of Japan is mountainous. Only about one-sixth (⅙) of the land can be farmed.

The islands of Japan are really the tops of a submerged mountain range. The range rises from one of the deepest known parts of the Pacific Ocean. Many of these peaks were or are volcanoes. As the map on page 476 shows, there are few large plains on any of the islands. The largest plain is the one surrounding Tokyo. The coastal plains, especially on the main island of Honshu, are where most of the people live. These areas are very crowded.

The most famous volcanic peak in Japan is Mount Fuji (*foo* jee). It is the highest mountain peak, rising 3776 meters (12,388 feet) above sea level. The upper part of Mount Fuji is covered with snow during the winter and spring months.

No part of Japan is more than 145 kilometers (90 miles) from the sea. As a result, the climate of the islands of Japan is milder than mainland regions at the same latitude. There are considerable differences in climate within the country of Japan, however. The island of Hokkaido, which is the farthest north, has long cold winters with a large amount of snow. Kyushu, the island farthest south, almost never has snow or freezing temperatures.

Typhoons. Rain falls in Japan all through the year. The heaviest rains come in the summer growing season. *Typhoons* are common in the fall months. **Typhoons** are huge storms with high winds which bring heavy rains. They are **tropical cyclones**, like the storms called hurricanes in the Caribbean. Often, because of the hilly and mountainous land, destructive floods and mud slides are caused by the typhoon's rain.

Earthquakes. The land areas of Japan could be called one of the shakiest places on the Earth. Hundreds of earthquakes are felt in Japan each year. Most of the quakes are faint tremors which cause little, if any, damage. Every once in a while, though, much damage is caused by a major quake. Modern high-rise buildings in cities like Tokyo have to be specially constructed to withstand earth tremors. Similar construction problems exist in west coast cities in the United States.

REMEMBER, THINK, DO

1. On a desk outline map of Eastern Asia, name and label the four main islands of Japan.
2. Why is the climate of Japan milder than that of most mainland regions of the same latitude?
3. See a film about Japan which shows the four main islands and Mount Fuji.
4. Why do so many people in Japan live on the coastal plains?
5. Why has the industrial development of Japan been considered a major achievement?

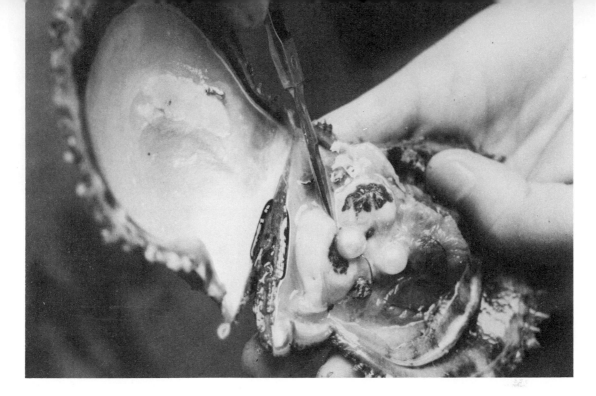

Home-Grown Pearls

One of the most interesting products of Japan is **pearls**. As you may already know, pearls grow in live oysters under certain conditions. The Japanese have learned how to make oysters develop pearls. They have even learned how to control the color of the pearl which the oyster grows.

Today the pearl-growing business is quite scientific. Divers bring oysters up from the bottom of the oyster beds. The live oysters are taken to a laboratory where a tiny bit of oyster shell is placed inside the oyster. No one knows exactly why, but the best pearls grow when a piece of oyster shell from Mississippi is placed inside the oyster. Then the oyster is returned to its place in the oyster beds. Three to five years later the shell is opened. Often, a beautiful, round pearl is found inside. Such pearls are called **cultured pearls**.

Agriculture in Japan

Most Japanese farmers have very small farms. The average size is about 1 hectare (about 2½ acres). Nevertheless, Japanese farmers raise a great deal of food on their land.

The combination of hard work, skill, irrigation, fertilizer, and newly developed seeds usually results in high crop yields. About one-third (⅓) of the working force of Japan is farmers. Much of the work is done by gasoline-powered, small machines.

One-sixth (⅙) of all the fish caught in the world are caught by Japanese fishers. Almost all the catch is eaten in Japan.

Of course, simple tools and hand labor are also needed. Many small tractors, about the size of a lawn mower, are used in the fields. There is such a shortage of good farmland that almost every inch of space where crops will grow is used. When climate permits, more than one crop is planted on the same land each year.

The Main Food Crop. As in the rest of Asia rice is the most important food crop. About half of the farmland is planted with this grain. The average Japanese farmer raises more rice per hectare or acre than do farmers anywhere in the world. Since 1968 Japanese farmers have been able to produce more than enough rice. The nation exports its surplus rice to other Asian countries. Japan imports much of its needed wheat, corn, and soybeans from the United States. About one-fifth (⅕) of

Japan's food needs to be imported.

Other crops raised on Japanese farms include vegetables, tea, fruits such as apples and mandarin oranges, and some grains other than rice. Mulberry trees are grown so their leaves may be used to feed silkworms.

Many Japanese farmers now use a method of growing plants called **hydroponic farming**. Instead of placing seeds in soil, plants are raised in water to which needed chemicals are added.

Animals. For many years, Japanese farmers have had very few animals on their farms. Using valuable cropland for animals was not considered a good way to use the land. After World War II, farmers began to raise some dairy and beef cattle, as well as hogs and chickens. Meat from these animals decreases the need for some food imports.

Fish. Protein is needed in everyone's diet. In the United States most people eat meat to get needed protein. Do you know how the Japanese add protein to their diet? The Japanese get most of their protein by eating fish. The average Japanese eats about 30 kilograms (64 pounds) of fish a year as compared with the yearly average of 5 kilograms (11 pounds) for most Americans.

The Japanese fishing industry, in terms of size of the catch, is first in the world. Japanese fishers not only fish near the islands of Japan. They also fish throughout the Pacific Ocean and wherever fishing is good in other oceans. Whaling ships from Japan search the Arctic and Antarctic continental waters for whales. In 1976, however, most nations established a fishing limit of 370 kilometers (200 nautical miles) from their shorelines. Fishers from other nations are not allowed to fish there. This zone was established to protect the fish reserves of individual nations. As a result Japanese fishers may have difficulty catching as much fish as they did in the past.

Many kinds of fish are caught by Japanese fleets. Tuna, sardines, crab meat, salmon, and oysters are canned or frozen for export. Herring, mackerel, cod, and pollack are caught in huge amounts and sold fresh to the Japanese people.

Forestry

More than half of the land of Japan is forested. The Japanese have given much attention to their forest resources. The government made a list of all the different kinds of trees that grew in the country. Then they decided which ones were most useful. The planting of trees has been controlled so that only the most useful ones would be grown first. Old trees must be replaced with young seedlings in order to maintain Japan's forest resources.

Much use is made of forest products in Japan. Most homes are built of wood. The textile industry uses wood pulp in making many synthetic fibers. Most fishing vessels until recently were made largely of wood. Mines need wood for supporting the sides of underground tunnels to prevent cave-ins. The railways need wood for ties upon which steel rails are laid. Some wood is also used for fuel. Almost nothing is wasted in the Japanese forests. Even the leaves, grass, and plants which grow in the forests are used as fuel or for enriching soils. Although much wood is harvested in Japan, some has to be imported. Much of the imported wood comes from Canada or the United States.

Other Natural Resources

Japan's mineral resources are rather limited for a major industrial nation. Large quantities of coal, copper, silver, sulfur, and zinc are mined. Gold and chromium are also found. Most other needed minerals and fuels have to be imported for the industries. Japan is able to mine about one-fourth (¼) of the iron ore it uses. The rest is imported. Petroleum needs are mainly met by imports.

Much of the power to run the many factories is produced by hydroelectric sources in the mountains. Most of these operate at full capacity only during periods of good rainfall. Japan is rapidly building power plants that can be operated by nuclear energy.

Industrial Achievements of Japan

The industrial development of Japan has provided its people with the highest standard of living of any nation in Asia. Manufacturing in Japan is carried on in much the same way as in nations of western Europe and North America. In fact, Japan's total of goods produced now exceeds every nation except two. Can you name them? Japan's industrial output per worker is now the world's highest.

Throughout Japan, factories turn out iron and steel, textiles, chemicals, cement, cameras, and electrical appliances. Almost anything you can name is probably made in Japan. Look at the appliances and tools in your home. How many of them were made in Japan?

Shipyards in Japan build more ships each year than any other nation in the world. Only two other nations manufacture more steel than Japan.

Japan is one of the leading textile manufacturing nations of the world. Most of the cotton used for textiles is imported from the United States, India, Pakistan, and Egypt. Imported wool is also used in fabrics. Silk is also a major textile made in Japan,

and the silk thread is produced there. A variety of new synthetic fabrics are produced by Japanese textile factories.

Japan leads the world in the manufacture of two-wheeled vehicles. It is second in the world in making passenger automobiles. Do you know which country makes the most automobiles?

The Japanese have become world famous for products which require skill and patience to make. Optical goods and time pieces have become valuable export items for Japan. Cameras made there are known everywhere as among the best made in the world. Electronic products have also become a major business for Japan. Television sets, radios, tape recorders, transistors, computers, electron microscopes, and automatic control systems are some of the major products exported by Japan.

Plastic toys and novelty items of all kinds are made by the millions in Japan. These are shipped all over the world. Factories also make locomotives, machine tools, pottery, china, and fertilizers.

Railways and Roads

Japan has one of the finest railway systems in the world. Major cities and rural areas are connected by many convenient railway lines. Japan is one of the few countries where trains carry more passengers than they do freight. It is also one of the few countries where trains almost always operate on time. One of

A worker inspects part of an electrical generator. The manufacturing of this kind of equipment is an important industry in Japan.

Japan's famous trains, the Osaka (oh *sah* kuh) Express, maintains a schedule that averages 162 kilometers (101 miles) per hour. At times the train goes as fast as 210 kilometers per hour (130 miles per hour). This train is called a "bullet train."

Japan was slow in building a good highway system. Until the 1960s very few Japanese people could afford to buy automobiles. Although many vehicles were being manufactured in Japan, they were built mostly for export. Such luxury items were beyond the income of the average Japanese worker. Thus, there was little road traffic and not much need for highways.

All that has changed. As wages were increased, workers began to have money to spend. They were soon able to buy luxury items for themselves. People who once rode a bicycle to work soon bought a motor scooter. It was not long before the scooter was replaced by a small car. As a result of this rapid change, highway construction came after the increased use of automobiles in Japan.

The auto boom caught the government unprepared. During rush hours in many cities of Japan, traffic is crowded and slow moving. At peak hours autos are often bumper to bumper. They rarely move more than 16 kilometers (10 miles) per hour. This condition is not too different from many cities in the United States.

Tokyo and Other Cities

Tokyo is one of the largest cities on earth. Huge cities like Shanghai, New York City, London, and Tokyo are so large that it is very difficult to know which is the largest. Tokyo is a very crowded and modern city. It has many high-rise buildings, elevated railways, subways, streetcars, buses, and taxicabs. Its shops, amusements, and restaurants serve the people working in or visiting this urban area.

Yokohama (*yoh* kuh *hahm* uh) used to be about 32 kilometers (20 miles) from Tokyo. It is now practically a part of Tokyo. It is the port city for Japan's capital. It is a center for foreign trade as well as having many large factories.

Osaka is the second largest city in Japan. Find it on the map on page 464. Osaka is a center for the huge textile industry.

Not very far from Osaka are two other large cities, Kobe (*koh* bee) and Kyoto (kee *oht* oh). Both have slightly more than a million people. Kobe serves as a port for both Osaka and Kyoto. Kyoto is a religious center. It was the ancient capital of Japan. During most of its history, emperors of Japan were crowned in the Imperial Palace there. Many famous handicraft factories are located in Kyoto.

Population Growth—a Major Problem Just About Solved

While Japan was making rapid progress as a commercial and industrial nation, its population was growing rapidly. Japan's population almost doubled in 50 years. Since 1960, population growth has slowed due mostly to a decrease in the birth rate. By 1977 Japan's birth rate was the lowest in Asia. It is still higher than most European countries, however. It appears that the Japanese may be solving one of their major problems.

REMEMBER, THINK, DO

1. Why do most Japanese eat more fish than meat?
2. What is hydroponic farming?
3. Do you think fishers of the world should be permitted to fish closer than 370 kilometers (200 nautical miles) from a nation's shores?
4. Visit local shops and appliance centers in your neighborhood. Make a list of items for sale that were manufactured in Japan. Compare the prices of Japanese-made items and American-made items.
5. Do a mini-report on silk production, from worm to fabric. Make a chart showing the major steps in the process. A sample display of silk and synthetic fabrics might also be arranged.

The People's Republic of China

China is an old nation that is changing rapidly. This nation has more people than any other nation. No one knows exactly how many people live in China. However, it is believed that the population is more than 900 million. At current rates of growth the population of China will double in about 58 years. The population in the United States will double in about 115 years.

The government of China is attempting to decrease the growth rate. Young people are encouraged to wait until they are a little older to marry. Family planning has been introduced. The government is also trying to move people from the crowded eastern region. They are moving them to the less populated parts of the country.

Look at the population map on page 484. You will see that most Chinese live in the eastern part of the country sometimes called China Proper. That part of China is an area about twice the size of the United States east of the Mississippi River.

Look at China on the map on page 464. Notice that China extends north and south about 35 degrees of latitude. The southern part of China, Hainan (*hy nahn*) Island, is at about the same latitude as Hawaii. The northern parts of China are at about the same latitude as southern Alaska. What differences do you think there will be in the climate of northern and southern China?

Since the revolution in 1949, the Chinese people have been able to increase greatly the amount of land used for farming.

A Brief History of Mainland China

Long before Columbus visited America the Chinese people had invented paper, printing, and gunpowder. They had developed a civilization rich in art and literature. The people also had adopted the religion of Buddhism.

At times during its history, China has been invaded and ruled by people from other countries. The easiest way for armies to invade China was from the north. For that reason, to protect their land, the Chinese built the **Great Wall**. This wall is about 8 meters (25 feet) high. It is about 2400 kilometers (1,500 miles) long.

In spite of the Great Wall, China was conquered in the 1200s by Kublai Khan (*koo* bly *kahn*) from Mongolia. Do you remember what other large country the Mongols conquered? The Mongols were driven out of China in 1368.

In 1644 China was conquered by the Manchus from Manchuria. Most ways of living in China changed very slowly under the Manchus. At this same period, Europeans were exploring other lands, developing the steam engine, and building factories. The Chinese, however, continued their old ways of doing things.

During the 19th century, however, some of China's ports were opened to foreign trade. Ideas from Europe and America began to influence the Chinese. Some Chinese visited other countries. They became dissatisfied with their government and their ways of living. These Chinese, under the leadership of Sun Yat-sen, organized a revolution. In 1912 the Republic of China was started. For a dozen years, military men known as **war lords** governed most of the provinces in the Republic. They became rich by setting high taxes. In 1928 General Chiang Kai-shek (*chang ky shehk*) defeated the warlords. He tried to unite all of China.

In 1931, however, armies from Japan invaded Manchuria. The war between Japan and China continued until it became a part of World War II. In 1928 a civil war began in China. It continued after World War II. It was a war between the government of Chiang Kai-shek who led a group known as the Nationalist Party, and the Chinese Communist Party led by Mao Tse-tung (*mahoh* zeh *doong*). By 1949 the Communists won control of all of the mainland of China and the island of Hainan. The Nationalist army fled to the island of Taiwan. That island off China's coast is sometimes called Formosa (for *moh* suh). Find it on the map on page 464.

The official name of mainland China now is the People's Republic of China. The country's land south of the Great Wall is known as China Proper. The People's Republic also includes Manchuria, Inner Mongolia, Kansu (*kan soo*) Province, Sinkiang (*shihn* jee *ahng*), and Tibet (tuh *beht*). Find all these areas on the map on page 464. The People's Republic of China claims Taiwan as part of their land. The Nationalist government on Taiwan, however, still claims that it is the rightful government of all China. The United States recognizes the People's Republic of China as the real government. The Communist government has made Peking (*pee kihng*) the capital of China.

The Chinese Communists, with aid from the U.S.S.R., brought about a great many changes in China during the 1950s. During the 1960s the governments of China and the U.S.S.R. gradually grew apart. They disagreed openly on many policies. Some clashes occurred between the Chinese and Soviets along their common border. During the 1970s China improved its relationships with many other countries.

Buddhism

Before the introduction of communism to China, the Buddhist religion was a major religion in China. However, Buddhism had its origin thousands of miles away, near India. The man who became known as the Buddha (*bood* uh) was originally a wealthy prince. His family had provided Prince Siddhartha (sihd *dahr* tuh) with all the possible luxuries of life. Like other people of that time and place, the family practiced the Hindu religion.

As Prince Siddhartha grew into manhood he began to travel outside the family's lands. He became familiar with the poverty suffered by many people. The prince became dissatisfied with many Hindu religious practices. These practices cost poor people a great deal of money. The prince believed that religious ceremonies should be simple and helpful. Prince Siddhartha gave up the family fortune and his title.

For many years he traveled about India. He developed ideas about how people might live in harmony with nature and with one another. He began to teach his ideas to others. Many people were attracted to this man's teachings. He was called "the Buddha" by his followers. The Buddha was against acts of violence between people. He objected to the killing of animals. He was also against the caste system that operated in India then.

Some places were started where followers of the Buddha could spend their lives doing good works and praying. Many people went to these places, called **monasteries**, to learn the teachings of Buddha.

Actually, the Buddha himself remained a Hindu all his life. Followers of his teachings became known as Buddhists. Many Buddhist monks traveled through Asia spreading this new faith. It was not long before the religion spread into Tibet, China, and most of southern and southeastern Asia.

The Land of China

As the map on this page shows, much of China is mountainous. There are low mountains along the coast, high mountains in the west, and very high basins and mountains in Tibet. The Tsinling Shan (*chin lihng shahn*) range extends eastward from southern Kansu province. This range divides China Proper into two main parts—northern China and southern China.

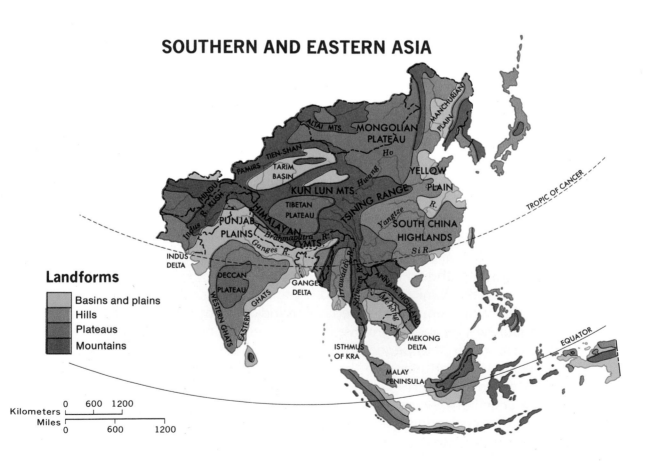

SOUTHERN AND EASTERN ASIA

Landforms
- Basins and plains
- Hills
- Plateaus
- Mountains

Kilometers 0 600 1200
Miles 0 600 1200

China Proper. Much of the land in central and southeastern China is hilly and mountainous. It is drained by three important rivers, the Hwang Ho (*hwahng hoh*), the Yangtze (*yang see*) and the Si (shee). The most important croplands in China are in the valleys and deltas of these rivers. Most of the large cities of China Proper are located either on these rivers or near them.

The Hwang Ho has been called "China's Sorrow" because of its many destructive floods. It flows for many miles in northern China through an area of loessland. Do you remember what loess is? It is fertile, fine soil that has been deposited on the land by winds. Because few forests and little grass grow on the hillsides in this part of China, the loessland erodes easily. So much of the topsoil is carried downstream that the river is named for it. Hwang Ho means "Yellow River." For thousands of years the Hwang Ho has been carrying soil downstream. It has built a huge flat plain near its mouth which is sometimes called the "Yellow Plain." Peking is located in the northern part of this plain.

To prevent floods along the lower Hwang Ho, levees were built along its banks. In many places the river is now higher than the land on either side. Canals carry water to areas needing it for irrigation. This water helps make the soil more fertile because it leaves rich topsoil behind. Grass and trees have been planted along much of the Hwang Ho to help stop erosion, too.

The Yangtze is China's largest and longest river. It is also one of the largest rivers on the Earth. The Yangtze flows eastward from the Tsinling Shan range. The Yangtze is about 5500 kilometers (3,400 miles) long. Small, powerful ships can go upstream more than half its length. Sea-going ships can sail upriver as far as Wuhan (*woo hahn*), about 1100 kilometers (680 miles) from the sea. Shanghai, China's largest city and probably the world's largest, is built near the mouth of the Yangtze.

The Si River in southern China also has formed a fertile delta plain where many people live. "*Si*," in Chinese, means "west." Do you see why the river was named Si? Canton (kan *tahn*) is the main city near its mouth.

The northern part of China Proper has cold winters and hot summers. Barely enough rain falls most years for crops. Main food crops are millet and wheat. On irrigated land, cotton, rice, and fruits are grown. During droughts, not enough food can be raised to feed the people. The Yellow Plain area is dry and dusty except during the spring and early summer months. Main crops grown there include wheat, cotton, soybeans, millet, sweet potatoes, peanuts, and sorghum.

The southern part of China Proper has a warmer climate and plenty of rainfall. In the basins of the Yangtze and Si rivers, rice is the main crop. Crops can be grown almost all during the year, but much of the land is hilly or mountainous. Terraces have been built on many hillsides in order to raise more rice.

REMEMBER, THINK, DO

---

1. Locate and label on an outline map of China the approximate location of the Hwang Ho, Yangtze, and Si rivers of China Proper.
2. Using a globe, check to see whether any part of China is as far north as Juneau, Alaska. Then check to see whether any part of mainland China is as far south as Honolulu, Hawaii.
3. Why did the ancient Chinese build the Great Wall?
4. What have the Chinese done to prevent flooding along the Hwang Ho River?
5. Prepare a brief oral report on one of the following:
   a. The Great Wall of China
   b. Contributions of China to the Western World
   c. Buddhism
   Or choose another topic for a short report.

Manchuria. Northeastern China is known as Manchuria. The land is mainly a rolling plain surrounded by mountains. The Sungari (*soong* guh ree) River drains much of the northern part of the central plain. It flows into the Amur (ah *moor*) River which forms the border between Manchuria and the U.S.S.R.

Because Manchuria is farther north than China Proper, the climate is colder during winter months. The growing season is shorter there, too. Most rainfall comes during summer months when monsoon winds blow from the ocean across the land. Enough rain falls throughout most of Manchuria for grains to grow. The rainfall decreases as one goes west, however. Some of the land along the border with Mongolia is desert. The crops most suited to the climate and soil of Manchuria are spring wheat, millet, corn, barley, sorghum, and soybeans. These are Manchuria's main crops. Fruits, such as apples, pears, cherries, grapes, and peaches, are grown in southern Manchuria. Sugar beets, tobacco, hemp, and some rice are also grown.

Inner Mongolia. The land north and west of the Great Wall was, until recently, an area where few people lived. Inner Mongolia is mainly an arid region through which the Hwang Ho flows. The southern part of Inner Mongolia is grassland. The northern area is part of the Gobi (*goh* bee) Desert. Until about 1960, most of the people who lived in this region were nomadic herders. These people raised sheep, camels, and horses. Since the 1960s, however, railroads have been built along and near the Hwang Ho River. One of them extends northward to the Trans-Siberia Railroad near Lake Baikal. New iron and coal mines have been developed and steel mills have been built.

Kansu Province. Located between Inner Mongolia and Sinkiang, is Kansu Province which has had rapid growth. Lanchow (*lahn joh*), the capital city of the province, is one of the fastest growing cities in the world. Large hydroelectric and irrigation projects have been built nearby. A railroad has been built to the northwest. Petroleum fields were discovered in northern Sinkiang. That is why Lanchow grew so rapidly. It now has more than a million people.

Enough rain usually falls around Lanchow to grow grain. In irrigated orchards farmers can grow excellent fruits and vegetables. To protect the hillsides from erosion and help control dust, grazing on the hills around Lanchow is no longer permitted. Many trees are being planted.

Sinkiang. The province farthest northwest in China, Sinkiang, is almost surrounded by mountains. The high Tien Shan (tee *ehn shahn*) mountain chain, with peaks reaching more than 6100 meters (20,000 feet), extends eastward through Sinkiang. This range divides the province into two basins. The southern basin extends eastward into Kansu Province. Sinkiang Province is about the same size as the states of Idaho, Montana, North Dakota, South Dakota, Wyoming, Utah, and Nevada combined. It has a few more people than these states have. Much of the land is very sparsely settled. That is true of the states mentioned, too. There are a number of quite large cities in Sinkiang. The cities are located near the edge of the basins

Modern industrial development began in China with the manufacture of cotton and silk textiles. Most textile manufacturing takes place in China Proper.

where water is available. The southern basin, called the Tarim (*dah reem*) Basin is a very dry desert. One area in this basin is 154 meters (505 feet) *below* sea level.

Two famous old caravan routes, now roads, lead through the Tarim Basin. In the west the two routes meet near a pass leading into the Soviet Union. The southwestern border of Sinkiang and the western border of Tibet are the places where the border with India is disputed. Most of the land near the border is about 6100 meters (20,000 feet) high.

Tibet. Most of the land of Tibet is more than 3000 meters (9,800 feet) above sea level. Although sometimes called a plateau, Tibet can be better described as a land of high basins separated by much higher mountains. As in the rest of arid

China has been able to mechanize portions of its agriculture since the revolution. However, much of the labor is still performed by hand.

China, little rain falls in Tibet. Summer monsoons bring some moisture from the Indian Ocean. Most of the year, however, Tibet is cold and dry. High altitudes, very cold temperatures, mountains more than 7600 meters (25,000 feet) high, and deep valleys make traveling across Tibet difficult.

For many years, little was known about this isolated area. Until the Chinese built a road into Tibet from Sinkiang, the only way to get into or out of Tibet was over mountain trails. **Yaks** (long-haired oxen), mules, and donkeys were used on the trails. Nevertheless, many small items, such as fountain pens and Swiss watches, could be bought in Lhasa (*lah* suh), the capital city. Now trucks take people and goods to and from Lhasa.

Lhasa was, until 1959, the home of the *Dalai Lama* (dah *ly lah* muh). The **Dalai Lama** is the leader of one branch of the Buddhist religion. According to his followers he is believed to be the Buddha born again. The Dalai Lama was also the ruler of the Tibetan government. When the Chinese Communists invaded Tibet, they agreed that the Dalai Lama should remain as head of the government. In 1959, however, the Communists tried to get him to do what they wanted. He refused and fled across the high mountains into India where he now lives in exile.

In Tibet the yak provides for many of the Tibetans' needs. It is one of the main sources of fuel, meat, leather, fiber, and milk for most Tibetans. A little barley and some fruit are raised in river valleys. The main diet of the people is a meal made from barley, yak butter, yak meat, and tea.

Agriculture in China

When the civil war in China ended in 1949, the agricultural conditions of the country were poor. The years of war and occasional drought had left both the farmer and the farmer's land exhausted.

Farming methods were still very old-fashioned. The farmers had always worked their land by hand. Most of the plots were small, as in India. Few power tools were available. The muscle power of animals was all the help most farmers had.

One of the major problems facing the Chinese Communist government in 1949 was feeding the millions of people. Only about one-eighth (⅛) of the land of China can be used at present to raise crops. That is slightly less than the amount used in the United States. At times food has had to be imported. Canada, Australia, France, and the United States have sold food to China.

To improve the agricultural system, the government first began to develop a series of cooperatives among the small farmers. Some machinery was brought to the farming communities. Farmers were expected to share the equipment, and help each other with their work. However, this did not rapidly increase farm production.

Then the government took the land away from the landowners and organized large farms called **communes**. From 200 to 4,000 families lived on a commune.

The communes were started to change the people's way of thinking, as well as their ways of working. No longer was the family the main economic unit. On the commune, families did not live together. The adults lived in dormitories. Farm workers were in the fields before sunrise and worked until dark. Children were not brought up by their parents. Youth centers and schools were responsible for both preschool and school-age youngsters. Elderly people, unable to work, were placed in homes.

In the communes the farmers learned how to raise more crops on the land by using fertilizer and newer methods of farming. They built dams on the rivers so that more land could be irrigated. The amount of food produced by these methods was improved.

The government, however, was not happy with the commune system. Often the people complained about the huge size of the farms. Sometimes it was difficult for people to organize their duties properly. During the 1970s the communes were reduced in size. A smaller group of families is now organized into what is called a **production team**. From 60 to 80 families are assigned an area of land to farm. The Chinese hope that these smaller farms will be managed better so that food production will increase.

Droughts still plague farming production. In 1977 one of the biggest droughts in 25 years threatened the entire winter wheat crop of China. Such natural disasters do not help any country in its stuggle to improve its agricultural output.

The Industrialization of China

The Communist government has been trying to make China a modern industrial nation. Under a series of Five-Year Plans, many changes have already occurred throughout the land. Much money has been spent to find new mineral resources and build new factories.

Minerals. Exploration for mineral resources in China has produced limited results. However, there is an abundant supply of coal and iron ore. China has increased its production of steel for use in its heavy industries. It is probably in sixth place among the world's steel producers. Steel is also used in the manufacture of farm, railroad, and military equipment; and machine tools.

New petroleum fields have been discovered. Hopefully, China is now self-sufficient in its petroleum output. Some petroleum is exported to Japan. Gold is found in eastern Manchuria. Uranium, copper, silver, lead, and zinc have also been found in the southwestern regions. Tin is mined in southern China.

Manufacturing. Although many of the goods used by people in their homes are still made by hand in China, the development of factories is rapidly changing this. Many textiles, ready-made clothing, and household items are machine-made and are available in local markets.

As a rising world power, much of China's new industrial growth is in the area of military weapons. Factories in China now build military aircraft. Both merchant and naval vessels have also been constructed. China exploded its own atomic bomb in 1964. Since then the development of nuclear weapons and missiles has also been pushed.

The bicycle is the most popular form of transportation for the people of China.

Trade and Transportation. In the early days after the end of the revolution in China, most outside trade was conducted with the U.S.S.R. and Eastern European countries. After the disagreement with the Soviet Union in the 1960s, China decreased its trade with the U.S.S.R. China then turned to Western European nations and Japan for trade. Japan is now one of the main trading partners of China.

China has been building railways into the northern and western regions of the nation. As new industrial sites are developed, both roads and railways are needed. Many new airfields have also been constructed. The **Grand Canal**, perhaps the oldest and longest canal in existence, is being improved. This canal connects the Hwang Ho and Yangtze rivers. In spite of these many improvements, bicycles, two-wheeled carts, and "foot power" still remain the main means of transport for many people.

Language

Many different dialects of the Chinese language were spoken in China when the Communist Party seized power. People from the north could not understand people from the south. Today Chinese children are learning to speak, read, and write in a common language.

Throughout history written Chinese has used very complicated drawings which stand for words. Most of these drawings are called **ideographs**. Each symbol stands for an idea rather than a sound. The Chinese did not have a phonetic alphabet. Now the government has developed a phonetic alphabet for the language. It is being taught to all the children in school.

Education in China

The government of China has placed great emphasis upon adult education. The Communist government wants to change the ideas and attitudes of all the people. In their educational programs there is no difference between education, propaganda, and indoctrination. Not only schools, but radio, newspapers, books, theater, and movies all present Communist ideas.

Young children are required to attend primary school. It is reported that nine out of ten primary age children now attend government schools. However, only eight million out of 38 million youths between 13 and 16 years of age attend school. Many people of this age group are full-time workers. After work they may join others their own age and adults in the adult education programs that are available.

Work-Study Schools are one type of adult education programs in China. Students attend classes part of the time, and work in fields or factories part of the day.

Another type of school is the Spare-Time School. People work full time, and attend classes in their free time. For farmers, such schools may be opened between growing seasons when they are not needed in the fields.

Worker-Peasant Schools offer a program to teach adults how to read and write.

SOUTHERN AND EASTERN ASIA

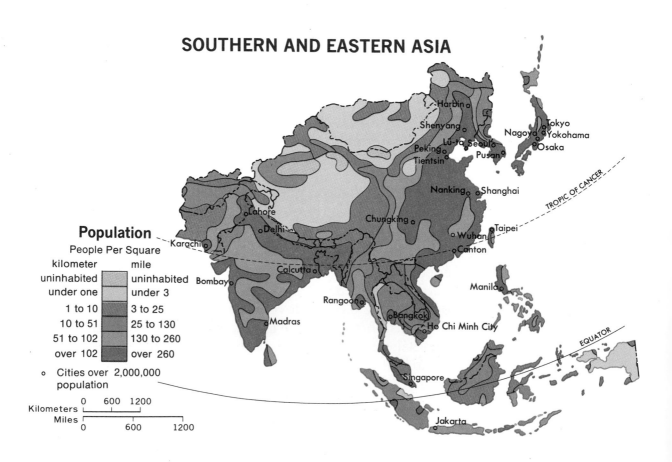

Population

People Per Square

kilometer	mile
uninhabited	uninhabited
under one	under 3
1 to 10	3 to 25
10 to 51	25 to 130
51 to 102	130 to 260
over 102	over 260

o Cities over 2,000,000 population

Kilometers 0 600 1200

Miles 0 600 1200

All of the education programs are designed to teach people usable work skills. The programs also teach them about the government of the People's Republic of China.

There are about 400 universities in China. As in the Soviet Union, much of the university training is in science. The government hopes that university graduates will help the country to improve its industries and agricultural practices rapidly.

The Great Cultural Revolution

The political struggles of China have at times upset China's plans for progress. From 1966 to 1968 the "Great Cultural Revolution" occurred. Chairman Mao Tse-tung wanted to make sure that the people of China, particularly the young people, would not forget his ideas. He wanted to inspire the young people with the communistic ideas he had developed.

During the Cultural Revolution secondary schools and universities were closed. This was done so that the students could join what was called the **Red Guard**. Members of the Red Guard were encouraged to help Mao Tse-tung maintain his power over the party.

Many workers and Communist Party leaders resisted the Red Guard activities and the Cultural Revolution. At times rioting and public disturbances became battles. Many industrial and agricultural projects were halted by the violence.

Towards the end of 1967 the leadership of the country urged the students to return to their school or work. By the end of 1968 the Cultural Revolution was over.

Chairman Mao died in 1976. His death opened the way for more power stuggles within China. A month after Mao's death Hua Kuo-feng (*hwah* gwoh fuhng) became the new chairman of the Communist Party as well as the premier of the country. He officially ended the Cultural Revolution.

Major Cities

Since China has more people than any other country, it is not surprising that it has some very large cities. On the map on page 464 there are shown 31 cities that have more than a million people. Shanghai is probably the largest city on Earth. It has many cotton textile mills and other kinds of factories. Before the Communists seized power, Shanghai was an important international trade center. Many people from other lands who had business interests in China lived there. Since the Communist government now controls all business, most of the foreigners have left China.

Other large cities near the Yangtze River include Nanking (*nan king*), Wuhan, and Chungking (*choong king*). Nanking has been the capital of China several times in the past. It is a major manufacturing center. Chungking was the capital of Nationalist China during the war with Japan. Wuhan is an important industrial center, with many steel mills.

Peking, to the north of the Yellow Plain, is the capital and second largest city of China. Tientsin (tee *ehn tsin*), nearby on the coast of the Yellow Sea, serves as the port city for Peking. Tientsin is about as large as Boston and is a major textile-making center.

Mukden (*mook dehn*) is one of the largest cities in Manchuria. It is a major rail center. Anshan (*ahn shahn*), nearby, is a major center for steel production. Metropolitan Luta (*lyoo dah*) is the main port city for Manchuria and has about three million people.

The Future of China

The Communist government of China promised the people that their children would lead better lives than their grandparents. In many ways that promise has been kept. No one in China goes hungry any more. There is no doubt that more Chinese are being educated than ever before in history. All children as well as adults are taught to work. There is little doubt that the country is becoming an important industrial nation. China has again become a major power in Asia and the world.

In 1971 member countries of the United Nations voted to admit the People's Republic of China. They also voted to expel the Republic of China which had held the UN seat for many years. The United States was not in favor of expelling the Republic of China.

Since the mid-1970s the United States and China have discussed ways of improving their diplomatic relationships. A number of United States leaders have visited Peking. On January 1, 1979, the governments of the United States and the People's Republic of China renewed official diplomatic relations. With this action—the recognition of China—perhaps the future will see an era of friendlier relations between these two nations.

REMEMBER, THINK, DO

1. Locate and label on an outline map of China these major cities:
   a. Shanghai   c. Nanking   e. Mukden
   b. Peking   d. Tientsin   f. Wuhan
2. See a film showing life in the People's Republic of China. See if your library has some recent information on China.
3. Why do you think the government of the People's Republic of China is manufacturing military equipment?
4. Make a mineral/industry map of mainland China. Show where major ores are found and industries are located.
5. What is the difference between a Work-Study School and a Spare-Time School in China?

LEARN MORE ABOUT EASTERN ASIA

Eastern Asia contains many very important nations. The Nationalist Chinese have created the Republic of China on the island of Taiwan. The countries of North and South Korea occupy a peninsula between the Yellow Sea and the Sea of Japan. Mongolia occupies a vast area of land between the People's Republic of China and the U.S.S.R. Two colonies still exist along mainland China—Macau and Hong Kong. You may wish to learn more about these lands. Use the Study Guide on page 31 to help you plan additional work.

The Republic of China (Taiwan)

This government was established on Taiwan by the Nationalist Chinese after they were forced off mainland China. The island is located only 145 kilometers (90 miles) from the southeastern coast of the People's Republic of China. Both the Communist and Nationalist Chinese claim each other's territories. Despite such difficulties, the Republic of China has grown into one of the most developed nations of Asia. One should learn more about this nation to understand the problems of more than one China in the world.

North and South Korea

The countries of North and South Korea are on the Korean Peninsula. This land was the scene of a major war during the early 1950s. Military

Hong Kong has a humid continental climate. The summer monsoon provides the colony with most of its 150 to 200 centimeters (60 to 80 inches) of rain each year.

forces of the United Nations helped defend South Korea from an invasion by North Korean forces. American military forces and supplies made up a major part of this UN aid. Military forces of the UN still guard a cease-fire line along the 38th parallel to maintain peace between the two Koreas.

North Korea is called the Democratic People's Republic of Korea. Its government is controlled by the Communist Korean Labor Party. It has close ties with the People's Republic of China and the U.S.S.R. North Korea has developed heavy industry because it has a good supply of iron ore, coal, and hydroelectric power.

South Korea is called the Republic of Korea. It has many factories that make consumer goods. Much clothing worn in the United States, for instance, is made in South Korea. The standard of living for the people in South Korea is one of the best in Eastern Asia.

Mongolia

North of central China and south of the U.S.S.R. is the landlocked Mongolian People's Republic. About a million people live in this high, cold, dry land. In land area this country is a little larger than the state of Alaska. About one-fifth (⅕) of the people live in the capital city of Ulan Bator (yoo lahn *bah* tohr). It is the largest city in Mongolia.

Since 1924 Mongolia has been largely under the influence of the U.S.S.R. However, the Chinese have also been gaining influence there. A railroad has been built between Ulan Bator and the Trans-Siberian Railway. A railroad has also been built connecting Mongolia to China.

Hong Kong

The British Crown Colony of Hong Kong is located on an island and a small area of the mainland in South China. It is near the mouth of the Si River and has long been a trading port. British goods move to and from China through Hong Kong. It is the only safe anchorage for large ships between Shanghai and Hanoi (hah *noy*), Vietnam. Because Hong Kong is a free port many goods can be purchased there at bargain prices.

For many years Chinese people have been fleeing from China to Hong Kong. They have done so to escape from living under the Communist government. The population of Hong Kong has grown very rapidly. It has about 4.5 million people in its small land area. Many of the refugees arrive without money or food. The people of Hong Kong have tried to find them jobs and places to live. A number of nations and groups throughout the world have given money to help. Many factories in Hong Kong produce toys, electric appliances, clothing, and other goods. Do you have anything made in Hong Kong?

Macau

About 65 kilometers (40 miles) west of Hong Kong is the small Portuguese colony of Macau. It was once known as a center for illegal drugs. Now it is mainly a tourist attraction for people who visit Hong Kong.

CHAPTER 25 REVIEW

WHAT HAVE YOU LEARNED?

1. What is the official name for mainland China?

2. What are the four main islands of Japan?

3. Why did the shoguns of Japan stop most trade with Europeans before Commodore Perry came to Japan?

4. Why is Japan's industrial development since World War II seen as a remarkable achievement?

5. How is the climate of Manchuria different from that of China Proper?

6. What are some of the major mineral resources of China?

BE A GEOGRAPHER

1. Is Japan located within low latitudes, middle latitudes, or high latitudes?

2. Which of the following cities is farther north than Peking?
   a. Rome, Italy
   b. Boston, Massachusetts
   c. San Francisco, California
   d. Tokyo, Japan

3. Look at a globe and decide how many time zones China probably has. Then check your guess with the map on page 27.

4. About how far is it from Canton to the equator? About how far is it from Canton to the South Pole? How did you figure out your answers?

QUESTIONS TO THINK ABOUT

1. Why do you think the government of the People's Republic of China is trying to develop friendly relations with other countries?

2. How do you think life on a Chinese farm commune might differ from life on a wheat farm in the United States?

3. How do you think the Japanese people might have felt about Commodore Perry's visit and treaty demands?

4. For how long should the United Nations maintain peace-keeping forces between North Korea and South Korea? Give reasons for your answer.

5. Do you think atomic weapons should be used for warfare in the future? Why or why not?

Chapter Review

Chapter 26

─────SOUTHEAST ASIA─────

The last major part of Asia is the region known as Southeast Asia. Countries in Southeast Asia are both on the mainland and on islands nearby. On the Indochina Peninsula are the nations of Burma, Thailand, Laos (*lah* ohs), Vietnam, and Kampuchea (kam poo *chee* uh). On the narrow Malay (muh *lay*) Peninsula that extends farther southward are parts of Burma, Thailand, and Malaysia (muh *lay* shuh). At the southern tip of this peninsula is the city-nation of Singapore. Find all these nations on the map on page 493. Note on the map, too, that most of the northern coast of Kalimantan (*kal* uh *man tan*) also is part of Malaysia. The island nation of Indonesia (*in* doh *nee* zhuh) separates the Indian Ocean and the Pacific Ocean. North of Indonesia in the Pacific Ocean is the Republic of the Philippines.

All of these countries, except Thailand, were controlled by European nations or the United States during the first half of the 20th century. All achieved independence in the 1940s, 1950s, and 1960s. These nations, however, are quite different in their development, culture, and government. Many Americans knew little about this part of the world before World War II. Millions of American military personnel fought in this area in World War II. Many more thousands were sent to Vietnam and Thailand during the 1960s and 1970s. As a result, Southeast Asia is now well-known by millions of Americans.

More people live in the nations of Southeast Asia than live in the United States. Many of the people are poor. A majority of them are farmers who raise enough rice for their own needs. Some of them work on plantations raising coconuts, sugar cane, rubber, coffee, and tea. Many of them live and work in several large cities, too.

Most of Southeast Asia is mountainous. Therefore, parts of the area, like Japan, are very crowded with people. Look at the population map on page 484. A majority of the people live in river valleys, plains, and deltas.

Throughout Southeast Asia at low altitudes the climate is hot and humid most of the year. At higher altitudes, of course, the temperatures are cooler. In the northern part of Southeast Asia the climate is milder in the winter months. Freezing weather and snow are unknown in Southeast Asia, however, except on the highest mountain peaks. Most of Southeast Asia has a great deal of rain. Both summer and winter monsoons bring moisture to the land. In some areas because of the location of mountains, a great deal more rain falls at one season than at another. Some places have droughts which last for several months. Then such places have a rainy season that lasts for several months at a time.

Some Problems in Southeast Asia

Life in Southeast Asia is much as it has always been for many people. However, in some areas, the way of life has changed rapidly as a result of agricultural and industrial developments since independence. Vast differences exist in the standard of living from one nation to another in Southeast Asia. Differences also exist within nations, of course.

People from India and China through the centuries have visited and settled in Southeast Asia. The immigrants brought with them their religions, languages, ways of farming, building homes, and conducting business. In some of these nations this mixture of ethnic groups poses a problem. No ethnic group is a majority in language, culture, or religion. Sometimes distrust and lack of cooperation exists among these ethnic groups. This fact makes it difficult for governments to introduce new programs.

Since independence in 1965 Singapore has built many new buildings. The country has prospered because of manufacturing and trading.

Many of these new countries had a one-crop economy at independence. This made it hard for government leaders to improve the nation's economy rapidly.

Recent political changes in Southeast Asia have caused more strains. The Communist countries of Indochina are a growing political force. The governments of many neighboring nations wish to avoid joining with the Communists. They also wish to avoid taking sides against the Communists. It is difficult to predict what changes will occur in Southeast Asia in the future. Some of the changes may affect us in the United States.

REMEMBER, THINK, DO

1. How did the government of Thailand differ from neighboring nations during the first half of the 20th century?
2. Why are parts of Southeast Asia so well known to many adult Americans?
3. Arrange, if you can, to see a film about Thailand and its ancient temples and canals. Compare what you learn about life in Thailand with your own way of life.
4. Prepare a bulletin board display about current events in the countries of Laos, Kampuchea (Cambodia), and Vietnam.
5. What nations are located on the Indochina Peninsula? On the Malay Peninsula? What are the island nations of Southeast Asia?

THE COUNTRIES OF INDOCHINA

The broad peninsula that extends southward from western China is called Indochina. The French gained control of this area, except for Burma and Thailand, in the 19th century. However, after World War II the countries of Laos, Cambodia, and Vietnam were created. You may wish to study these nations in greater detail than is presented in this text. Use the Study Guide on page 31 to help you plan your work.

Burma

Burma has about 33 million people in an area about as large as the state of Texas. It is surrounded on three sides by high mountain ranges and on the fourth side by the sea. Notice that three main rivers flow southward through Burma. The Irrawaddy (*ihr* uh *wahd* ee) River flows through broad valleys in the central part of the country. The Sittang (*sih* tahng) and Salween (*sal* ween) rivers are farther east.

In the past, Burma has been a nation with very little interest in the outside world. In fact the name of the capital, Rangoon (ran *goon*), means "the center of the universe." Burma has been described as an inward-looking nation. It prefers to be left alone.

For more than 100 years the British controlled Burma. The nation became independent in 1948. A socialist form of government was established. In recent years Burma has been caught in the struggles of Southeast Asia. However, the people of Burma are trying to develop their own system of government without outside interference.

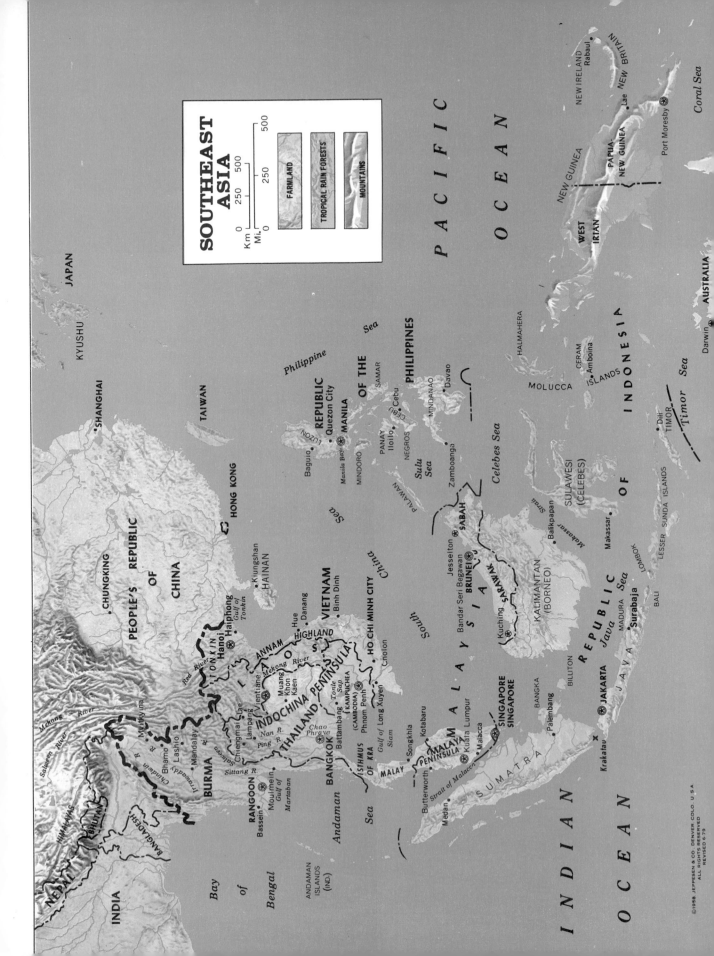

SOUTHEAST ASIA

| Km | 0 | 250 | 500 |
| Mi. | 0 | 250 | 500 |

FARMLAND

TROPICAL RAIN FORESTS

MOUNTAINS

JAPAN

KYUSHU

SHANGHAI

TAIWAN

HONG KONG

PEOPLE'S REPUBLIC OF CHINA

CHUNGKING

HAINAN

Kiungshan

Gulf of Tonkin

Haiphong

Hanoi

TONKIN

Red River

ANNAM

HIGHLAND

Hue

Danang

Binh Dinh

VIETNAM

L A O S

Vientiane

Muang Khon Kaen

Lampang

Chiengmai

Lae

Nan R.

Ping R.

Mekong River

Tonle Sap

Battambang

KAMPUCHEA (CAMBODIA)

Phnom Penh

HO CHI MINH CITY

Cholon

Gulf of Long Xuyen

INDOCHINA PENINSULA

THAILAND

BANGKOK

Chao Phraya

Sittang R.

ISTHMUS OF KRA

Gulf of Siam

Songkhla

Kotabaru

MALAY PENINSULA

MALAYA PENINSULA

Butterworth

Kuala Lumpur

Malacca

Strait of Malacca

SINGAPORE

SINGAPORE

M A L A Y S I A

Bandar Seri Begawan

BRUNEI

SARAWAK

Kuching

SABAH

Jesselton

KALIMANTAN (BORNEO)

Balikpapan

Makassar

Makassar Strait

SULAWESI (CELEBES)

CERAM

Amboina

MOLUCCA ISLANDS

HALMAHERA

PHILIPPINES

REPUBLIC OF THE

Quezon City

MANILA

Manila Bay

LUZON

Baguio

MINDORO

PANAY

Iloilo

CEBU

Cebu

NEGROS

SAMAR

Zamboanga

MINDANAO

Davao

Sulu Sea

PALAWAN

Celebes Sea

Philippine Sea

South China Sea

Sea

Philippine Sea

PACIFIC OCEAN

NEW IRELAND

Rabaul

NEW BRITAIN

NEW GUINEA

Lae

PAPUA NEW GUINEA

Port Moresby

WEST IRIAN

Coral Sea

AUSTRALIA

Darwin

Timor Sea

TIMOR

Dili

LESSER SUNDA ISLANDS

LOMBOK

BALI

Surabaja

MADURA

J A V A

Java Sea

REPUBLIC OF INDONESIA

JAKARTA

BANGKA

Palembang

BILLITON

S U M A T R A

Medan

Krakatau

INDIAN OCEAN

Bay of Bengal

ANDAMAN ISLANDS (IND)

Andaman Sea

Bassein

RANGOON

Moulmein

Gulf of Martaban

BURMA

Mandalay

Bhamo

Lashio

Irrawaddy

Chindwin

Sittang R.

Salween River

Mekong River

HIMALAYAS

NEPAL

BHUTAN

BANGLADESH

INDIA

©1958 JEPPESEN & CO. DENVER COLO. U.S.A.
ALL RIGHTS RESERVED
REVISED 6-79

Just east of Thailand is the landlocked, mountainous nation of Laos. This nation is about as big as the states of Pennsylvania and New York. It contains about 3½ million people. Most of this country is drained by the Mekong (*may* kawng) River. It flows along the western border of the country for more than 800 kilometers (500 miles). Most of the cities of Laos are located along this river. Vientiane (*vyehn tyahn*) is the capital of Laos. The Communists gained control of the government of Laos in 1975. Since then many conditions have changed.

Located south of Laos is Kampuchea (Cambodia). Kampuchea is about as large as the state of Oklahoma. It has about 8½ million people. The central part of the country is a low basin through which the Mekong River flows. During the rainy season, much of this basin is flooded. This area is an excellent one for rice cultivation.

The capital of Kampuchea is Phnom Penh (p'*nawm pehn*). It is located on the Mekong River in the southern part of the nation. As in Laos, this nation has recently been taken over by Communists.

SOUTHERN AND EASTERN ASIA

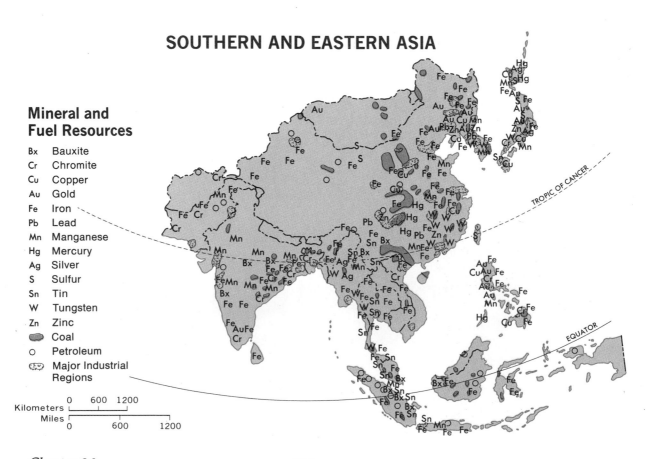

Mineral and Fuel Resources

Bx	Bauxite
Cr	Chromite
Cu	Copper
Au	Gold
Fe	Iron
Pb	Lead
Mn	Manganese
Hg	Mercury
Ag	Silver
S	Sulfur
Sn	Tin
W	Tungsten
Zn	Zinc
	Coal
o	Petroleum
	Major Industrial Regions

Kilometers 0   600   1200

Miles 0   600   1200

The canals of Bangkok, Thailand, are used like streets. Merchants operate food markets along the banks. Customers arrive and shop from their boats.

Malaysia

The Federation of Malaysia is one of the most prosperous nations in Southeast Asia. It is the leading country of the world in the production of tin and natural rubber. The former British colonies of Sarawak (suh *rah* wahk) and Sabah (*sahb* uh) joined with Malaya to form this new nation in 1963. The capital of Malaysia is the city of Kuala Lumpur (*kwahl* uh *loom* poor) in Malaya. Find these places on the map on page 493. You will notice that Malaya is on the southern end of the Malay Peninsula. It is about 645 kilometers (400 miles) from its eastern states on the island of Kalimantan.

Singapore

Singapore is the smallest and most prosperous of all nations in Southeast Asia. Its port is one of the busiest in the world. Look at the map or globe to locate Singapore. Can you guess why this port is so important? Ships sailing from the east or west almost have to pass this port on their round-the-world trips.

Singapore is built partly on the southern tip of the Malay Peninsula and partly on a small island just off the mainland. The island is only about 40 kilometers (25 miles) long by 25 kilometers (15 miles) wide. This island is about the size of Chicago. The island is connected to the mainland by a causeway. Singapore has a population of about 2½ million people. Three-fourths (¾) of these are of Chinese ancestry. Most citizens of Singapore work either at the port or in some business connected with the shipping of goods. Since Singapore is a free port, many goods are transferred from one ship to another in this busy port area.

Many years of fighting in Vietnam have resulted in the death of thousands.
Many people have had to leave their homes as refugees.

The Vietnam Conflict—a Brief Review

The French colonized Vietnam during the late 19th century. They established many plantations on the land. The French remained in Vietnam even after the Second World War. However, Communist forces were actively at work in Indochina during the early 1950s. The French tried to stop the Communist advances in Vietnam but were defeated by them in 1954.

At that time, Vietnam was divided into two parts. North Vietnam was controlled by a Communist government. The South Vietnam government tried to establish a non-Communist government. It was not long before fighting occurred between North and South Vietnam. The United States became deeply involved on the side of South Vietnam in the war. Many thousands of Americans were involved in the fighting in Vietnam. The entire world was concerned about the war in Southeast Asia.

Finally, in 1973, the United States withdrew from the war. The government of South Vietnam quickly fell to the North Vietnamese. Vietnam was reunited as one nation. The capital of Hanoi in the North became the capital for the entire nation. The city of Saigon (sy *gahn*), which had been the capital of the South, was renamed "Ho Chi Minh City." Ho Chi Minh (hoh chee mihn) was for many years the leader of North Vietnam. He died of old age during the war.

Thailand

The word *Thailand* means "land of the free." Although in time of wars Thailand has suffered invasions, it has always regained its independence.

Thailand is about the size of the states of Colorado and Wyoming combined. About 46 million people live in this nation. Thailand is not yet over-crowded. However, its population is growing rapidly.

The most densely populated part of Thailand is the basin of the Chao Phraya (chow *pry* uh) River. This river flows through a wide, low, delta area, and is frequently criss-crossed by canals and small streams.

Bangkok (*bang* kahk), the capital of Thailand, is a city on the river with many canals. It is often called "the Venice of the East."

The government of Thailand has tried to protect itself from a Communist take-over. Communists have caused trouble for people in parts of Thailand during the past few years. The government in Thailand has been taken over by the military. They hope to improve conditions within the nation.

Vietnam

The country of Vietnam has often been described as "two rice baskets at the end of a long bamboo pole." The bamboo pole is the Annam (a *nam*) Highlands near the coast. The two rice baskets are river deltas. In the north the Red River forms one delta. In the south the Mekong River forms the second delta near Ho Chi Minh City (Saigon). Find these areas on the map on page 493. Most of the 47 million people of Vietnam live along these rivers or in the deltas. Much rice is grown in these areas.

REMEMBER, THINK, DO

1. Where on the land do most people tend to live in Southeastern Asia?
2. Identify the following places on an outline map of Southeast Asia:
   a. Indochina Peninsula
   b. Irrawaddy River
   c. Malay Peninsula
   d. Mekong River
   e. Phnom Penh, Kampuchea
   f. Vientiane, Laos
   g. Hanoi, Vietnam
   h. Chao Phraya River
   i. Ho Chi Minh City, Vietnam
3. Why do you think some groups within a country in Southeast Asia might distrust other groups in the country?
4. Should the United States have diplomatic relations with the government of Vietnam? Why or why not?
5. What is the difference between a free port like Singapore and a port that is not free, like New York City?

Southeast Asia

SOUTHERN AND EASTERN ASIA

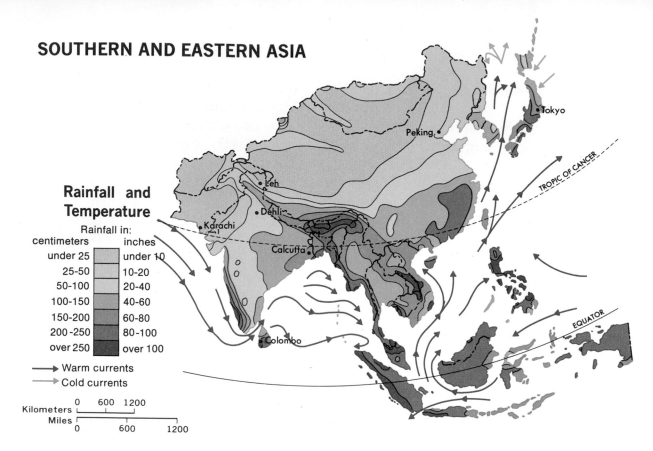

Rainfall and Temperature

Rainfall in:

centimeters	inches
under 25	under 10
25-50	10-20
50-100	20-40
100-150	40-60
150-200	60-80
200-250	80-100
over 250	over 100

→ Warm currents
→ Cold currents

Kilometers 0 600 1200
Miles 0 600 1200

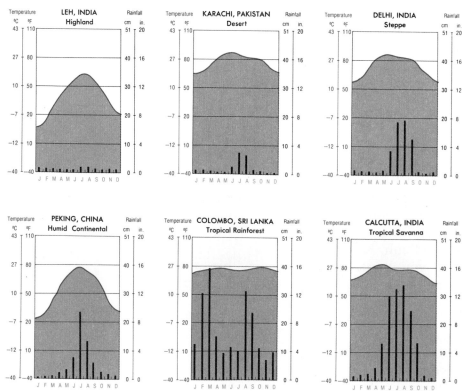

LEH, INDIA — Highland

KARACHI, PAKISTAN — Desert

DELHI, INDIA — Steppe

TOKYO, JAPAN — Mid-latitude Marine

PEKING, CHINA — Humid Continental

COLOMBO, SRI LANKA — Tropical Rainforest

CALCUTTA, INDIA — Tropical Savanna

THE ISLAND NATIONS OF SOUTHEAST ASIA

Indonesia

More than 13,000 islands form the Republic of Indonesia. These islands extend about 5500 kilometers (3,400 miles) along the equator south and east of the Malay Peninsula. Many of Indonesia's islands are very small. Others are among the largest on Earth. The largest and best known islands are Sumatra (soo *mah* truh), Java (*jahv* uh), New Guinea (*gihn* ee), and Kalimantan. On some maps Kalimantan is shown as Borneo (*bohr* nee oh).

Only the southern portion of the island of Kalimantan is part of Indonesia. In the north, two states on this island belong to Malaysia, while a third area is the sultanate of Brunei (*broo* ny).

New Guinea is the second largest island in the world. The western half of this island is part of Indonesia. It is called West Irian (*ihr* ee *ahn*). The eastern half became the independent nation of Papua (*pap* yuh wuh)— New Guinea in 1975. Find all these islands on the map on page 493. Can you also locate the island of Timor (*tee* mawr)? Until 1976 half of Timor was controlled by Portugal. Now all the island is part of Indonesia.

The Land and the People

Indonesia is Southeast Asia's largest country, both in area and in population. The combined area of the 13,000 islands is about as large as Alaska and California together. Only four nations in the world have more people. Can you name them? Indonesia has about 143 million people. About three-fifths (³⁄₅) of the people live on the island of Java. It is a very crowded area, with more than 650 people per km² (1,700 per sq. mi.). A majority of the people of Indonesia have ancestors who were Malayan. Except for those who live on the island of Bali (*bahl* ee), almost all Indonesians are Muslims. The Balinese are Hindus.

No one knows for certain, but most scientists believe that at one time Australia and Asia were connected by land. Then part of the land sank, or the ocean level rose, and water covered much of the land. Thousands of islands were left scattered here and there in the ocean. The scientists believe that these islands were once part of a single land mass because of the kinds of plants and animals found on them. Some scientists say that Asia stops and Australia begins on the small island of Lombok (*lahm* bahk). It is located in the long chain of islands, known as the Lesser Sunda (*syoon* duh) Islands, near Bali. Plants and animals found to the west of this island are similar to those of the Malay Peninsula. The islands east of Lombok have plants and animals like those found in Australia.

Many of the islands have become larger through the centuries because of volcanoes. In 1883 about 36,000 people were killed as a result of one of the most famous eruptions in history. The whole top of the small island of Krakatau (krak uh *toh* uh) was blown off. This island is located between Sumatra and Java. Large waves washed ashore on nearby islands drowning many people. Such great waves were caused by earthquakes that occurred below the surface of the ocean.

All of the islands have a central range of mountains. Some hilly areas are scattered here and there in the mountainous areas. The plains are wider and more fertile on the western islands of Sumatra, Java, Kalimantan, and Bali. As in other lands with volcanoes, the volcanic ash left behind after eruptions forms fertile soil. Kalimantan, however, has no volcanoes.

The Climate

Throughout Indonesia the climate in the lowlands is always hot and humid. Nights are not much cooler than days. At Jakarta (juh *kahrt* uh), the capital of Indonesia, the temperature has not fallen below 19° C (66° F) nor risen higher than 35° C (96° F) in a hundred years!

The western islands have slightly more rainfall than the eastern islands. The rainy season lasts mainly from November through February. Some rain usually falls each day from early September through late April. The amount of rainfall at any one place depends largely upon its location with respect to nearby mountains. The humidity is high almost always during the rainy season. The dry season in most places lasts from May to September. Even

then, however, it is not uncommon to have brief rain showers from time to time.

Agricultural Products

Rubber and spice plantations in Indonesia were started by the Dutch during colonial times. Most of the rubber plantations are on the island of Sumatra, southern Kalimantan, and eastern Java. Like Malaysia, Indonesia is a major exporter of natural rubber.

Rice is the main food crop grown on the islands. Other crops include corn, cassava, vegetables, tobacco, tapioca, sugar cane, sweet potatoes, coconuts, soybeans, peanuts, cacao, tea, and coffee. Among the spices grown, pepper, nutmeg, and cinnamon are most important. Almost two-thirds (⅔) of the people of Indonesia work in some phase of agriculture.

Java is almost covered with rice paddies. About two-thirds (⅔) of this island is cultivated. Since much of it is hilly, terraces have been built by the farmers to increase rice production. Until the 1960s farmers were able to raise enough food for all the people of Indonesia. Two crops a year are possible because of the even climate, rainfall, and fertile soil. However, because of rapid population growth, some food has to be imported today.

Forest Products

About one-fourth (¼) of the land of Indonesia is forested. To protect and conserve the wise use of these resources, the government now owns all the forest lands. More than half of the forests are kept as national reserves. Much of the forested land is open to commercial use.

Along many of the coastal waters are coconut palm and mangrove trees. Farther inland, many of the islands have tropical rain forests. But these forests are often difficult to work profitably because marketable trees are scattered. However, from the more accessible trees come wild rubber, resin, gum, camphor, rattan, and palm oil. Many of these products are exported to other nations. Bamboo, a popular tree in much of Asia, is used for furniture making and home construction. Hardwoods such as teak and ebony are very valuable export items.

Mineral Products

Indonesia has some valuable mineral deposits. At present, tin, petroleum, and bauxite are the most important mineral resources. These have been found mainly on the islands of Sumatra, and the two small islands of Bangka (*bang* kuh) and Belitung (buh *leet* ung), east of Sumatra. About one-seventh (1/7) of all the tin produced in the world comes from these two small islands.

Until China became an exporter of oil, Indonesia was the only large producer of petroleum between the Middle East and California. Much of the petroleum is exported to other nearby countries. The main oil fields are on the island of Sumatra. Recent discoveries of petroleum on the east coast of Kalimantan make this area the second largest petroleum region of Indonesia. Some oil production occurs on the islands of Java and Ceram (*say* rahm), but not much by world standards. Indonesia's oil production is much smaller than that of the Middle East, the U.S.S.R., and the United States.

Other minerals to be found in this nation are gold, silver, diamonds, low-grade coal, iron ore, and nickel. Nickel mines are located on the island of Sulawesi (*soo* lah *weh* see). On some maps the island's name is sometimes shown as Celebes (*sehl* uh *beez*). These mines are expected to yield more than 45,000 tons of ore yearly by 1979. Exploration for new minerals continues on the other islands of Indonesia.

Latex, the sap from which rubber is manufactured, drips into buckets from cuts made in the bark. The latex is often collected by hand.

Geometric designs often decorate the houses on Sakawesi, an island in Indonesia.

Industry in Indonesia

Getting manufacturing started in Indonesia has been a slow process. For several years after independence, most of the business trade was still controlled by the Dutch or by the Chinese. When the Dutch finally left Indonesia, they took with them much of their money and their knowledge of manufacturing.

Many young Indonesians were sent to other nations to learn how these countries were making use of their own resources. The young people then returned to Indonesia to help develop manufacturing.

Also, under colonial rule, most of the raw materials were exported to The Netherlands. The economy of the islands depended mainly on exports of plantation crops and ores. When sudden drops in world prices of rubber and tin occurred the effect upon the economy of Indonesia was terrible. It became necessary for the people of Indonesia to develop their own local industries.

At present, most of the industrial development has occurred on the island of Java. In fact, more than three-fourths (¾) of all industrial workers are on this one island. Jakarta, the capital, is the main trading center of Indonesia. Near Jakarta one finds textile factories and rubber processing plants. Chemical industries there also manufacture drug supplies and soap products. Printing plants, tobacco factories, and food processing plants also operate there.

A problem facing Indonesia in its industrial expansion, even with petroleum, is good energy sources. Hydroelectric power sources are limited. Java has very limited hydroelectrical potential. The island of Sumatra, however, has considerable energy potential, but little industrial development.

Education and Government

Education has also been a big problem for the government. At independence, only about one-tenth (1/10) of the people could read and write. Schooling is now required of young children. By 1977 about three-fifths (3/5) of the people could read and write.

Good political leadership is a great need in every country. This is especially important to a developing nation. Unfortunately, the first president of Indonesia was not effective. He had been a major hero to the nation in its push for independence. However, vast amounts of money were wasted on poorly planned projects during his administration. A great deal of money was spent on the military. Much of that was caused by strained relationships with neighboring Malaysia. Living costs in Indonesia skyrocketed. When Communists attempted to take over the government of Indonesia in 1965 a bloody civil war erupted. More than 300,000 people were killed in the battles. After the government defeated the Communist rebels, Sukarno (soo *kahr* noh), Indonesia's first president, was replaced. General Suharto (soo *hahr* toh) became the new leader.

Under Suharto the country has made rapid improvements. The country rejoined the United Nations, and signed a peace agreement with neighboring Malaysia. Loans from the World Bank have helped finance many new industrial projects. The future looks brighter for Indonesia now.

REMEMBER, THINK, DO

1. Use an outline map of Southeast Asia to complete these map questions:
   a. Locate the main islands of Indonesia: Sumatra, Java, Kalimantan, Sulawesi, Timor, and West Irian.
   b. Mark the equator on your outline map.
   c. Which islands of Indonesia are entirely north of the equator?
   d. Which islands are entirely south of the equator?
   e. Which islands are crossed by the equator?
2. Prepare a mini-report on the independent sultanate of Brunei.
3. As a member of a group prepare a news report about the explosion of the volcano on Krakatau in 1883. It might be a "You Were There!" type of report done with drawings and narration.
4. Compare the yearly high and low temperatures of your community with those of Jakarta, Indonesia.
5. Prepare a mineral and agricultural map for the main islands of Indonesia.

The Philippines—"The Pearl of the Orient"

Northeast of Indonesia and south of the island of Taiwan are about 7,000 islands which make up the Republic of the Philippines. Most of the islands are very small. Only about 500 of them have more than 2.5 square kilometers (about 1 square mile) of land. About 4,000 of the islands have no official name. Together the islands have about as much land area as the state of Arizona. About two-thirds (⅔) of all the land area is on the two large islands of Luzon (loo *zahn*) and Mindanao (*mihn* duh *nah* oh). More than 46 million people live in the Philippines. About half of them live on the island of Luzon. Manila (muh *nihl* uh), the largest city and capital, is located on a beautiful bay in central Luzon. Find these islands and Manila on the map on page 493.

A History of the Philippines

No one knows how long people have lived on the islands of the Philippines. In 1521 the first European explorer reached the Philippines after sailing across the Pacific. He was a Portuguese explorer named Magellan. This man was exploring for the King of Spain. When he reached these islands, he claimed the land for King Philip of Spain. He named the islands in honor of this Spanish king.

Magellan and a few of his sailors were killed in a battle with some of the people of these islands. Only one ship of the five that had started the journey finally reached Spain by sailing home around Africa. It was the first time a ship had ever sailed "around the world."

Not long after the news of these islands reached Spain, ships were sent there to settle the islands. Priests and Spanish settlers arrived. For about 300 years the Spanish ruled over the islands. A rebellion against Spanish rule took place in 1896.

Two years later, the United States went to war with Spain to free Cuba from Spanish rule. At the end of that war, Spain also turned over control of the Philippines to the United States for $20 million. The Filipinos who had been fighting the Spanish for their independence then began to resist the Americans.

It was not until 1934 that the Congress of the United States permitted the people of the Philippines to elect their first president. Complete independence was scheduled to take place 12 years later.

In December, 1941, Japanese armies invaded the islands of the Philippines. These armies occupied the Philippines until 1944. Many Filipinos went to the mountains, forming guerilla bands, to fight against the invaders. These guerilla forces were an important help to the American troops in finally defeating the Japanese armies in 1945.

In 1946, as promised, the Philippines became an independent nation.

The United States assisted the

About half the working population of the Philippines is engaged in agriculture.

Philippines in their recovery from the damages of World War II. The United States also supported many projects to establish a sound economy. In the years since independence, friendly relations have generally been maintained between the governments of the Philippines and the United States.

The Land and the Climate

Like Indonesia, most of the land of the Philippines is mountainous. There are about 50 volcanoes on the islands. Twenty are active from time to time. Large plains are found on Luzon. Most of the other islands have much narrower plains. These plains tend to have fertile soil. They receive adequate rain during the rainy season. About half of the land that could be farmed is currently under cultivation. During the rainy seasons, the land that is not cultivated or forested is covered with a tough, tall grass called **cogon** (*koh guhn*) grass.

Because the Philippines are in low latitudes, the climate is always warm at lower altitudes. The average daytime temperature is about 27° C (80° F). Northern Luzon has slightly cooler winter months than southern Mindanao. However, the temperature rarely drops below 15° C (60° F) except in the mountains.

There are really only two seasons —the rainy and the not quite so wet. Some rain tends to fall almost every month. The heaviest rains come with the monsoon winds. During June, July, and August, places on the western side of the islands have their rainy season. The moisture is brought by the southwest monsoons. During the winter months the monsoons blow from the northeast bringing more rain to eastern portions of the islands. Typhoons often strike the Philippines during the fall months.

Southeast Asia

Forests

Forests cover about half of the total area of the Philippines. Some of these forests are among the finest found anywhere on the Earth. At one time probably almost all the land was forested. For many years the local people cleared off land by burning trees. Much valuable timber was destroyed. Some of that still happens, but not as much as in the past. Today the government owns all the forests. Now the government controls the amount of lumber which is cut. This valuable resource would last for many years to come if it were managed wisely. But the forests are being cut at a very fast rate because the government needs money. Not much of the land where trees are cut has been reforested. Much of the lumber exported from the Philippines is shipped to the United States.

Agriculture in the Philippines

The main food crop in the Philippines, as in the rest of Southeast Asia, is rice. About half of the cultivated land is planted in rice. Much of the farm work in the paddies is done by hand. However, more and more machines are being used. Increased knowledge of fertilizers has

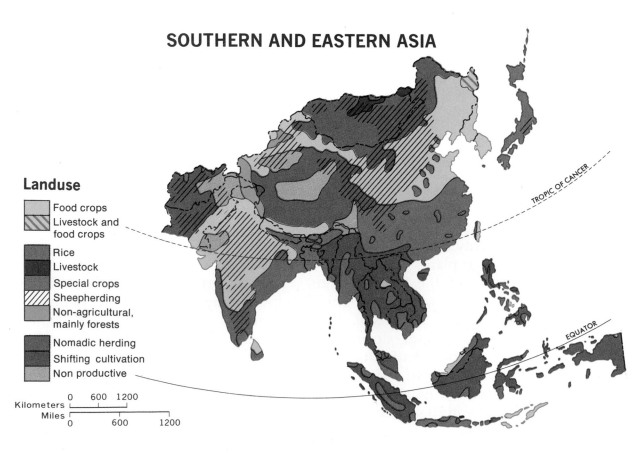

SOUTHERN AND EASTERN ASIA

Landuse

- Food crops
- Livestock and food crops
- Rice
- Livestock
- Special crops
- Sheepherding
- Non-agricultural, mainly forests
- Nomadic herding
- Shifting cultivation
- Non productive

Kilometers 0 600 1200

Miles 0 600 1200

TROPIC OF CANCER

EQUATOR

also been helpful to the farmers. New seeds, developed by agricultural scientists from the University of the Philippines, have also increased yields greatly.

The main cash crops of the Philippines are sugar cane, copra, pineapples, and *abaca*. **Abaca** is a plant which looks very much like a banana plant. The fibers from this plant are extremely strong. For many years the abaca fibers have been woven into strong rope and twine. It is commonly called **Manila hemp**.

The sugar cane plantations now use machines in plowing, planting, and cultivating the cane fields. Sugar cane is still harvested mainly by hand methods. Long knives called **machetes** are used for cutting.

The **coconut palm tree** provides many things to people who live in warm climates. The trunks are used in many ways where strong wood is needed. Some people use the trunk for part of the foundations of new houses. The leaves of the palm tree are often used as thatch on roofs. Many people claim the thatch is very good insulating roofing material. The tough fiber of the husk which surrounds the coconut, called **coir** (*koy* uhr), may be used in making mats. Nevertheless, it is the meat of the coconut that is the most valuable part.

The dried meat of the coconut is called **copra** (*koh* pruh). Large quantities of coconut oil can be pressed from the meat of the coconut. About a gallon of oil can be pressed from the copra of thirty coconuts. Coconut oil is used in the making of soap, medicine, and oleomargarine. Dried coconut is also used in cooking, especially in making candies and cakes. Some people enjoy the sweet taste of the coconut milk obtained from a ripe, fresh coconut.

Pineapples and pineapple juice are canned in the Philippines. Other fruits raised by the farmers include bananas, papayas, citrus fruits, and mangoes. Many of these are exported to other lands.

On Cebu (say *boo*) and other central Philippine islands, corn is the main food staple. Some root crops such as yams and *gabi* (*gah* beh) are also grown and eaten by the local people. **Gabi** is a starchy root that is like the taro plant. Other crops raised in the Philippines include tobacco, rubber, and coffee.

REMEMBER, THINK, DO

1. How did the United States acquire the Philippines as a colony?
2. Why was the trip around the world by one of Magellan's ships considered to be so important?
3. Invite a Filipino to talk to your class about the Philippine Islands. If such a person is not available, see if you can locate an up-to-date film about the Philippine Islands.
4. What is copra? What are some ways it is used?

A Tasaday and his son return to their village after gathering food.

Stone Age People in the 1970s

In 1971 scientists discovered a tribe of about 25 people living in caves on the island of Mindanao. They are called *Tasadays* (*tahs* uh days). They were found living much as people probably lived about 10,000 years ago. Everyone was surprised. So far as anyone can tell, the Tasadays have never had any contact with people outside their small tribe.

The Tasadays had no weapons. They ate mainly food that they could gather in the forests—roots and flowers. They did not kill animals for food. Their language had no word in it for anger. And Tasaday mothers never spank their children. How would you like to live with them?

The government has set aside an area of 20,000 hectares (50,000 acres) as a protected reserve for the Tasadays. Other people will not be permitted in the area. That way the gentle Tasadays may continue to live much as they always have.

Mineral Wealth and Manufacturing

A number of valuable mineral ores have been found in the Philippines. Gold, chrome, copper, and iron ore are of greatest importance at the present time. Many of these deposits are in the mountainous areas, at considerable distances from the ports. As a result the cost of mining and transporting the ores is fairly high. Thus, some of the mining operations have been rather small in size. A large mountain of high-grade iron ore has been discovered on the northern side of the island of Mindanao. Unfortunately, the ore contains nickel which makes it expensive to smelt. Once cheaper smelting processes are developed, it is likely the Philippines may become a major exporter of iron ore.

Many kinds of factories have been built since the Philippines became an independent nation. Oil refineries have been built to process crude oil which is imported by tankers from Indonesia. A number of factories assemble automobiles and trucks from parts imported from the United States. A variety of electrical appliances is also assembled in many plants as well. Rope, furniture, and cement plants have been in operation for many years. Many sugar refineries, copra-pressing plants, and cigar factories make products for home use as well as export. Other factories make matches, soap, drugs, tires, and other goods.

The government of the United States has helped the Philippines to make rapid progress in building many of these factories. Some hydroelectric power plants have been developed to answer some of the requirements for energy. One of the great needs of the growing industrial area is power.

REMEMBER, THINK, DO

1. Why are mining operations in the Philippines rather small?
2. Ask a librarian to help you find more information about the Tasadays.
3. When did the Philippines get their complete independence?
4. On an outline map of the Philippines, label the main islands of Luzon, Mindanao, Cebu, and Samar. Also, label the two cities which serve as capital cities.
5. What feelings do you think the people of the Philippines might have about the United States? Remember that the United States was the colonial power there for almost 50 years.

A SUMMARY OF ASIA

Asia is the largest continent on earth. More than half of the people of the world live in Asia. The world's highest and lowest points are found in Asia. Many differences exist among people of Asia—differences in governments, religion, language, customs, and development. Some of the world's most developed nations are in Asia. So, too, are some of the world's poorest nations.

China and India are the two most populated nations of Asia and the world. These two nations share a small, common border along the Himalaya Mountains near Tibet and the province of Kashmir. These nations also share problems of a rapidly growing population, and limited food supply. Problems of health and education are also major concerns of both governments.

Japan is one of the most industrialized nations of Asia. Its standard of living rivals that of most nations of the Western Hemisphere. Although their land is small in area, the Japanese people have used available space to produce much of their food. Japan, moreover, is a world leader in producing manufactured goods.

Indonesia and the Philippines are also island nations near Asia. Both have developed rapidly since World War II. Both nations are comprised of thousands of islands within the Pacific Ocean.

During the long period they were colonies, the islands were used for large plantations. Rubber and spices were—and still are—major crops grown in Indonesia. The Philippines have supplied large quantities of sugar cane, pineapples, and copra. Both are also capable of growing much rice.

Petroleum and iron ore from these nations have added to their wealth and enabled the growth of industries. Other minerals are also tapped for use. Small refineries and manufacturing plants have added to the progress of these independent nations in the Pacific.

WHAT HAVE YOU LEARNED?

1. In what ways are relationships between China and its neighbors like relationships between the U.S.S.R. and its neighbors? Are there differences?

2. What common problems do landlocked nations in Asia, like Nepal and Mongolia, share with African nations like Lesotho, Swaziland, and Malawi?

3. Identify the major landform characteristics of the continents of Asia and Africa. Use a globe or wall map to note similar and/or different characteristics.

4. Compare the basic food staples and mineral resources of Asian nations with those of African nations. What major similarities and differences exist in food and mineral resources between these two huge continents?

5. On the continents of Asia and Africa, there are rather limited transcontinental transportation systems compared with Europe and North America. What political, geographical, and economic factors may be hindering development of such cross-continental transportation systems?

BE A GEOGRAPHER

1. Between what meridians of longitude are the countries of Southern and Eastern Asia located?

2. Which is farther north:
   a. Manila or Singapore?
   b. New Delhi or Katmandu?
   c. Ho Chi Minh City or Jakarta?
   d. Tokyo or Shanghai?
   e. Calcutta or Bombay?

3. Would the sun ever be seen directly overhead at:
   a. Hong Kong?
   b. Canton?
   c. Kyoto?
   d. Manila?
   e. Kiungshan on Hainan Island?
   f. Shanghai?
   g. Calcutta?
   h. Taipei?

4. If you flew west around the world on 40° N. from northern Honshu:
   a. What continents would you fly over?
   b. What seas would you fly over?
   c. Would you fly over any of the U.S.S.R.?

5. Calculate the following:
   a. Use a flexible ruler or a piece of string and measure the distance from the northwestern tip of Sumatra to the eastern border of West Irian in Indonesia. Then, using the scale of kilometers on the globe, figure out the number of kilometers that this distance on the globe represents.
   b. Figure the same distance using the map on Southeast Asia on page 493.
   c. Figure the same distance using a large wall map. If different answers are obtained, be prepared to tell why.

6. Using the globe, figure the distances by air from Singapore to each of the following cities:
   a. Manila
   b. Rangoon
   c. Ho Chi Minh City
   d. Bangkok
   e. Jakarta
   f. Hong Kong

7. Find all the following on the map on page 493 and be prepared to locate them on the wall map when requested to do so:
   a. Bali
   b. Hanoi
   c. Baguio
   d. Bangkok
   e. Sarawak
   f. Jakarta
   g. Rangoon
   h. Singapore
   i. Phnom Penh
   j. Ho Chi Minh City
   k. Strait of Malacca
   l. Gulf of Martaban
   m. South China Sea
   n. Isthmus of Kra
   o. Gulf of Tonkin
   p. Gulf of Siam

8. What direction is it from:
   a. Hong Kong to Manila?
   b. Bangkok to Hanoi?
   c. Jakarta to Manila?
   d. Ho Chi Minh City to Singapore?

9. Determine the population of the following:
   a. What cities in Southeast Asia have a population of more than 500,000?
   b. What cities have a population of more than 2,000,000?

QUESTIONS TO THINK ABOUT

1. If you were the Prime Minister of India, what would you attempt to do to help India move ahead economically?

2. What are the advantages and the disadvantages of being a little country along the borders of two large countries? (Example: Nepal between India and China.)

3. Why have the governments of China and the U.S.S.R. been less friendly with each other in the 1970s than they were in the 1950s? What would you predict for the future?

4. What advantages and disadvantages does Japan have over China as far as developing a high standard of living is concerned?

5. Many village people in Southeast Asia have little or no feeling of national pride or identity. Why should this be so? Is love of country important? How can love of country be developed?

6. What advantages and what dis- advantages are there to being an island nation (such as Indonesia or the Philippines)?

7. What advantages or disadvantages are there to being a land-locked nation (such as Laos or Nepal)?

8. The population of Asia is already more than half the people on Earth. Unless conditions change, in 35 years there will be as many people on Asia as there are now on the whole planet! Does it matter? What should be done?

9. How might farmers in Java increase their food crop production in order to satisfy food demands of its island population?

10. When Indonesia first became independent, many students were sent to other countries to study. What reasons might a graduate have for returning home or staying in the United States after completing a study program?

Unit Review

Tree ferns surround Russell Falls on the island of Tasmania. Because there is 200-300 centimeters (80-120 inches) of rain a year the tree fern can grow to be 6 meters (20 feet) tall.

Unit

AUSTRALIA AND ISLANDS OF THE PACIFIC

Unit Preview

- What bodies of water surround the continent of Australia?

- Why are most of Australia's major cities near the coast?

- Is New Zealand north or south of the Tropic of Capricorn?

- What difference is there in time between your home and Sydney, Australia?

- What season of the year is it in New Zealand now?

Where would you rather live, on the island of Fiji or in Sydney, Australia?

DO YOU KNOW?

- Who the first inhabitants of Australia were?

- How the Great Barrier Reef was formed?

- How many islands make up the country of New Zealand?

- What unusual wildlife can be found in Australia?

- How many people earn a living on the many small islands of the Pacific?

Unit Introduction

Australia and Islands of the Pacific

South of the Philippines and Indonesia is the smallest continent on Earth, Australia. It is one of the two continents completely south of the equator. It is also one of the two not connected with any other continent. About 1900 kilometers (1,200 miles) southeast of Australia is New Zealand. Australia and New Zealand are the two main countries to be studied in this unit.

These countries are much alike—yet, they are very different. Both countries are members of the Commonwealth of Nations. Both nations were once British colonies. Australia and New Zealand both are younger nations than the United States. English is the official language of both countries. Most people in Australia and New Zealand have a high stan-

dard of living. Their way of life is much like that in Great Britain and the United States.

These countries are quite different, too. Australia is a continent, while New Zealand is an island nation. Much of Australia is arid, while most of New Zealand has lots of rain. There are few mountains in Australia, but there are many in New Zealand.

East and northeast of Australia are thousands of small islands in the Pacific Ocean. A majority of these islands are south of the equator. The islands west of 180° Longitude are considered to be in the Eastern Hemisphere. In this book the whole region is called Australia and Islands of the Pacific. You will remember that the terms Eastern Hemisphere and Western Hemisphere are terms of convenience. They are not hard and fast dividing lines.

No one knows exactly how many islands there are east and northeast of Australia. They have never been counted. Many are tiny, nameless, and uninhabited. Some islands were built by volcanoes from the bottom of the sea. The few people living on some of the islands are mostly farmers, fishers, a few government officials, traders, and missionary groups.

Things You Might Like To Do As You Study About Australia and Islands of the Pacific

You might wish to begin some activities to help you learn more about Australia and Islands of the Pacific.

1. Use an opaque or overhead projector to make several outline maps of Australia. Using legends to explain your symbols, place the following types of information on separate maps:

a. the amount of rainfall;

b. the population per km² (or per sq mi.);

c. the types of farming;

d. the temperatures in January and June;

e. railroads and major highways;

f. major mineral deposits;

g. the major cities.

Use atlases, encyclopedias, and the maps in this book to help you locate the information you need. Place the maps on the chalk tray or on the bulletin board. Study them and see if you can figure out any relationships among the maps.

2. Find out all you can about the Davis Cup. How is the Cup won, who has it at present, and what does it represent?

3. Visit a zoo and see as many marsupials (mahr *soo* pee uhls) as possible. If a zoo is not available, wildlife picture books in your school and community library may provide an imaginary trip. Look also at articles in encyclopedias under "animal" and "marsupial." List the interesting characteristics of your favorite marsupial.

4. Prepare a written report or an oral illustrated report on one of the following subjects:

a. Micronesia

b. Melanesia

c. Polynesia

d. Living on Sheep Stations in Australia

e. Australian Aborigines

f. The Great Barrier Reef

g. Sir Edmund Hillary, Mountain-climber and Explorer

h. Sports in Australia and New Zealand

i. Schools in Australia and New Zealand

5. Begin a current events bulletin board for news about Australia and New Zealand. Share important items with your classmates each day.

AUSTRALIA

Kilometers
0   250   500

Miles
0   250   500

DESERT

SAVANNA

MOUNTAINS

PRAIRIE

DRY GRASSLAND

SCRUB FORESTS AND GRASSLAND

EQUATOR

PACIFIC OCEAN

INDIAN OCEAN

REPUBLIC OF INDONESIA

JAVA

Timor Sea

NEW GUINEA

NEW IRELAND

NEW BRITAIN

PAPUA-NEW GUINEA

⊕ Port Moresby

Gulf of Papua

Coral Sea

Great Barrier Reef

NEW HEBRIDES

NEW CALEDONIA

TROPIC OF CAPRICORN

PACIFIC OCEAN

PACIFIC OCEAN

⊕ Darwin

CAPE YORK PENINSULA

Gulf of Carpentaria

NORTHERN TERRITORY

CARPENTARIA BASIN

COASTAL EASTERN

PLAINS HIGHLANDS

GREAT DIVIDING RANGE

QUEENSLAND

⊕ Brisbane

GREAT SANDY DESERT

Alice Springs

GREAT VICTORIA DESERT

CENTRAL BASINS

GREAT ARTESIAN BASIN

Darling River

SYDNEY ⊛

Canberra ⊕   2230m

NEW SOUTH WALES

Mt. Kosciusko, 7,316′   SNOWY MTS.

GREAT DIVIDING RANGE

WESTERN AUSTRALIA

WESTERN PLATEAU

• Kalgoorlie

SOUTH AUSTRALIA

Broken Hill •

Murray   River

River

VICTORIA

MELBOURNE ⊛

Tasman Sea

NEW ZEALAND

• Auckland

Wellington ⊛

NORTH ISLAND

Mt. Cook 12,319′
SOUTH 3755m
ISLAND   ALPS

Christchurch •

Tasman Glacier

SOUTHERN

Dunedin •

STEWART IS.

Perth ⊛

Adelaide ⊛

BARRON ISLAND

Bass Strait

TASMANIA

Hobart •

⊕ Hobart

0°   10°   20°   30°   40°   180°   170°   160°   150°   140°   130°   120°   110°   100°   90°

Chapter

───── AUSTRALIA ─────

The Lonely Continent

Australia is isolated from other continents by vast expanses of water. Because of this, some people call Australia "the lonely continent." The country of Australia also includes the island of Tasmania (taz *may* nee uh), just south of the mainland. Many nearby smaller islands are also part of Australia.

Australia's mainland extends about 4000 kilometers (2,500 miles) from west to east. It is about 3200 kilometers (2,000 miles) from north to south. This is almost as much land as the 48 touching states in the United States. About 14 million people live in Australia. Unlike the continent of Asia, Australia has very few people per square kilometer. Australia has encouraged immigrants from Europe and North America to settle the land. Even so, there are still only about 2 persons per square kilometer (5 per square mile).

Look at Australia on the map on page 519. Note the major bodies of water that surround the continent. To the north lies the Timor Sea. Northeast is the Coral Sea. The Pacific Ocean touches the east coast of Australia. The Indian Ocean is south and west of the continent.

More than nine-tenths (⁹/₁₀) of Australia's land does not rise above 600 meters (2,000 feet) above sea level. It has the lowest average elevation of any continent. The highest mountain is Mount Kosciusko (*kahs ee uhs* koh). This peak is only 2228 meters (7,310 feet) above sea level. A large part of the land is quite level. The western half of Australia is a low plateau. As the map on page 519 shows, much of the land is also desert.

The Climate of Australia

In general, Australia has a warm, dry climate. In fact, it is the driest of all continents. The summers tend to be warm and hot. Winters are mild with light to moderate rainfall. Since

Australia lies within the Southern Hemisphere, seasons are opposite to those in the United States. For example, when we are having winter, they are having summer. When they have spring, we have fall. But as you would expect, Australia's climate is quite different from region to region.

The Major Regions of Australia

Australia can be divided into three main regions. One is the Western Plateau. Another is the Central Lowlands. The third is the Eastern Highlands and Coastal Plains.

The Western Plateau. The Western Plateau region takes up about three-fifths (⅗) of the continent. It includes most of the land of Western Australia, the Northern Territory, and South Australia. See map on page 532. Much of the land in the plateau region is about 300 meters (1,000 feet) above sea level. The land is fairly level except for a few hills. Many of the lakes and stream beds on the Western Plateau are dry most of the year. A few near the southwest coast contain water all year long. Much of the interior land is desert. The Great Sandy Desert and the Great Victoria Desert extend through much of west and central Australia.

On the western edge of this plateau in southwestern Australia, dairy farming is quite important.

Sheep are a major product of Australia. The wool and meat of the sheep, as well as live animals, are shipped all over the world.

There, enough rain falls for good pastures. Wherever enough rain falls, cattle and sheep are raised. Some trees grow on the hills. Much of the interior plateau has less than 25 centimeters (10 inches) of rain a year. Very little grows on this land.

The Central Lowlands. Lowlands stretch from the Gulf of Carpentaria (*kahr* puhn *tehr* ee uh) in the north to the mouth of the Murray River in the south. The lowlands are located between the Western Plateau and the Eastern Highlands. In the south the lowland area is drained by the Murray River. This is the longest and most important river in Australia. Irrigated crops are grown at several places in the Murray River basin.

The central part of the lowland area is known as the **Great Artesian** (ahr *tee* zhun) **Basin**. Water drains into this basin from the higher eastern rim. It enters a spongelike rock layer between two hard layers of clay or rock. The clay or rock neither absorbs water, nor lets it go up or down. When a well is drilled through the top layer of clay or rock, water flows out of the well. Pressure built up from the underground area forces the water up and out. Such a well from which water flows without pumping is called an **artesian** well. Often these wells are drilled as deep as 1200 meters (4,000 feet) or more.

Water from most artesian wells in this area is filled with minerals. Much of it cannot be used for irrigation. It is not good for people to drink. However, cattle and sheep can drink it. As the map on page 528

shows, this area is mainly grassland. Many cattle and sheep are raised there. On the eastern edges of the lowland region much wheat is grown. More rain falls there.

The northern part of the lowland area is called the Carpentaria Basin. It is an area of poor soils. Summer rains cause tall grass to grow, but long periods of drought follow the rainy season. The few people who live in this part of Australia raise cattle.

The Eastern Highlands and Coastal Plains. This region extends all along the eastern coast of the continent. It also includes the island-state of Tasmania. East of the Great Dividing Range are narrow coastal plains. These extend inland from the coast for about 80 to 300 kilometers (50 to 185 miles). As the winds blow moist air from the southeast across the land, the air rises and cools. Moisture in the air condenses to form water droplets that form clouds. With further cooling, the droplets become larger. These droplets fall as rain. The eastern slopes of the Great Dividing Range and the coastal plains have more rainfall than any other part of Australia. About 150 centimeters (60 inches) of rain fall here each year. The winds blow from the ocean all during the year so this region has fine pastureland and forests. High in the Snowy Mountains of southeastern Australia heavy snows occur during the winter season. About two-thirds (⅔) of all the people in Australia live in this Eastern Highlands and Coastal Plains region.

Koala bears are difficult to raise outside of Australia because they will eat only the leaves of the eucalyptus tree.

Australia's Unusual Wildlife

Separation from other continents has helped make Australia a very interesting land. Many of its plants and wildlife are not found on any other continent.

More than half of the **mammals** (animals that nurse their young) that are native to Australia are *marsupials*. Animals that carry their young in a pouch are called **marsupials**. The kangaroo, opossum, koala (koh *ahl* uh), and wombat (*wahm* bat) are all marsupials. Kangaroos and koalas are two of Australia's most interesting animals. Kangaroos can jump about 8 meters (25 feet) in one

bound. They use their back legs and strong tail to make these jumps. The little koalas look like teddy bears. Koalas never drink water. Instead, they eat leaves of eucalyptus (*yoo* kuh *lihp* tuhs) trees for moisture.

Perhaps the most interesting animal found in Australia is the *platypus* (*plat* ih puhs). The **platypus** is one of the few egg-laying mammals on the Earth. The platypus has a bill like a duck, a tail like a beaver, and fur over most of its body. After its eggs hatch, the platypus nurses its young.

Many interesting birds live in Australia, too. The emu (*ee* myoo) is a large bird like an ostrich. It does

Australia

AUSTRALIA AND NEW ZEALAND

EQUATOR

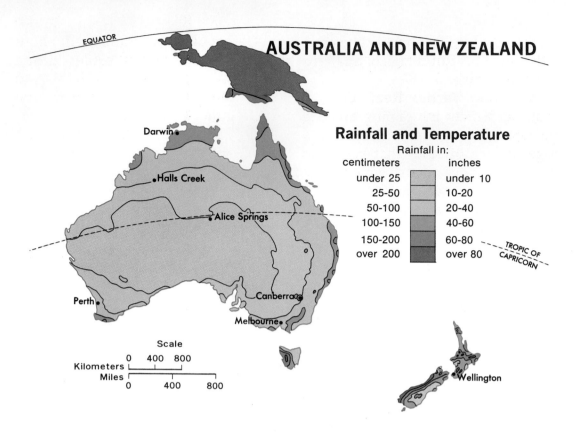

Rainfall and Temperature
Rainfall in:

centimeters		inches
under 25		under 10
25-50		10-20
50-100		20-40
100-150		40-60
150-200		60-80
over 200		over 80

TROPIC OF CAPRICORN

Scale

Kilometers 0 400 800

Miles 0 400 800

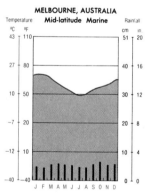

MELBOURNE, AUSTRALIA
Mid-latitude Marine

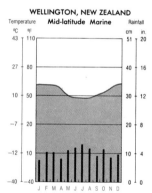

WELLINGTON, NEW ZEALAND
Mid-latitude Marine

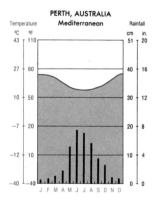

PERTH, AUSTRALIA
Mediterranean

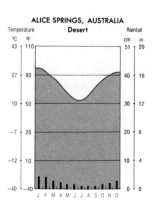

ALICE SPRINGS, AUSTRALIA
Desert

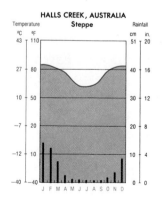

HALLS CREEK, AUSTRALIA
Steppe

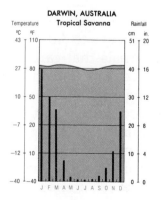

DARWIN, AUSTRALIA
Tropical Savanna

not fly. The lyre (*ly* uhr) bird has one of the most beautiful tails of any bird found on Earth.

The Great Barrier Reef. One of Australia's most interesting areas is the *Great Barrier Reef*. A **reef** is a ridge of rock or sand that lies at or near the surface of the water. This reef extends for about 2000 kilometers (1,250 miles) along the northeast coast of Australia. The reef has been built slowly during the years by tiny underwater animals called *coral polyps*. The tiny **coral polyps** deposit limestone a little at a time in warm ocean waters. This reef is of special interest to people who study tropical fish, ocean life, and coral.

Australian Natives and Immigrants

Most of the people living in Australia today have ancestors who came from Europe. However, they are not the native Australians. When British settlers first came to this continent they found some small, dark-skinned people living on the land. The settlers called them **aborigines** (*ab* uh *rihj* uh nees) because they were the first people to live on this land. The Australian aborigines lived in a very primitive way at that time. They used simple tools made of stone and wood. They were a nomadic people. They moved from place to place, eating fruits and berries, and meat. Animals were killed with spears and *boomerangs*. A **boomerang** is a curved weapon which returns close to the spot from which it is thrown, if it misses the target.

The aborigines numbered about 300,000 when the first settlers began to colonize the land in 1788. Up to that time, the aborigines were the undisputed owners of the land.

The British and Irish immigrant population grew slowly. By 1820 it was estimated that only about 150,000 colonists were in Australia. When gold was discovered in Victoria in 1851 many more European immigrants came. In the late 1800s the European population had increased to about 400,000. Little by little the aborigines retreated farther inland, losing their lands on the way.

The Australian government has set aside lands in various sections of the country as **reservations** for the aborigines. Most of these reservations are in the north. Some of the aborigines still live much as they did when the first British settlers arrived. Others tend cattle and sheep on ranches. However, only about 40,000 pure aborigines are left in Australia. In many ways the treatment of the aborigines in Australia is much like that of Native Americans in the Untied States.

During the early years people of Anglo-Saxon background were invited to settle in Australia. However, people from Asia were not permitted to enter the country. This policy is no longer followed. Following both World Wars many different Europeans came to live in Australia. Although the population of the country is quite mixed today there are still few people of African or Oriental ancestry living in Australia now.

The desire for quick wealth led many people to Australia. Even today people, like this man, look for precious metals and gems, such as sapphires.

The Modern Immigrants—"Project Down-Under"

Most of Australia's first immigrants came from England and Ireland in the early 19th century. Those immigrants had been poor and unemployed in the "old country." Some had been in prison. They hoped to make a better life for themselves in Australia. The grazing lands of Australia provided an opportunity for many ambitious settlers. Others tried farming. Some people opened small shops and businesses. The immigrant's life demanded hard work. For many, their efforts were rewarded.

Gold was discovered in 1851. Prospectors came to Australia. The miners hoped to strike it rich in this land. Some did. Other miners missed the gold, but discovered other ores. Silver, copper, and tin were found. Settlements began to grow in the west, south, and central regions of Australia. Through the years more immigrants came. Coal, iron, and lead were discovered. In the 1960s an oil boom brought even more immigrants to this prospering nation. Recently, uranium was discovered. The resources of Australia have helped many people to make a good life for themselves.

As cities and towns developed schools were built for the children. Australia's school-age population was very large. The country soon discovered that there were not enough trained teachers to staff the schools. A teacher shortage had occurred in many parts of the continent. Where could Australia locate English-speaking teachers willing to immigrate to this continent? The government created "Project Down-Under." The Project's task was to locate needed teachers.

During this same time the United States had many extra trained teachers. So people from Australia's Project Down-Under came to the United States to hire some of these teachers.

Interested teachers were interviewed in major cities all over the United States. The Australians explained about life in their country. Selected teachers would be assigned to schools in remote areas. Conditions in some of the places might be difficult for new immigrants. An interested teacher had to consider many facts before making a final decision to go. Project Down-Under teachers were then flown to Australia. There are still some teaching jobs available "Down-Under." Maybe some users of this text will some day migrate to Australia!

REMEMBER, THINK, DO

1. Do you know why Australia is called "the lonely continent"?
2. What continent, other than Australia, is completely south of the equator?
3. Prepare a chart comparing the similarities and differences of the Western Plateau, the Central Lowlands, and the Eastern Highland and Coastal Plains regions of Australia.
4. Why are the seasons of the Southern Hemisphere opposite to those of the Northern Hemisphere?
5. Find out what you can about relationships between Australians who came from Europe and the aborigines.
6. One of the world's best women's tennis players is from Australia. Find out what you can about Evonne Goolagong.

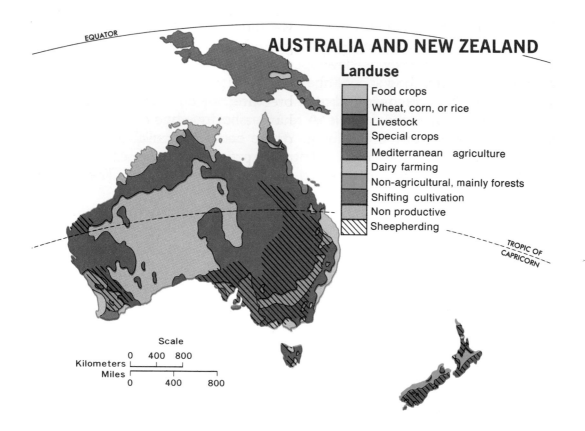

EQUATOR

AUSTRALIA AND NEW ZEALAND

Landuse

Food crops
Wheat, corn, or rice
Livestock
Special crops
Mediterranean agriculture
Dairy farming
Non-agricultural, mainly forests
Shifting cultivation
Non productive
Sheepherding

TROPIC OF CAPRICORN

Scale

Kilometers 0   400   800

Miles 0   400   800

Farming and Ranching

The grazing and farming lands of Australia are not extremely fertile. In fact, large areas lack at least one or more necessary mineral elements. These weaknesses, however, can be cured by adding minerals to the soil. Ranchers and farmers have learned how to enrich the soil. Certain types of clovers have also been planted to improve ranch lands. By applying scientific farming methods in this way, Australian farmers and ranchers have become among the world's best.

Australian Wool. Australia has become one of the greatest sheep-raising countries of the world. There are about 11 times more sheep than

people! Australian herders graze about 150 million sheep a year. Shearings from the sheep provide more than one-fourth (¼) of all the wool used in the world. Wool is Australia's most valuable export.

Australian wool is known for its high quality. Most sheep raised for wool are a breed named **merino** (muh *ree* noh). These sheep were first developed in Spain. For several centuries the Spanish had laws against taking these sheep out of the country. A small flock of merino sheep was taken to South Africa, however. A few of them were then taken to Australia. The climate and the grass in southern Australia seemed perfect for the sheep. The merinos began to produce a better quality of wool—and more of it—than they did in Spain. As a result of selective breeding, Australian sheep now grow twice as much wool as did their ancestors. The average production of wool is about four kilograms (nine pounds) per sheep per year.

In addition to wool, much mutton is exported. Australian farmers have learned that they can make more money by raising sheep that are good for both wool and mutton. Though the merino sheep is the best breed for wool, the meat of that animal is not tasty to eat. Farmers therefore developed other breeds that are good producers of both wool and meat.

Beef and Dairy Cattle. Cattle raising and dairy farming are also important in Australia. Most of the dairy farms are located on the good pastures of the eastern or south-western coasts. Many beef cattle are grazed on the northern grasslands where the climate is too warm for sheep to produce wool. By selective breeding, the Australian farmers have also improved their beef and dairy herds. Australian beef and dairy products are exported to many countries. Great Britain, Japan, and the United States are the leading buyers of these products.

Wheat and Other Crops. Next to wool, wheat is Australia's most valuable agricultural export. Millions of bushels of wheat are harvested annually. About two-thirds (⅔) of the wheat crop is exported. Most of the wheat is grown in the fairly dry areas west of the Great Dividing Range. Only about 30 to 60 centimeters (12 to 15 inches) of rain fall in the wheat-growing regions every year.

Since much of Australia is desert or semi-desert, water must be stored and used carefully.

Oats and barley are also grown in the same area. These two grains are used mainly as feed for livestock. Corn and rice are grown where rain or irrigation is available for their cultivation.

Sugar cane is a leading crop in northeastern Australia. Farmers in Queensland also grow cotton, tobacco, pineapples, bananas, and papayas. In southern Australia the weather is too cool for such tropical fruits to grow well. There, citrus fruits and grapes are grown at the edge of the desert areas where irrigation is possible. Along the cooler southeastern coast and in Tasmania, fruits such as apples, pears, plums, peaches, cherries, and berries are grown.

Mining and Manufacturing

Although Australia is best known for its agricultural exports, the majority of people live in cities and work in industries. Manufacturing represents the major single source of wealth for most people in Australia. Almost half of the people live in six cities, all of which are state capitals. More than one-third (⅓) of all the nation's people live in the two largest cities, Sydney and Melbourne.

Australia is fortunate to have vast coal fields. Large iron ore deposits have been discovered, too, in Western Australia and in northwestern Tasmania. These deposits are believed to have enough iron ore to supply the country's needs for 50

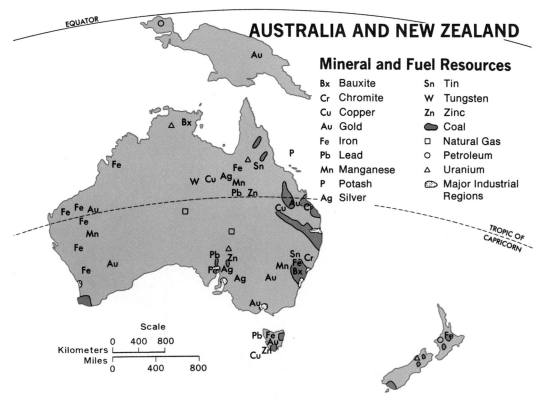

AUSTRALIA AND NEW ZEALAND

Mineral and Fuel Resources

Bx	Bauxite	Sn	Tin
Cr	Chromite	W	Tungsten
Cu	Copper	Zn	Zinc
Au	Gold		Coal
Fe	Iron		Natural Gas
Pb	Lead		Petroleum
Mn	Manganese		Uranium
P	Potash		Major Industrial Regions
Ag	Silver		

Scale
Kilometers 0 400 800
Miles 0 400 800

years. The iron ore is located fairly close to the coast. That makes it possible to move it easily to steel manufacturing plants and to other countries. Australian steel mills make high-quality steel at low cost. Many countries in Asia buy steel from Australia. Iron ore is also exported.

Large bauxite deposits are found in northern Australia. Gold, lead, zinc, and silver are mined in important amounts in southern and southwestern Australia. Broken Hill and Kalgoorlie (kal *guhr* lee) are important mining towns. (See map, page 519.) Uranium is found in large amounts near Darwin in northern Australia.

Important oil and gas fields have been found on Barron Island off the northwest coast, and in Bass Strait between the mainland and Tasmania. Other fields have been located near the center of the continent. Unfortunately these fields are far from population centers, except for the Bass Strait underwater field. And they are expensive to develop. Australia has to import much of the petroleum used at present.

Many kinds of goods are manufactured in Australia. Heavy industrial firms produce locomotives, autos, farm machinery, trucks, and structural steel for buildings. Smaller goods such as textiles, clothing, electrical appliances, drugs, paints, paper, glass, leather goods, and furniture are also produced. Despite the numbers of local industries, Australia imports many goods from Japan, Great Britain, Canada, and the United States. At times the demand for imports has had a bad effect upon the economy of Australia. Too much money spent for imports can produce an imbalance in trade. This tends to weaken an economy.

The Snowy Mountain Project

The land east of the Great Dividing Range receives enough rain for agricultural needs. However, to the west of this range, much of the land was not used due to a lack of water. To provide water for irrigation in the Murray River Basin west of the Range, a huge project was planned. Several large dams were built on streams in the mountains. Long tunnels were made through the mountains to the western slopes. Some of these tunnels were as long as 15 to 45 kilometers (10 to 30 miles). Tunnels were cut so the water would flow rapidly from the lakes through the tunnels. Seventeen underground power stations were built to make electricity from this water. Finally the water is used to irrigate fertile farmland in the Murray Basin.

Government

The Commonwealth of Australia is a union of states, like the United States. There are six states in the nation. They are: Queensland, New South Wales, Victoria, Tasmania, South Australia, and Western Australia. Find them all on the map on page 519. The Northern Territory has not yet been made a state since so few people live there.

As in the United States, the people of Australia have built a special city for the capital. Canberra (*kan* buhr uh) is a federal district like the District of Columbia.

Australia has a parliamentary form of government. Members of Parliament are elected by the people. Australia recognizes the monarch of Great Britain as its chief of state. The monarch is represented by a governor-general. But this official has little real power, and acts only with the consent of the Australian government. The most powerful leadership position in Australia is the prime minister. This person is the head of the government.

Education

Each state of Australia has its own education programs. Many parts of the nation are remote and difficult to reach. It has often been hard to provide schools for all sections of the country. In some very isolated regions a "School of the Air" has been developed. Children study at home. They receive their lessons in the mail and send papers to their teachers the same way. They also have two-way radios by which the teacher can talk to them. That way they can get help when they need it. In the cities, schools are like the best schools of any nation in the world. Many fine universities are also located in Australia.

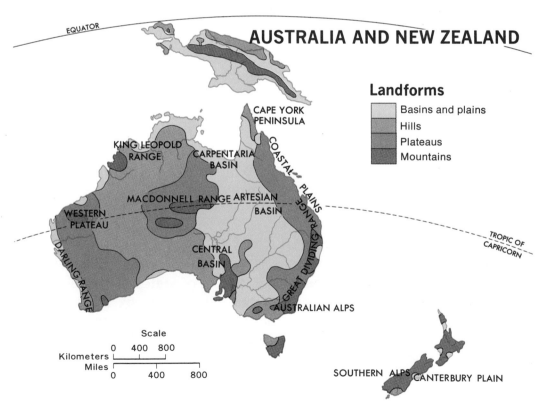

AUSTRALIA AND NEW ZEALAND

EQUATOR

CAPE YORK PENINSULA

KING LEOPOLD RANGE

CARPENTARIA BASIN

COASTAL PLAINS

MACDONNELL RANGE ARTESIAN BASIN

WESTERN PLATEAU

CENTRAL BASIN

DARLING RANGE

GREAT DIVIDING RANGE

AUSTRALIAN ALPS

TROPIC OF CAPRICORN

Landforms
- Basins and plains
- Hills
- Plateaus
- Mountains

Scale
Kilometers 0 400 800
Miles 0 400 800

SOUTHERN ALPS CANTERBURY PLAIN

WHAT HAVE YOU LEARNED?

1. Why are the seasons in Australia opposite those of the United States?

2. What does "marsupial" mean? What marsupials live in Australia?

3. What are some reasons why immigrants have settled in Australia?

4. What have Australian farmers and ranchers done to improve the soil and make it more productive?

BE A GEOGRAPHER

1. On an outline map of Australia label the six states and territory, as well as the capital city.

2. Is southern Australia closer to Antarctica than the mainland of any other continent?

3. Make an agricultural and product map of Australia. Try to identify major items grown and produced in each state and territory.

4. Where in Australia do you think winter sports are popular?

QUESTIONS TO THINK ABOUT

1. What do you think might be done to improve conditions for aborigines in Australia?

2. If you were a sheep rancher and depended on the world market price of wool, how do you think the use of machine-made fibers might affect wood prices?

3. Why do you think it is sometimes harmful to the economy of a nation to import more goods than it exports?

4. Which do you think is more important to the economy of Australia, agriculture or manufacturing? Why?

28
Chapter

NEW ZEALAND AND OTHER ISLANDS OF THE PACIFIC

New Zealand

New Zealand is made up of two large islands and one small island. They are located about 2900 kilometers (1,200 miles) southeast of Australia. The large islands are simply named North Island and South Island. Stewart Island is the small island just south of South Island. A few more than 3 million people live in New Zealand.

Notice the position of New Zealand on a globe or the map on pages 542-543. Except for the southern tip of South America, no other land is as close to Antarctica as New Zealand is. Explorers bound for Antarctica have often used either Christchurch or Dunedin (duh *need* uhn) on South Island as their last port of call.

Land and Climate of New Zealand

New Zealand is a beautiful land extending more than 1600 kilometers (1,000 miles) from north to south. High snow-capped mountains called the Southern Alps reach to more than 3700 meters (12,000 feet) on

South Island. Huge glaciers move down these mountains almost to sea level. Tasman Glacier near Mt. Cook, for instance, is almost 30 kilometers (18 miles) long. Broad, fertile plains stretch east of the mountains to the sea.

Much of North Island is rolling grassland, with high wooded hills rising here and there above the plains. Four volcanoes, three of them active, extend from east to west across the island near its widest point. Throughout much of the year, the volcanoes have snow-capped peaks. Near their bases are many beautiful geysers, bubbling hot pools, and mud pots.

When the first European settlers came to the islands, the forests were full of huge trees. Some were said to be as large as the giant redwoods of California. The larger trees in New Zealand have long since been cut. However, about one-fourth (¼) of the land is still forested. Forests are now being carefully managed. Fast growing trees are planted as older ones are cut. The wood is used for the pulp and paper industry.

Most of New Zealand has plenty of rainfall throughout the year for good pastureland. Temperatures are mild enough on North Island for pastures to stay green throughout the year. The southeast winds bring rain to North Island in the summer. Winds from the west bring rain in winter. Along the mountainous coast of South Island the rainfall is very heavy. More than 250 centimeters (100 inches) fall a year. The rain falls as snow on the high peaks. Tons of snow pack into ice, forming the glaciers. On the eastern side of the mountains on South Island are plains. Less rainfall occurs on this land. The plains provide excellent pastureland for sheep. It is the best region in New Zealand for growing crops such as wheat, barley, and oats.

Sheep and Cattle

New Zealand is one of the world's leading exporters of wool, meat, and butter. Most exported meat is either mutton or beef.

New Zealand has more sheep per square kilometer of land and more sheep per person than any other nation. More than 60 million sheep are grazed on the pasturelands. The New Zealanders, like the Australians, have developed a breed of sheep that produces both superior wool and good mutton. Great Britain buys most of New Zealand's mutton. Wool is shipped to many nations, including the United States.

The farmers of New Zealand have almost a million cattle. About half of these are dairy cattle. The others are raised for beef. The mild climate and rich pastures favor dairy farming.

Minerals and Manufacturing

Few minerals have been found in New Zealand except for gold and some coal. Almost enough coal is mined each year to meet the needs of the people. However, coal is imported from time to time. Electricity, made at hydroelectric power plants on the rivers, provides abundant power. Steam trapped below the earth's surface also provides a

Herding sheep with a motorcycle is quicker and easier than doing it on foot.

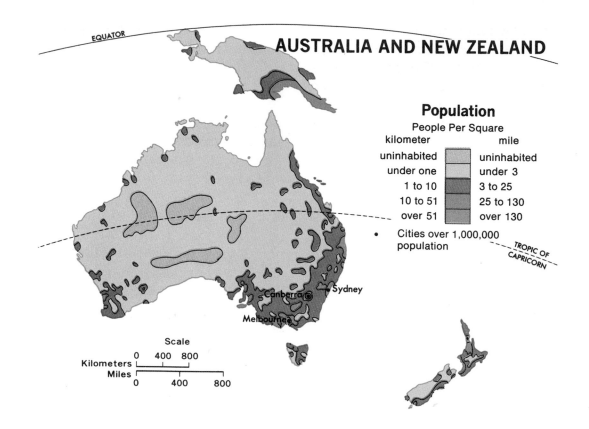

Population

People Per Square

kilometer		mile
uninhabited		uninhabited
under one		under 3
1 to 10		3 to 25
10 to 51		25 to 130
over 51		over 130

• Cities over 1,000,000 population

EQUATOR

TROPIC OF CAPRICORN

Canberra

Sydney

Melbourne

Scale

Kilometers 0 400 800

Miles 0 400 800

source of power. As a result electricity is plentiful in New Zealand.

Among the consumer products made in New Zealand are woolen textiles and blankets. Shoes, cigarettes, electrical appliances, and kitchen utensils are also made. Food processing is one of the leading industries, of course. Almost all of New Zealand's food needs are produced locally. Tea and sugar have to be imported.

Transportation

Building railroads and roadways across the hilly land of New Zealand was somewhat difficult. Road beds on the steep hills required skillful engineering. Bridges also had to be constructed over the many waterways. The railroads are owned by the government. Many of them are powered by electricity. The railway is one of the largest businesses of New Zealand.

Good roads have been built throughout the islands. Many people own autos. Farm produce is moved rapidly to towns by trucks. The four large cities, Auckland (*aw kluhnd*), Wellington, Christchurch, and Dunedin have airports. International air service is available from Auckland to all parts of the world.

Government and Education

New Zealand is an independent nation and a member of the Commonwealth of Nations. Much of its trade is with Great Britain and other Commonwealth countries. New Zealanders pride themselves on being "the most British of the Dominions." They think of Queen Elizabeth II of England as their queen, and speak of Britain as "home." The queen, as in Australia, is represented by a governor-general. The governor-general acts according to the advice of the Executive Council. The duties of this council are similar to the cabinet's duties in the United States. New Zealand has only one law-making body, the House of Representatives. Members of the House are elected by the people every three years. All meetings of the House are broadcast by radio to the entire country.

New Zealand's lawmakers have passed many laws to protect workers, care for elderly people, and help those unable to work. Many houses have been built by the government and rented to workers at low cost. Medical care and treatment in public hospitals are free to all. School children are given free dental care and free milk. New Zealand has free public schools. All children between 7 and 15 years of age must attend school. These services, of course, are

Spectacular mountain scenery is common in New Zealand.

not really free. High taxes are necessary to pay for all these government benefits. High taxes also make it difficult for anyone to become very rich. On the other hand, there are very few people in the country who go hungry and do not live in a good house.

REMEMBER, THINK, DO

1. What are the main export items of New Zealand?
2. Why do many explorers to Antarctica make New Zealand their last port of call?
3. You might be interested in seeing a film which describes how glaciers are formed and how they move.
4. New Zealand's government provides many services to children and adults. Find out what services are provided in your community and who pays for them.
5. About how many farm animals per person are there in New Zealand? In Australia? In the United States?

LEARN MORE ABOUT THE ISLANDS OF THE PACIFIC

North and east of New Zealand and Australia are many, many islands. The people of these islands have much in common. Most depend in part on the coconut tree for food, timber, and thatch. They catch fish from the waters near their islands. Some carry on farming on a small scale, shifting from one plot of land to another, every few years. On some of these islands, crops of coconut, rubber, cacao, and coffee are grown on large plantations. Food items such as sweet potatoes, cassava, beans, bananas, and peanuts are raised. Some of the people work on large plantations.

For the last 200 years many Europeans, Asians, and Americans have visited these islands. Some stayed on as business people, missionaries, and farmers. Some of these islands have been managed as United Nations trust territories for many years. Some are currently independent, or working towards that goal. In the years of the 21st century, however, life may be quite different for the people living in these Pacific islands. You may wish to enrich your study of the world by learning more about the islands of the Pacific. Use the Study Guide on page 31 to help plan your activities.

Sugar cane is a major crop on many islands in the Pacific Ocean. It grows well in warm, moist climates.

New Guinea

North of Australia is the second largest island in the world, New Guinea. The eastern half of New Guinea plus a number of nearby islands are part of the nation of Papua-New Guinea. Papua-New Guinea became an independent nation in 1975. It has almost 3 million people. It includes the islands of New Britain, New Ireland, and Bougainville (*boog* uhn *vihl*). Most of the people living in Papua New Guinea are *Melanesians*. Other island groups located in the Western Pacific south of the equator are the New Hebrides (*hehb* ruh deez) Islands, the Solomon Islands, and New Caledonia (*kal* uh *doh* nyuh).

Micronesia and Polynesia

Most of the islands north of the equator in the Western Pacific, and a few south of it, are grouped together and called **Micronesia**. This is a Greek term meaning small islands.

The islands in the central part of the Pacific are usually considered to be part of **Polynesia**. This is a Greek term meaning many islands. Polynesia extends from Hawaii to New Zealand. Try to locate all these island groups on a globe.

A SUMMARY OF AUSTRALIA
AND ISLANDS OF THE PACIFIC

Australia and New Zealand have an economy and way of life much like that of European countries and the United States. These two nations, however, are thousands of miles away from Europe. They are also isolated from other continents by vast bodies of water.

Australia's industries and businesses provide employment for many people. Many other people, as in New Zealand, work on farms and ranches. The climates in these nations tend to be moderate. The seasons are opposite to lands of the Northern Hemisphere.

Thousands of islands are located in the Pacific Ocean. Many of these islands seem to create a link between the lands of the Eastern and Western Hemisphere.

———— WHAT HAVE YOU LEARNED? ————

1. How is life for people in Australia and New Zealand similar to and different from that of people living in the island nations of Asia?

2. What other countries of the Eastern Hemisphere, besides Australia, have irrigation projects to help improve farming?

3. How are the mineral resources of Australia and Africa similar? How are they different?

4. British settlers developed colonies in parts of the Middle East, Africa, Asia, and Australia and New Zealand. What major contributions and problems have resulted from such colonial settlements?

———— BE A GEOGRAPHER ————

1. Compare the population map of Australia and New Zealand on page 537 with the rainfall map on page 524. What relationships, if any, do you note between rainfall and population?

2. Locate the following on a globe or the map on pages 542-543 and be prepared to point out their locations on a wall map:
   a. Gulf of Carpentaria
   b. Great Barrier Reef
   c. New Hebrides
   d. Papua New Guinea
   e. Mount Cook
   f. Botany Bay
   g. Murray River
   h. Southern Alps
   i. New Caledonia
   j. Great Artesian Basin
   k. Stewart Island
   l. Gulf of Papua
   m. Cook Strait
   n. Cape York Peninsula
   o. Bass Strait

3. Prepare a list of the highest peaks and their elevations in each of the following countries:
   a. Australia

   b. China
   c. India
   d. Indonesia
   e. Japan
   f. New Zealand
   g. Nepal
   h. The Philippines
   i. Switzerland
   j. France

4. Using the map on page 519, compute the approximate distance between the following places:
   a. Darwin and Melbourne
   b. Perth and Sydney
   c. Cape York and Hobart
   d. Adelaide and Brisbane

5. What direction is it from:
   a. Darwin to Melbourne?
   b. Port Moresby to Perth?
   c. Adelaide to Brisbane?
   d. Alice Springs to Sydney?
   e. Sydney to Wellington?

QUESTIONS TO THINK ABOUT

1. What things do you think a person should consider before deciding to migrate to another country?

2. Why do you think the original inhabitants of Australia did not develop methods of farming, such as the native people did in Africa and Asia?

3. If you had your choice of three places in the Eastern Hemisphere to visit, where would you choose to go? Why?

4. If you could choose a place to live in the Eastern Hemisphere for two years, where would you choose to live? Why?

Unit Review

Political Map of the World

International Boundaries

kilometers
miles

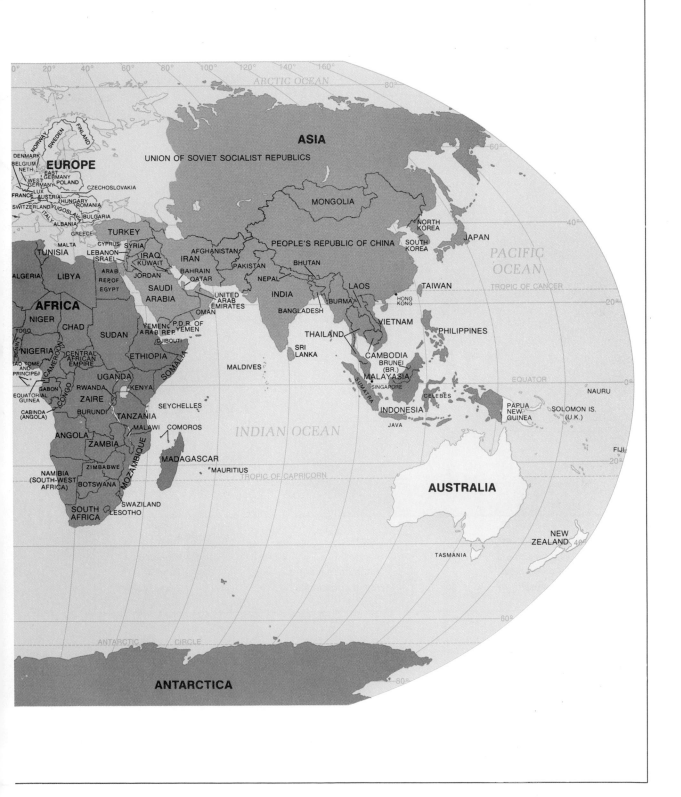

543

EASTERN HEMISPHERE: CLIMATE REGIONS

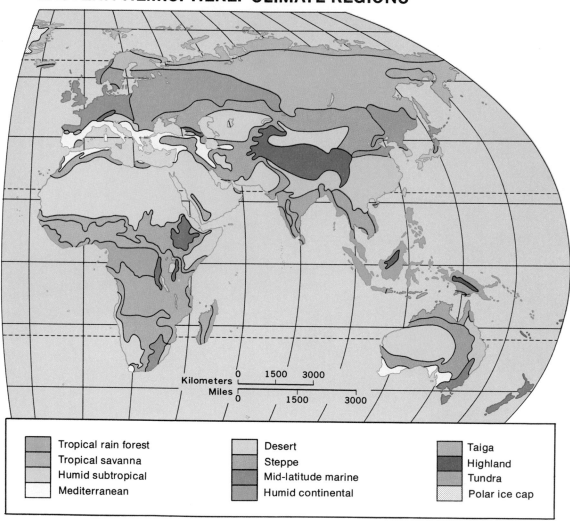

Kilometers 0 1500 3000
Miles 0 1500 3000

Tropical rain forest
Tropical savanna
Humid subtropical
Mediterranean

Desert
Steppe
Mid-latitude marine
Humid continental

Taiga
Highland
Tundra
Polar ice cap

EASTERN HEMISPHERE: ELEVATIONS

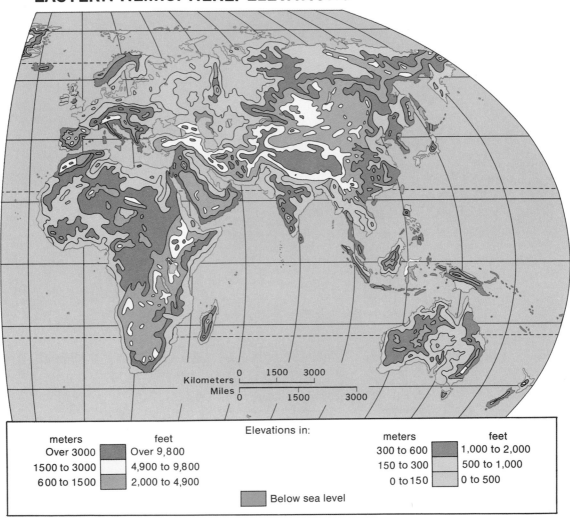

Kilometers
0   1500   3000

Miles
0   1500   3000

Elevations in:

meters	feet
Over 3000	Over 9,800
1500 to 3000	4,900 to 9,800
600 to 1500	2,000 to 4,900

meters	feet
300 to 600	1,000 to 2,000
150 to 300	500 to 1,000
0 to 150	0 to 500

Below sea level

APPENDIX

TABLE I   THE SIZE OF THE EARTH

Diameter of the Earth at the equator	12 756 km	7,926 mi.
Distance around the Earth at the equator	40 075 km	24,902 mi.
Land area	148 350 000 km²	57,280,000 sq. mi.
Water area	361 563 400 km²	139,660,400 sq. mi.
Total area	509 917 870 km²	196,940,400 sq. mi.

Source: *The World Book Encyclopedia (1978)*

TABLE II   SIZE AND POPULATION OF MAJOR AREAS OF THE WORLD

Continent	Estimated Population in Thousands	Approximate Area in Thousands of Km²	Approximate Area in Thousands of Sq. Mi.	Population Per Km²	Population Per Sq. Mi.
Africa	437,000	30 320	11,707	14	36
Antarctica	Uninhabited	13 209	5,100	—	—
Asia, Excluding area of U.S.S.R.	2,377,149	27 145	10,480	87	226
Australia	14,213	7 687	2,968	2	5
Europe, Excluding area of U.S.S.R.	487,286	4 952	1,912	98	254
New Zealand and Pacific Islands*	8,998	1 246	481	7	19
North America	358,000	24 390	9,417	15	39
South America	233,000	17 829	6,884	13	34
U.S.S.R.	262,000	22 402	8,650	12	30
World	4,178,000	148 300	57,259	28	73

Source: *The World Book Encyclopedia (1978)*   *Excluding Hawaii, which is included with North America

TABLE III   HIGHEST MOUNTAIN PEAKS OF THE EASTERN HEMISPHERE

Mountain	Location	Height Meters	Feet	Mountain	Location	Height Meters	Feet
Europe				**Asia**			
Mont Blanc	France-Italy	4807	15,771	Mount Everest	Nepal-Tibet	8848	29,028
Monte Rosa	Switzerland	4634	15,203	K2 (Godwin			
Dom	Switzerland	4545	14,911	Austen)	Kashmir	8611	28,250
Liskamm	Italy-Switzerland	4527	14,852	Kanchenjunga	India-Nepal	8598	28,208
Weisshorn	Switzerland	4506	14,782	Lhotse I (Everest)	Nepal-Tibet	8511	27,923
				Makalu I	Nepal-Tibet	8481	27,824
U.S.S.R.				Lhotse II (Everest)	Nepal-Tibet	8400	27,560
Communism Peak	U.S.S.R.	7495	24,590	Dhaulagiri	Nepal	8172	26,810
Pobedy Peak	Sinkiang-U.S.S.R.	7439	24,406	Manaslu I	Nepal	8156	26,760
Lenin Peak	U.S.S.R.	7134	23,405	Cho Oyu	Nepal-Tibet	8153	26,750
Elbrus	U.S.S.R.	5642	18,510				
Shkara	U.S.S.R.	5201	17,064				
Africa				**Australia and New Zealand**			
Kilimanjaro	Tanzania	5895	19,340	Cook	New Zealand	3764	12,349
Kenya	Kenya	5199	17,058	Kosciusko	Australia	2228	7,310
Margherita	Uganda-Zaire	5109	16,763				
Ras Dashan	Ethiopia	4620	15,158				
Meru	Tanzania	4566	14,979				

Source: *World Almanac 1978*

TABLE IV   MAJOR RIVERS OF THE EASTERN HEMISPHERE

River	Approximate Length Km	Mi.	Outflow	River	Approximate Length Km	Mi.	Outflow
Europe				**Asia**			
Danube	2858	1,776	Black Sea	Yangtze	5472	3,400	East China Sea
Rhine	1320	820	North Sea	Hwang Ho	4667	2,900	Yellow Sea
Elbe	1165	724	North Sea	Mekong	4184	2,600	South China Sea
Loire	1020	634	Bay of Biscay	Euphrates	3597	2,235	Persian Gulf
Tisza	966	600	Danube River	Indus	2897	1,800	Arabian Sea
				Brahmaputra	2897	1,800	Bay of Bengal
U.S.S.R.							
Ob-Irtysh	5568	3,460	Gulf of Ob				
Amur	4345	2,700	Tatar Strait				
Lena	4313	2,680	Laptev Sea				
Yenisey	4130	2,566	Kara Sea	**Australia**			
Volga	3685	2,290	Caspian Sea	Murray-Darling	3718	2,310	Indian Ocean
Africa							
Nile	6671	4,145	Mediterranean Sea				
Zaire (Congo)	4374	2,718	Atlantic Ocean				
Niger	4184	2,600	Gulf of Guinea				
Zambesi	2736	1,700	Indian Ocean				

Source: *World Almanac 1978*

TABLE V AREA AND POPULATION OF THE EASTERN HEMISPHERE

Country	Estimated Population in Thousands 1978	Area Km²	Area Sq. Mi.	Population Km²	Population Sq. Mi.	Capital	Estimated Population in Thousands*
Northwestern Europe							
Belgium	9,962	30 513	11,781	326	844	Brussels	153 (1,051)
Denmark	5,167	43 069	16,629	120	311	Copenhagen	626 (1,384)
Finland	4,757	337 009	130,120	14	36	Helsinki	513 (817)
France	54,208	547 026	211,208	99	256	Paris	2,300 (8,550)
Iceland	226	103 000	39,769	2	5	Reykjavik	81 (94)
Ireland	3,224	70 283	27,136	46	119	Dublin	568 (680)
Luxembourg	372	2 586	998	144	373	Luxembourg	76
Monaco	25	1	.5	16 779	43,457	Monaco	n/a
The Netherlands	14,091	41 160	15,892	342	886	Amsterdam	758 (991)
Norway	4,100	324 219	125,182	13	34	Oslo	477 (556)
Sweden	8,292	449 964	173,732	18	47	Stockholm	724 (1,352)
United Kingdom	56,555	244 035	94,223	232	601	London (Greater)	7,281
Central Europe							
Austria	7,680	83 849	32,374	92	238	Vienna	1,615
Czechoslovakia	15,042	127 869	49,371	118	306	Prague	1,078
Germany, East	17,304	108 178	41,768	160	414	Berlin (East)	1,086
Germany, West	63,544	248 468	95,934	256	663	Bonn	299
Hungary	10,678	93 030	35,919	115	298	Budapest	2,038
Liechtenstein	23	157	61	146	378	Vaduz	4
Poland	34,920	312 677	120,725	112	290	Warsaw	1,317
Romania	21,796	237 500	91,699	92	238	Bucharest	1,488 (1,575)
Switzerland	6,798	41 288	15,941	165	427	Bern	153 (288)
Southern Europe							
Albania	2,730	28 748	11,100	95	246	Tiranë	175
Andorra	27	453	175	59	153	Andorra	11
Bulgaria	8,889	110 912	42,823	80	207	Sofia	966 (1,065)
Greece	9,143	131 944	50,944	69	179	Athens	867 (2,540)
Italy	57,154	301 253	116,314	190	492	Rome	2,868
Malta	320	316	122	1 013	2,624	Valletta	15
Portugal	8,805	92 082	35,553	96	249	Lisbon	758 (1,635)
San Marino	19	61	24	311	792	San Marino	4,400
Spain	36,818	504 750	194,885	73	189	Madrid	3,520
Yugoslavia	21,925	255 804	98,766	86	223	Belgrade	746
U.S.S.R.	261,565	22 402 000	8,649,500	12	31	Moscow	6,942 (7,528)
Middle East							
Bahrain	275	622	240	442	1,145	Manama	110
Cyprus	649	9 251	3,572	70	181	Nicosia	116
Iran	35,687	1 648 000	636,296	22	57	Tehran	4,716
Iraq	12,258	434 924	167,925	28	73	Baghdad	2,184
Israel	3,742	20 700	7,992	181	469	Jerusalem	326
Jordan	3,036	97 740	37,738	31	80	Amman	500
Kuwait	1,155	20 118	7,768	57	148	Kuwait	80 (218)
Lebanon	3,321	10 400	4,015	319	826	Beirut	475 (939)
Oman	840	212 457	82,030	4	10	Muscat	10
People's Democratic Republic of Yemen	1,733	332 968	128,560	5	13	Aden	264 (285)
Qatar	101	11 000	4,247	9	23	Doha	80
Saudi Arabia	9,799	2 153 090	831,313	5	13	Riyadh	300

TABLE V   AREA AND POPULATION OF THE EASTERN HEMISPHERE (cont.)

Country	Estimated Population in Thousands 1978	Area Km²	Area Sq. Mi.	Population Km²	Population Sq. Mi.	Capital	Estimated Population in Thousands*
Syria	8,109	185 180	71,498	44	114	Damascus	837
Turkey	43,009	780 576	301,382	55	142	Ankara	1,522 (1,554)
United Arab Emirates	243	86 600	32,278	3	8	Abu Dhabi	95
Yemen Arab Republic	7,262	195 000	75,290	37	96	Sana	125
North Africa							
Algeria	18,460	2 381 741	919,595	8	21	Algiers	1,504
Egypt	39,729	1 001 449	386,662	40	104	Cairo	5,715
Libya	2,766	1 759 540	679,362	2	5	Tripoli	551
Morocco	19,290	446 550	172,414	43	111	Rabat	368
Tunisia	6,202	163 610	63,170	38	98	Tunis	469 (648)
Southern Sahara							
Chad	4,291	1 284 000	495,755	3	8	N'Djamena	150
The Gambia	563	11 295	4,361	50	129	Banjul	39
Mali	6,138	1 240 000	478,767	5	13	Bamako	170
Mauritania	1,435	1 030 700	397,956	1	3	Nouakchott	50
Niger	4,979	1 267 000	489,191	4	10	Niamey	130
Senegal	4,744	196 192	75,750	24	62	Dakar	580
Sudan	19,122	2 505 813	967,500	8	21	Khartoum	852
Upper Volta	6,459	274 200	105,869	24	62	Ouagadougou	115
Western Equatorial Africa							
Benin	3,422	112 622	43,484	30	78	Porto-Novo	97
Cameroon	6,773	475 442	183,569	14	36	Yaoundé	178
Central African Republic	2,870	622 984	240,535	5	13	Bangui	187
Congo, Peoples Republic of the	1,480	342 000	132,000	4	10	Brazzville	289
Equatorial Guinea	321	28 051	10,830	11	28	Malabo	37
Gabon	1,029	267 667	103,347	4	10	Libreville	n/a (112)
Ghana	10,687	238 537	92,100	45	117	Accra	564 (738)
Guinea	4,741	245 857	94,926	19	49	Conakry	46 (197)
Guinea-Bissau	549	36 125	13,948	15	39	Bissau	71
Ivory Coast	5,260	322 463	124,504	16	41	Abidjan	500
Liberia	1,831	111 369	43,000	16	41	Monrovia	96
Nigeria	68,160	923 768	356,669	74	192	Lagos	1,149 (1,477)
Sierra Leone	2,873	71 740	27,699	40	104	Freetown	179
Togo	2,406	56 000	21,622	43	111	Lomé	95 (141)
Zaire	27,051	2 345 409	905,568	12	31	Kinshasa	2,202
Eastern Africa							
Burundi	4,070	27 834	10,747	146	378	Bujumbura	79
Comoro Islands	337	2 171	838	155	401	Moroni	18
Djibouti	114	22 000	8,494	5	13	Djibouti	100
Ethiopia	30,067	1 221 900	471,778	25	65	Addis Ababa	1,161
Kenya	14,874	582 646	224,961	26	67	Nairobi	535
Madagascar	7,971	587 041	226,658	14	36	Tananarive	347
Mauritius	911	2 045	790	445	1,151	Port Louis	141
Rwanda	4,622	26 338	10,169	175	453	Kigali	30
Seychelles Islands	65	376	145	173	448	Victoria	14
Somalia	3,424	637 657	246,201	5	13	Mogadiscio	445
Tanzania	16,408	945 087	364,900	17	44	Dar es Salaam	273
Uganda	12,721	236 036	91,134	54	140	Kampala	80 (331)

TABLE V   AREA AND POPULATION OF THE EASTERN HEMISPHERE (cont.)

Country	Estimated Population in Thousands 1978	Area		Population		Capital	Estimated Population in Thousands*
		Km²	Sq. Mi.	Km²	Sq. Mi.		
Southern Africa							
Angola	6,265	1 246 700	481,354	5	13	Luanda	600
Botswana	756	600 372	231,805	1	3	Gaborone	18
Lesotho	1,108	30 355	11,720	37	96	Maseru	17 (29)
Malawi	5,409	118 484	45,747	46	119	Lilongwe	70
Mozambique	9,889	783 030	302,330	13	34	Maputo	355 (750)
Namibia	1,027	824 292	318,261	1	3	Windhoek	64
Swaziland	540	17 363	6,704	31	80	Mbabane	18
South Africa	27,615	1 221 037	471,445	23	60	Cape Town	691 (1,097)
Zambia	5,410	752 614	290,586	7	18	Lusaka	348
Zimbabwe	7,000	390 580	150,804	18	47	Salisbury	557
Southern Asia							
Afghanistan	20,666	647 497	250,000	32	83	Kabul	597
Bangladesh	82,453	143 998	55,598	573	1,484	Dacca	1,680
Bhutan	1,245	47 000	18,147	26	67	Thimphu	12
India	637,082	3 287 590	1,269,346	194	502	New Delhi	302
Nepal	13,545	140 797	54,362	96	249	Kathmandu	333
Pakistan	77,674	803 943	310,404	97	251	Islamabad	125
Sri Lanka	14,982	65 610	25,332	228	591	Colombo	562
Eastern Asia							
Hong Kong	4,563	2 916	1,126	1 565	4,052	Victoria	521
Japan	115,486	377 389	145,711	306	793	Tokyo	8,640 (11,623)
Korea, North	17,175	120 538	46,540	142	368	Pyongyang	2,500
Korea, South	35,793	98 484	38,025	363	940	Seoul	6,889
Macau	279	16	6	17 438	46,500	Macao	n/a
Mongolia	1,579	1 565 000	604,250	1	3	Ulan Bator	267
People's Republic of China	865,493	9 561 000	3,691,523	91	236	Peking	7,570
Republic of China	17,009	35 916	13,885	473	1,225	Taipei	1,840
Southeast Asia							
Burma	33,326	676 552	261 218	49	127	Rangoon	1,352 (2,055)
Kampuchea	8,809	181 035	69,898	49	127	Phnom Penh	200
Indonesia	143,732	1 919 270	741,034	75	194	Jakarta	5,490
Laos	3,581	236 800	91,429	15	39	Vientiane	130
Malaysia	12,830	329 749	127,316	39	101	Kuala Lumpur	452
Philippines	46,660	300 000	115,831	156	404	Manila	1,455 (4,904)
Singapore	2,374	581	224	4 086	10,583	Singapore	2,075
Thailand	46,532	514 000	16,408	91	236	Bangkok	1,867
Vietnam	47,588	332 559	128,402	143	370	Hanoi	415 (644)
Australia and Islands of the Pacific							
Australia	14,213	7 686 848	2,967,909	2	5	Canberra	156
Melanesia	2,126	502 432	193,990	4	11		
Micronesia	283	3 155	1,218	90	232		
New Zealand	3,264	268 704	103,747	12	31	Wellington	185
Papua-New Guinea	2,826	461 691	178,260	6	16	Port Moresby	77
Polynesia	501	9 479	3,659	53	137		

Source: *The World Book Encyclopedia* (1978)   *Metropolitan areas are in parentheses.

GLOSSARY

Alluvial soil   Soil that has been carried by streams from mountains or hills. It has been dropped by the streams in valleys or near their mouths.

Altitude   How high above sea level, or how much below sea level, a place is.

Arid   Dry; having too little rainfall for agriculture without irrigation.

Atmosphere   The air surrounding the Earth.

Basin   A large area of land sloping inward and surrounded by higher land.

Bay   A part of a sea or ocean that extends into the land and is partly surrounded by land. (*See also* Gulf.)

Belt   An area or region in which climate and soils are much alike. Therefore, an area in which almost the same crops may be grown.

Brass   An alloy made out of copper and zinc. Brass resists rust.

Canal   A waterway constructed across land to connect two bodies of water.

Cape   A point of land that extends out into the water.

Capital   The city where the government of a state or nation is located; also money used as an investment in the future—to start new industries, operate and enlarge industries already started, search for mineral ores or fuels, buy goods for resale, etc.

Cash crops   Crops such as cotton, cacao, peanuts, etc., that are raised mainly for sale.

Climate   The condition of the atmosphere at any given place (or places or regions) in an average year. This average is figured by using data gathered for many years.

Coast   Land next to or touching a sea or ocean.

Colony   An area of land that is ruled by a government in another land, usually a long distance away.

Cooperative   A marketing agency, usually made up of farmers, to control the quality of products and to market them. Members receive profits based upon their share of production.

Coral reef   A ridge of hard material deposited by coral polyps. The ridge extends upward from the ocean floor to, almost to, or above the surface of the water.

Degrees   Parts into which a circle or sphere is divided. There are 360 degrees (°) in a full circle or on the globe.

Delta   Land that is formed at the mouth or mouths of a river as the river drops the silt it has been carrying.

Desert   Dry wasteland where little will grow. Most deserts are rocky.

Dictator   A person who rules a government as an all-powerful authority.

Direction   Any way that one may face or point. There are six main directions: north, south, east, west, up, and down. There are sixteen points on a compass.

Drought   A long period of time with little or no rainfall; common in arid areas.

Earthquake   A shaking or trembling of the Earth's crust caused by slippage along a fault or by the folding of rock strata.

Economy   A country's way of producing, distributing, and consuming goods.

Equator   A great circle drawn east and west around a globe halfway between the North and South poles.

Equinox   The two days each year when the sun's rays shine directly on the equator, so that day and night are of equal length everywhere on the Earth. The Vernal Equinox occurs about March 21, and the Autumnal Equinox about September 22.

Erosion   The wearing away of the Earth's crust by water, wind, frost, and other forces.

Estuary   The very wide mouth of a river that has tides as oceans and seas do, and where fresh water and salt water mix.

Exports   Products grown or made in a country that are sent out of the country for sale.

Fault   A break or crack in the Earth's crust.

Federal   Of or pertaining to a union of states under a central government. The states retain certain responsibilities and the central government has other responsibilities.

Fiord   A long, narrow inlet of the sea, usually with hills or mountains rising steeply from the water on both sides.

Free Trade   Trade without taxes levied by one state or country on goods produced and imported from another state or country.

Glacier A large mass of ice which, because of its weight, moves slowly down a mountain or valley.

Great Circle A line drawn on a globe which divides the Earth into two equal parts (as does the equator).

Grid A system of crossing lines used on globes and maps to help locate places on the Earth. The crossing parallels of latitude and meridians of longitude form the "squares" in the grid.

Gulf A part of a sea or ocean that extends into the land and is partly surrounded by land. (*See also* Bay.)

Heavy industry Industry that is basic to the growth of other industries (steel, machine tools, aluminum, chemicals, etc.).

Hills Land higher than the area nearby, but not as high as mountains.

Horizon Where the land and the sky seem to meet. The sun drops below the horizon in the west each evening, and rises above the horizon in the east each morning.

Humidity The amount of moisture in the atmosphere.

Hurricane A tropical cyclone (a huge, whirling low-pressure cell) about 150-500 kilometers (100-300 miles) in diameter. Hurricanes have winds of more than 120 kilometers (75 miles) per hour.

Imports Products grown, mined, or made in another country which are brought into a land for sale.

Inflation A rapid rise in the cost of goods or services and, therefore, in the cost of living. As inflation occurs, money drops in value.

Irrigation Taking water from a lake or river, or pumping it from under the ground, to water crops.

Isthmus A narrow strip of land with water on both sides that connects two larger bodies of land.

Landforms The form or contour of the land in an area; the topography; the location of mountains, rivers, lakes, cities, etc.

Latitude Distance north or south of the equator, measured in degrees. Each degree of latitude equals about 111 kilometers (69 statute miles). Lines of latitude are drawn on globes and maps parallel to the equator and are called parallels. Parallels are numbered from 0° (the equator) to 90° North or South (the poles). Low latitudes are between the equator and the Tropics of Cancer and Capricorn (0° to about 23½°); middle latitudes are between the Tropic of Cancer and the Arctic Circle and the Tropic of Capricorn and the Antarctic Circle (about 23½° to 66½°); high latitudes are north of the Arctic Circle and south of the Antarctic Circle (66½° to 90°).

Legend A key or explanation of what the symbols mean that are used on maps, charts, graphs.

Legislature The lawmaking body of a republic or a state.

Longitude Distance east or west of the prime meridian, measured in degrees. At the equator, a degree of longitude equals about 111 kilometers (69 statute miles). Near the poles, it equals only a few meters. Meridians of longitude (north-south lines) are drawn on globes and maps so that they cross the equator at right angles. Each meridian is half of a great circle. On globes, meridians meet at both poles and are farthest apart at the equator.

Lowland A part of a country or region in which most of the land has little altitude or is considerably lower than the land that surrounds it.

Metropolitan area A main city and the smaller cities around its edges, or a main city and its suburbs.

Mineral fuels Coal, petroleum, and other matter taken from the Earth's surface or below it that can be used to produce heat and power.

Minerals Ores and other materials taken from the surface of the Earth or below it for use by people.

Monsoon Seasonal winds that blow in one direction during summer months and in the reverse direction during winter months.

Mountain A very high and usually steep part of the land that is higher than the land surrounding it.

Natural resources The animals, plants, minerals, and mineral fuels found on the Earth that people can use for food, clothing, shelter, fuel, and to make other needed things.

Navigable river A river that has a channel or passageway deep and wide enough for the safe passage of ships. Some rivers are navigable all through the year, others only during rainy or flood seasons.

Oasis A green and fertile area in a desert where water is found.

Ocean The entire body of salt water that covers nearly three-fourths of the Earth's surface, and that surrounds the continents of the Earth.

Peninsula A peninsula is land that is almost surrounded by water. Usually a peninsula seems to stick out into a sea or ocean. A peninsula can be large or small. An example of a large peninsula is the continent of Europe. An example of a much smaller peninsula is southwestern England.

Plain Rather level land, sometimes with a few rolling hills; usually of fairly limited altitude and of considerable extent. A coastal plain is low, nearly level land near an ocean or sea.

Plateau An upland area of considerable size which may be crossed and cut by streams or canyons.

Port A city located on an ocean, sea, lake, or river from which goods and people are taken by ship and to which people and goods are brought by ship.

Prime Meridian The north-south line (meridian) which is numbered 0° and from which all east and west locations are figured.

Province An administrative part of a country, usually with some voice in government; similar to a state.

Race Persons who have fairly definite, inherited physical traits. Anthropologists recognize these main races: Caucasoid (white); Mongoloid (yellow); and Negroid (black).

Rainfall The amount of rain that falls on the land. Rainfall is measured by the total number of centimeters (or inches) that fall in one year. Snowfall at a place is also included in measured rainfall. About 25 centimeters (10 inches) of snow equals 2.5 centimeters (one inch) of rainfall.

Range A long row of mountains or hills that are connected.

Raw materials Substances that are still in their natural state that can be used in producing other goods.

Refining Process of taking impurities from an ore or other materials. In the oil industry, the refining process also separates different parts of the petroleum. A refined product is pure or has a higher degree of purity.

Region Part of a continent or country that is alike in some way. Usually a region is known for a certain kind of land, climate, or crops.

Republic A state or nation in which citizens elect persons to represent them as law-makers. Some countries that use the name "Republic," however, no longer have that type of government; military dictatorships have taken over many such governments.

River basin The land area that is drained by a river and its tributaries.

Rural In the country—not in a city. In the United States, towns with fewer than 2,500 people are considered rural.

Savanna Grassland that has some areas with trees.

Sea A smaller part of an ocean, especially if partly surrounded by land.

Sea level The level of the oceans of the world from which measures of altitude are figured.

Season A part of the year in which the weather is much the same, caused by the way the Earth's axis is tipped as it moves around the sun.

Smelting Refining by using heat to melt an ore. Chemicals or other ores are often added during the smelting process to purify or bring about changes in the product.

Soil The layer of loose earth covering much of the land areas of the world. Some soils are fertile or productive (capable of growing good crops) while, for various reasons, others are not.

Solstice The times each year when the direct rays of the sun reach farthest north and south and, thus, are directly over either the Tropic of Cancer or the Tropic of Capricorn. On the day of the Summer Solstice, the sun's rays shine directly down on the Tropic of Cancer and the entire area within the Arctic Circle has 24 hours of sunlight. The entire area within the Antarctic Circle has 24 hours without direct sunlight that same day.

Steppe A large, semi-arid region where short grasses cover the land and grazing usually is the main agricultural pursuit. Farming in steppe areas is hazardous unless irrigation is possible. Some years enough rain falls at the right time for successful grain crops. Other years, harvests are poor because of long periods of drought.

Strait A natural narrow waterway that connects two larger bodies of water.

Swamp An area of wet or spongy land; marshy ground not suitable for farming.

Synthetic Made through chemistry. Materials from petroleum, for example, can be combined to make a synthetic product that is like natural rubber.

Tableland Fairly level land that is higher than lowland.

Tannin A liquid obtained from quebracho trees used in making leather from hides.

Tariffs Taxes placed on goods imported or exported.

Temperature How hot or cold it is in a place at a particular time. Temperature is measured in degrees (°) by a thermometer.

Territory Land owned and controlled by a country that is not yet self-governing. A Trust Territory is land controlled by a country (but not owned by it) under the supervision of the United Nations.

Topography The surface or land features of an area, including both location and altitude.

Trade winds Winds blowing from about 30° North and South latitudes almost continually toward the equator. Trade winds blow from the northeast in the Northern Hemisphere and from the southeast in the Southern Hemisphere.

Tribe A group of people including many families and generations that have similar beliefs and ways of living. Tribes may be large or small. They share a common language and have a feeling of kinship toward other members of the tribe even though they may not know them personally.

Tributaries Streams that flow into a river. Usually, but not always, tributaries are smaller streams than the main river.

Tropical rain forest A dense growth of tall, broadleaf, evergreen trees often with heavy vines growing up the trees. Small shrubs and grasses are rarely found in tropical rain forests because little sunlight reaches the ground.

Tundra Areas in high latitudes where mosses, lichens, bushes, grasses, and summer flowers are the main plants. Such areas are permanently frozen (permafrost), as deep as 3000 meters (1,000 feet), except during the summer when perhaps the top 60 centimeters (24 inches) may thaw.

Typhoon Large tropical cyclones in the Pacific and Indian oceans that may be from 160 to 480 kilometers (100 to 300 miles) in diameter and in which winds blow from 120 to 200 kilometers per hour (75 to 125 miles per hour). Heavy rains fall during the passage of a typhoon, and high tides are driven ashore by the winds. A typhoon is like a hurricane in the Caribbean Sea or Gulf of Mexico.

Upland Higher ground than most of the land in a region—not high enough to be called a highland.

Urban In the city—not in the country.

Valley A low area between hills or mountains that usually has a stream or river, flowing through it.

Volcano A mountain that has formed over an opening in the Earth's surface. Far below the volcano is a deposit of hot, melted rock called magma. Active volcanoes sometimes erupt, throwing ashes and rocks into the air and sometimes having the molten rock flow from openings. Once it reaches the surface, the molten rock is called lava.

Watershed The ridge that divides one drainage area from another. Also, the area or region that drains into or supplies a river or lake.

PHOTO ACKNOWLEDGMENTS

The abbreviations indicate the position of the picture on the page: T is top, M is middle, B is bottom, L is left, R is right, TR is top right, TL is top left, MR is middle right, ML is middle left, BR is bottom right, and BL is bottom left.

UNIT I
Page x: Jangdish Agarwal/Uniphoto.
5 Philip Sykes/Katharine Young. **6** Tony LaTona/Uniphoto. **10** Stephanie Fitzgerald/Peter Arnold. **11, 12** G. R. Roberts. **15** David O'Connor. **19** Susan Jane Lapides. **20** Taurus Photo. **26** Taylor Instrument of Sybron Corporation. **28** *(top to bottom)*: W. Luthy; Shostal; P.I.P.; Anonymous.

UNIT II
34 E & F Bernstein Productions. **36** Dan Porges/Peter Arnold. **38** A. L. Gonzales. **41** Brian Foster. **42** Carmine Fantasia. **46** Sven Simon/Katharine Young. **49** UPI.

UNIT III
50 Jacques Jargoux/Peter Arnold. **52** TL Talbot D. Lovering (Allyn and Bacon Staff Photographer). **52** TR G. R. Roberts. **52** MR George Guzzi. **52** BR G. R. Roberts. **52** BL Sven Samelius. **54** Richard V. Foster. **55** Brian Foster. **57** Central Office of Information/British Consulate. **63** Taurus Photo. **65** John Topham Picture Library. **67** British Tourist Authority. **69, 71** UPI. **73** E & F Bernstein Productions. **74** T, **74** B British Petroleum Company. **77** Aerofilms, Ltd. **78** G. R. Roberts. **79** Brian Foster. **80** Pictorial Parade. **83** Scottish Tourist Board. **85** Talbot D. Lovering (Allyn and Bacon Staff Photographer). **89** Vignoble d'Alsace, Collection CIVA F, 68003 Colmer. **93** Rene Delon/Roquefort Society. **95** G. R. Roberts. **97** Perfumerie Fragonard. **98** British Aircraft Association. **101** Steven Rosendahl/Uniphoto. **105** Sven Samelius. **107** Torbjorn Lovgren/Carl E. Ostman. **108** Swedish National Tourist Office. **111** Swedish Information Service. **112** Eivon Carlson. **115** Kay Honkanen/Carl E. Ostman. **117** L, **117** R Swedish Information Service. **118** G. R. Roberts. **120** Lennart/Eivon Carlson. **121** S. E. Tormi/EPA. **122** Frank Siteman. **123** E.P.A. **124** A. J. Gonzales.

UNIT IV
128 Lou Jones. **133** IN Press/Germany. **135** Gunther Brinkman/Peter Arnold. **137** YIVO Institute. **139** U.S. AIRFORCE Photo. **140** Sepp Seitz/Woodfin Camp. **142** George Guzzi. **146, 148** Talbot D. Lovering (Allyn and Bacon Staff Photographer). **153** John L. Lemkers. **154** Jaye R. Phillips. **156** Talbot D. Lovering (Allyn and Bacon

Staff Photographer). **158, 160** Swiss National Tourist Office. **161** John Topham Picture Library. **165** Pictorial Parade. **166** J. J. Foxx/Woodfin Camp. **168** Richard V. Foster. **169** R RCA Victor. **171** Sovfoto/Eastfoto. **172, 173** J. J. Foxx/Woodfin Camp. **177** W. H. Hodge/Peter Arnold. **178** Desjardin/Agence TOP. **180** Lon Jones. **185** L, **185** R Sven Simon/Katharine Young. **187** Wide World Photos Inc.

UNIT V
190 Carmine Fantasia. **192** TR Yugoslav Press and Cultural Services, **192** BR Nikos Kontos, **192** BL Jean Boughton, **192** TL Katharine Young. **194** Tony LaTona/Uniphoto. **197** Vance Henry/Taurus Photos. **198** UPI. **203** Peter Arnold. **204** Vance Henry/Taurus Photos. **206** Klaus Francke/Peter Arnold. **209** Talbot D. Lovering (Allyn and Bacon Staff Photographer). **212** George Guzzi. **214** Fototecha Unione. **215** Alinari. **217** Jean Boughton. **219** Vance Henry/Taurus Photos. **220** George Guzzi. **222** E.N.I.T. **227, 230** Tony LaTona/Uniphoto. **233** Gamma/Liaison. **236** M. Renaudeau/Agence TOP. **237** Robert Burke. **239** Rogers Fund/Metropolitan Museum of Art. **241** Richard Wiess/Peter Arnold. **242** Nikos Kontos. **244** H. Armstrong Roberts/E. P. Jones. **245** Taurus Photos. **247** Culver Pictures Inc. **249** A. Devaney/E. P. Jones. **250** Sovfoto/Eastfoto.

UNIT VI
254 Sovfoto. **256** T From "The Great Red Train Ride," by Eric Newby. Published by Weidenfeld and Nicholson, Ltd., **256** M, **256** B Sovfoto. **258** Sovfoto. **259** From "The Great Red Train Ride," by Eric Newby. Published by Weidenfeld and Nicholson, Ltd. **263** Culver Pictures Inc. **265, 269, 270** Sovfoto. **272** Richard V. Foster. **274** Vance Henry/Taurus Photos. **277** Sovfoto. **280** Vance Henry/Taurus Photos. **282** NASA. **285** Sovfoto. **289** Tass/Sovfoto.

UNIT VII
292 Kendall Dudley. **296** Jordan Information Bureau. **300** Talbot D. Lovering (Allyn and Bacon Staff Photographer). **304** Gamma Liaison. **307** UPI. **315, 316** Jordan Information Bureau. **319** Talbot D. Lovering (Allyn and Bacon Staff Photographer). **321** Jean Claude Francolon/Gamma-Liaison. **322** Natural Council of Tourism of Lebanon. **324** Mary V. Elliff/Kendall Dudley. **326** Boubat/Agence TOP. **329** Gamma-Liaison. **331** Marvin Ickow/Uniphoto. **332** Talbot D. Lovering (Allyn and Bacon Staff Photographer). **335** The Spencer Collection/New York Public Library. **338** ARAMCO. **340** Akher SAA & Hassam Nassif/

WHO. **342** Kuwait Ministry of Information. **343, 344** ARAMCO. **349** NASA. **351** John Topham Picture Library. **353** Talbot D. Lovering (Allyn and Bacon Staff Photographer). **354** Gamma-Liaison. **357** Talbot D. Lovering (Allyn and Bacon Staff Photographer). **359** Thomas P. Huf/Kendall Dudley.

UNIT VIII
364 Victor Englebert. **366** T *(left to right)*: Michael Laurent/Gamma-Liaison; Michael Evans/Liaison; G. Walker/Liaison; Peter Jordan/Liaison. **366** BL, **366** BR Marion Kaplan. **371** American Museum of Natural History, New York. **373** Victor Englebert. **374, 380, 383** Marion Kaplan. **386** Victor Englebert. **389** United Nations/Katharine Young. **393** Marion Kaplan. **394** Babs Yahaya and H. Godicke/UNICEF Photo. **397, 400** Marion Kaplan. **402** Paul Almasy. **405, 407, 409** Victor Englebert. **411** Dr. Edward R. Degginger. **412** Robert Frerck. **414** Marion Kaplan. **417** Robert Frerck. **420** Villalobos/Liaison. **423** Marion Kaplan. **426** Peter Jordan/Liaison. **429** Southern Rhodesian Department of Tourism.

UNIV IX
432 Lou Jones. **434** TL Karen Rubin/Liaison. **434** TR Martha C. Guthrie/EPA. **434** MR Talbot D. Lovering (Allyn and Bacon Staff Photographer). **434** BR Robert Frerck. **434** BL Air India. **434** ML Janice Blumberg/Uniphoto. **441** Chester Beatty Library, Ireland. **443** From the book, "Early India and Pakistan" by Sir Mortimer Wheeler © 1959, London, England. Published in the United States by Frederick A. Praeger Inc., Publishers. **448** David O'Connor. **449** Martha C. Guthrie/EPA. **452** David O'Connor. **455** Paul Almasy/WHO. **458** UPI. **460** David O'Connor. **463, 468** Martha C. Guthrie/EPA. **473** Marvin Ickow/Uniphoto. **475** British Museum. **479** Eric Kroll/Taurus Photos. **480** G. R. Roberts. **483** Marvin Ickow/Uniphoto. **487** Lou Jones. **491** G. H. Richardson/Taurus Photos. **495** Talbot D. Lovering (Allyn and Bacon Staff Photographer). **496** Wide World Photos. **501** Sandra Hedberg/Global Focus. **502** Frandeon/Agence TOP. **505** Martha C. Guthrie/EPA. **508** Wide World Photos.

UNIT X
514 Robert Frerck. **516** L G. R. Roberts. **516** R Robert Frerck. **521, 523** G. R. Roberts. **526, 529** Australian News & Information Service. **535** Robert Frerck. **537, 539** G. R. Roberts.

All work not otherwise credited is from the Allyn and Bacon Inc. Photo Collection.